AF817264

Climate Change Management

Series Editor

Walter Leal Filho, International Climate Change Information and Research Programme, Hamburg University of Applied Sciences, Hamburg, Germany

The aim of this book series is to provide an authoritative source of information on climate change management, with an emphasis on projects, case studies and practical initiatives – all of which may help to address a problem with a global scope, but the impacts of which are mostly local. As the world actively seeks ways to cope with the effects of climate change and global warming, such as floods, droughts, rising sea levels and landscape changes, there is a vital need for reliable information and data to support the efforts pursued by local governments, NGOs and other organizations to address the problems associated with climate change. This series welcomes monographs and contributed volumes written for an academic and professional audience, as well as peer-reviewed conference proceedings. Relevant topics include but are not limited to water conservation, disaster prevention and management, and agriculture, as well as regional studies and documentation of trends. Thanks to its interdisciplinary focus, the series aims to concretely contribute to a better understanding of the state-of-the-art of climate change adaptation, and of the tools with which it can be implemented on the ground.

Notes on the quality assurance and peer review of this publication

Prior to publication, the quality of the works published in this series is double blind reviewed by external referees appointed by the editor. The referees are not aware of the author's name when performing the review; the referees' names are not disclosed.

Abdulaziz I. Almulhim · Walter Leal Filho ·
Ayyoob Sharifi
Editors

Urbanization, Climate Change, and Health

Integrating Strategies for Sustainable and Resilient Cities

Springer

Editors
Abdulaziz I. Almulhim
Urban and Regional Planning Department
College of Architecture and Planning
Imam Abdulrahman Bin Faisal University
Dammam, Saudi Arabia

Walter Leal Filho
Research and Transfer Centre "Sustainable Development and Climate Change Management"
Hamburg University of Applied Sciences
Hamburg, Germany

Ayyoob Sharifi
The IDEC Institute and Network for Education and Research on Peace and Sustainability (NERPS)
Hiroshima University
Higashihiroshima, Japan

ISSN 1610-2002 ISSN 1610-2010 (electronic)
Climate Change Management
ISBN 978-3-031-99131-8 ISBN 978-3-031-99132-5 (eBook)
https://doi.org/10.1007/978-3-031-99132-5

© The Editor(s) (if applicable) and The Author(s), under exclusive license to Springer Nature Switzerland AG 2025, corrected publication 2025

This work is subject to copyright. All rights are solely and exclusively licensed by the Publisher, whether the whole or part of the material is concerned, specifically the rights of translation, reprinting, reuse of illustrations, recitation, broadcasting, reproduction on microfilms or in any other physical way, and transmission or information storage and retrieval, electronic adaptation, computer software, or by similar or dissimilar methodology now known or hereafter developed.
The use of general descriptive names, registered names, trademarks, service marks, etc. in this publication does not imply, even in the absence of a specific statement, that such names are exempt from the relevant protective laws and regulations and therefore free for general use.
The publisher, the authors and the editors are safe to assume that the advice and information in this book are believed to be true and accurate at the date of publication. Neither the publisher nor the authors or the editors give a warranty, expressed or implied, with respect to the material contained herein or for any errors or omissions that may have been made. The publisher remains neutral with regard to jurisdictional claims in published maps and institutional affiliations.

This Springer imprint is published by the registered company Springer Nature Switzerland AG
The registered company address is: Gewerbestrasse 11, 6330 Cham, Switzerland

If disposing of this product, please recycle the paper.

Preface

As cities continue to expand and global temperatures rise, the intersection of urbanization, climate change, and public health has emerged as one of the most urgent and complex challenges of our time. *Urbanization, Climate Change, and Health: Integrating Strategies for Sustainable and Resilient Cities* is a timely response to this challenge—offering a multidimensional exploration of how urban systems can be reimagined to promote both environmental sustainability and human well-being.

This book is grounded in the recognition that cities are both sites of vulnerability and powerful platforms for transformation. From rising heatwaves and deteriorating air quality to mental health strain and mobility inequities, the impacts of climate change are increasingly concentrated in urban areas. Yet, these areas also offer the infrastructure, innovation, and collective capacity needed to lead the way toward resilience.

Bringing together contributors from diverse disciplines—including urban planning, public health, environmental science, gender studies, and climate policy—this volume offers critical insights into the local and global dimensions of climate-resilient urban development. Through regionally grounded case studies from India, Indonesia, Spain, Senegal, and beyond, the chapters reflect both the common challenges and the unique opportunities faced by cities in different contexts.

Structured across three thematic parts, the book moves from foundational frameworks and regional perspectives to sectoral strategies and forward-looking solutions. Topics such as salutogenic urbanism, reproductive health and climate, green and blue infrastructure, climate-resilient housing, and mobility systems are explored through an integrative lens, emphasizing the interconnected nature of urban resilience and public health.

This book is not merely a compilation of academic thought—it is a call for integrated action. It urges decision-makers, practitioners, researchers, and community leaders to go beyond siloed planning and instead embrace inclusive, evidence-based, and cross-sectoral strategies for healthy and climate-smart cities. It also highlights the importance of context-sensitive planning that honors local needs, knowledge, and capacities while aligning with global sustainability goals.

We hope that this volume serves not only as a resource for those working at the intersection of urbanism, climate, and health but also as a source of inspiration to imagine cities that are livable, equitable, and resilient in the face of an uncertain future.

Dammam, Saudi Arabia — Abdulaziz I. Almulhim
Hamburg, Germany — Walter Leal Filho
Higashihiroshima, Japan — Ayyoob Sharifi

About This Book

Urbanization, Climate Change, and Health: Integrating Strategies for Sustainable and Resilient Cities presents a timely and interdisciplinary exploration of how urban development, environmental pressures, and public health intersect in the era of global climate change. As cities grow and climate-related risks intensify, the need to design integrated, human-centered solutions has never been more critical.

This comprehensive volume brings together diverse perspectives from across the globe, offering in-depth analyses, real-world case studies, and innovative strategies that bridge the fields of urban planning, climate adaptation, public health, and sustainability. From the heat-stressed coasts of Indonesia to transforming infrastructure in Spain and mobility challenges in India, the book highlights how cities—especially in the Global South—are navigating complex health and environmental dynamics.

Drawing on a mix of conceptual frameworks, empirical research, and applied solutions, this book examines pressing topics such as green and blue infrastructure, healthy housing, microclimate assessment, reproductive health in climate planning, and the role of resilience in shaping future-ready cities.

Designed for urban planners, public health professionals, researchers, and policymakers, this book offers actionable insights for creating cities that are not only sustainable and smart but also inclusive and health-promoting. *Urbanization, Climate Change, and Health* envisions a future where urban resilience is grounded in equity, sustainability, and strategic integration across disciplines.

Contents

Part I
Foundations and Regional Perspectives on Climate Change, Urbanization, and Health

Chapter 1
Urbanization Under Global Climate Change with Health Impacts and Strategies for Sustainable and Resilient Cities

Abdulaziz I. Almulhim

Abstract The speed at which urbanization has increased, in conjunction with the consequences of global climate change, has exacerbated public health concerns in cities around the world. This chapter aims to explore the interconnection between urbanization, climate change, and public health, highlighting key interventions and strategies for fostering resilient and sustainable cities. Climate-related threats to health—namely polluted air, excessive heat, and the growing number of vector-borne illnesses—have become more widespread in rapidly urbanizing nations. The chapter emphasizes the importance of integrated urban planning methods, which include both resilience to climate change and public health accomplishments. It also addresses significant thematic areas related to urban climate adaptation, mitigation, and urban resilience. It highlights the environmental footprint of cities and demonstrates how strategic planning can mitigate the adverse health impacts of climate change. By focusing on adaptation strategies and resilient urban systems, this chapter contributes to the ongoing discourse on urban health and sustainability in the face of global environmental challenges.

Keywords Urbanization · Climate change · Health · Urban sustainability · Urban resilience · Sustainable and resilient cities

---

A. I. Almulhim (✉)
Department of Urban and Regional Planning, College of Architecture and Planning, Imam Abdulrahman Bin Faisal University, Dammam, Saudi Arabia
e-mail: aialmulhim@iau.edu.sa

© The Author(s), under exclusive license to Springer Nature Switzerland AG 2025

A. I. Almulhim et al. (eds.), *Urbanization, Climate Change, and Health*, Climate Change Management, https://doi.org/10.1007/978-3-031-99132-5_1

Introduction

Urbanization and Climate Change Synergy

Both urbanization and climate change have become interwoven in a chain reaction that compounds both health and environmental crises. Growing urbanization, in particular for developing countries, is moving forward at an unimaginable level, aggravating climate change in multiple ways and heightening its adverse consequences. Compared to preceding decades, the population of cities has risen dramatically globally and continues to do so today. Approximately 57%, or 4.4 billion, of the global population are currently living in metropolitan areas, and this number is predicted to double in size by 2050, leading the world populace to experience an exceptional rate of expansion (Dyvik 2024). Following such expansion, climate change is directly proportional to urbanization, causing further urban expansion. Urban areas contribute considerably to climate change, as their daily operations are the primary generators of greenhouse gas emissions. Cities are estimated to account for 70% of worldwide carbon dioxide emissions, largely due to construction and transportation (IPCC 2022).

The synergy between urbanization and climate change explains how rapid urbanization actually exacerbates climate change impacts and intensifies public health challenges. For instance, greenhouse gas emissions increase due to both urban expansion and transportation (Khalili et al. 2019). As urban areas expand, they are frequently dependent on fossil fuels for power generation, transport, and manufacturing operations, and such dependence considerably amplifies greenhouse gas emissions (Yoro and Daramola 2020). The building of structures, utilization of numerous transport vehicles, and both industries and households contributing to pollution generate huge quantities of carbon dioxide as well as multiple contaminants, worsening the effects of global warming. For instance, India alone, with the majority of its cities including Guwahati, Mullanpur, Begusarai, and Delhi, serves as the most polluted country in 2023 worldwide, exceedingly more than 100 times the average air quality index of the WHO's safety guidelines, ranking it as the country with one of the worst air qualities globally (IQAir 2025; Fuller et al. 2022).

Urban heat islands also occur because of urbanization's potential for heat absorption. City centers usually remain warmer than comparable rural areas due to the extensive development of roads, structures, and other infrastructures that hold and absorb heat. This is called the urban heat island impact (Mohajerani et al. 2017). The temperature discrepancy amplifies the negative consequences of climate change, leading to greater heatwave severity. In such a manner, increased temperatures may result in heat exhaustion, heatstroke, and dehydration, especially among certain groups such as seniors, young children, and those with previously existing medical problems. The cumulative effect of heat and pollution in the air may aggravate underlying illnesses such as cardiovascular disease, asthma, diabetes, heatstroke, and mental problems. Zhao et al. (2021) found that around 489,000 deaths related to heat issues took place annually from 2000 until 2019, with 36% from Europe and 45% from Asia.

Urbanization causes increased vulnerability to extreme weather events. Climate change worsens both the intensity and frequency of severe weather conditions, including excessive rainfall causing catastrophic flooding (Clarke et al. 2022; Almulhim and Cobbinah 2023). Flooding imposes direct harm by means of fatalities and relocation, in addition to amplifying the spreading of waterborne infections such as dysentery and cholera and eventually overpowering medical facilities, thus provoking recurrent public health catastrophes. As a result of increased urbanization, ecosystem disruption also takes place, causing the loss of natural buffers; for instance, in most cities in China, which has had negative implications for the country's vegetation (Du et al. 2019). The rapid decline of these biological barriers stimulates urban receptivity to climatic repercussions, leaving populations increasingly vulnerable to natural disasters and negative implications for public health with the growth of emergent transmissible diseases (Jordan 2020).

The rapidly increasing urbanization, together with the compounding consequences caused by climate change, generates a potentially hazardous synergy that not only exacerbates the destruction of the environment but also intensifies dangers to human health. Nevertheless, it is important to note that cities are both key contributors to and vulnerable victims of global climate change. Along with contributing via major quantities of waste materials, consumption of electricity, emissions from transportation, and space-heating and cooling demand, which are all attributable to growing urbanization, city regions are also disproportionately faced with the adverse consequences of urbanization as it relates to climate change, thus necessitating mitigation strategies. Such a dual function emphasizes the complicated correlation between climate change and urbanization, whereby cities worsen the ecological catastrophe and make themselves more severely vulnerable to its ramifications.

There are several ways in which urban environments act as contributors to climate change, namely through land use and deforestation, greenhouse gas emissions, and waste generation, among others (Greenwalt et al. 2018). Urban expansion commonly invades natural environments, leading to wetland deterioration, deforestation, and a decline in important carbon storage systems, which impairs the ability of the planet to absorb carbon dioxide, thus intensifying climate change. In addition, urban areas, being economical and production centers, bring about substantial global greenhouse gas production resulting from heightened energy utilization in cooling and heating for manufacturing, construction, and transportation, which exacerbates airborne pollutants and congested roadways. Cities account for approximately 70% of worldwide carbon dioxide releases, though covering roughly 2% of the planet's land surface (IPCC 2022).

The transport industry represents the biggest carbon dioxide contributor, predominantly from fossil fuel-powered motor vehicles. Seoul, New York, and Guangzhou are the top three cities responsible for more than 25% of the overall carbon footprint worldwide (Charlton 2018). The electricity and gas needed by structures in cities for lights, heating, and cooling generate high emission levels. In addition, urban areas operate as locations for massive consumption, which leads to enormous volumes of garbage, including food scraps, industrial waste, and plastics. Such trash, specifically found in garbage landfills, releases methane, an extremely harmful greenhouse gas (Nižetić et al. 2019), which amplifies climate change and causes detrimental impacts on public health.

On the other hand, although urban environments function as the major contributors to climate change, they also present as the worst victims of climate change as they face all its repercussions, namely by suffering the effects of heat islands and rising temperatures, air pollution, rising sea level, extreme weather phenomena, and water scarcity, to name a few (Chapman et al. 2017; Saxena 2025). Urbanization and climate change further fuel air quality deterioration as cities frequently suffer from increased volumes of air contaminants such as sulfur dioxide, particulate matter, and nitrogen oxides, which can culminate in chronic obstructive pulmonary disease (COPD), pneumonia, stroke, lung cancer, and heart disease (Kaur and Pandey 2021). In addition, cities frequently encounter major strain on water supply systems because of their large populations. The adverse consequences of climate change, notably alterations in precipitation patterns, worsening droughts, and decreased snowpack, aggravate the shortage of safe and clean water across various urban regions and have resulted in more than one million deaths annually due to health impacts alone (Saxena 2025). Furthermore, urban areas inadequately prepared or insufficiently set up for resilience to extreme weather conditions are susceptible to substantial losses and exposure to injuries, diseases, and disorders.

Urbanization and climate change are often directly proportional and interrelated in a cycle of feedback that worsens the destruction of the environment and causes hardship for humans by way of resource strain, social inequity, and health and economic consequences. Enhanced urbanization imposes a burden on energy, water, and food supplies, hampering municipalities' resilience to changing climates. Informal groups within towns, occasionally lacking sufficient services and amenities, are at the most risk from climate change as they have a greater chance of suffering from bad air quality, excessive temperatures, floods, and epidemics of illnesses, thereby deepening social and economic disparities (Pasquini et al. 2020). The increasing incidence of climate-induced catastrophic events, in combination with the increased demand for urban healthcare systems, contributes to soaring public health spending and economic damage. Significant expenses unfairly harm less wealthy urban neighborhoods that are less ready to face the consequences of climate change. This demonstrates how urban environments function as both contributors to and victims of global climate change and how rapid urbanization exacerbates climate change impacts while intensifying public health challenges.

Climate-Induced Health Risks

Undoubtedly, climate change causes considerable crisis internationally, entailing serious dangers to human health. Prominent health issues brought about by climate change include air pollution, heat stress within urban locations, and the propagation of vector-borne illnesses. While temperatures around the world rise and environmental conditions shift, this disrupts multiple states of health-related variables, particularly air quality, food safety, availability of water, and accessibility to healthcare. Such constraints are especially acute in rapidly expanding urban areas, wherein the adverse consequences of climate change are frequently worsened due to scarce means for adaptability, growing population levels, and unsuitable facilities. Pollutants are further amplified by the changing climate from two fundamental sources: rising temperatures that boost the generation of ground-level ozone causing smog, and climate-induced events such as droughts and wildfires, which amplify the overall number of fine particles covering the atmosphere (Wang et al. 2016).

Pollution of the air has been associated with cardiovascular and respiratory disorders, notably COPD, strokes, asthma, and cardiac arrest, while prolonged exposure could boost the likelihood of cancer of the lungs along with early fatality (Shahriyari et al. 2022). Among speedily urbanizing domains, specifically in nations with low to medium incomes, air pollution resulting from rising transport emission levels, manufacturing operations, and the generation of power commonly represents a major challenge. Such areas typically exhibit suboptimal air quality along with lack of healthcare premises, keeping citizens at risk from pollution-related diseases. Cities in eastern, southern, and central Asia, along with central and eastern Europe, suffer a substantially increased rate of illness caused by ambient (outdoor) fine particle air pollution (PM2.5) exposure, as seen in Table 1.1 (Health Effects Institute 2022), which can result in disability and early death from several illnesses, notably respiratory infections, ischemic heart disease, COPD, stroke, lung cancer, diabetic complications, and poor birth results after prolonged exposure.

Soaring global temperatures rapidly bring about greater severity and frequency of heatwaves. City regions, notably those with heightened population density, are very susceptible to the urban heat island impact, whereby constructions, highways, and various other structures take in and store heat with greater efficiency compared to the natural environment, which causes drastically greater temperatures in comparison to countryside settings (Heaviside 2020). Heat stress within cities has been linked to numerous negative health effects, such as cardiovascular strain, heatstroke, dehydration, and heat exhaustion. Young children, senior citizens, and others with existing health problems constitute the most endangered, especially when long-term exposure to extremely high temperatures brings higher chances of fatalities. Among the strongest examples of a city facing regular heat stress caused by its climate is Phoenix, Arizona, since the very layout of the city may lead to hazardous heat traps, which can be confirmed by its record of having 197 heat-related fatalities in 2019 alone (Denny 2021). Increasingly, urbanizing populations have become more exposed to heat stress due to increasing population numbers, the shortage of open natural areas, and constrained infrastructures such as cooling facilities or air conditioning (Pereira et al. 2024).

Climate change has shifted the spread of vector-borne illnesses worldwide, accounting for 17% of all infectious diseases transmitted through species including ticks, mosquito larvae, and parasites, causing approximately 700,000 yearly deaths (WHO 2024). Infections, including dengue virus, plague, malaria, and tick-borne illnesses such as Lyme disease, continue to spread across fresh territories as an

Table 1.1 Death rates attributed to PM2.5 in 2019 per regional urban media with deaths per 100,000 (Health Effects Institute 2022)

Region	Global urban median
Asia, East	106
Asia, South	86
Asia, Central	78
Europe, Central	77
Europe, Eastern	77
Sub-Saharan Africa, West	50
Asia, Southeast	48
North Africa/Middle East	43
Asia Pacific, high income	36
Europe, Western	26
Sub-Saharan Africa, central	23
Latin America, Central	19
Sub-Saharan Africa, southern	19
Latin America, Andean	16
North America, high income	15
Latin America, tropical	15
Latin America, Southern	14
Caribbean	14
Sub-Saharan Africa, East	11
Australasia	10
Oceania	1

aftermath of amplified rainfall, raised temperatures, and exacerbated weather-related issues, which has allowed a greater range of environments for vectors and elongated their reproduction cycle (Yadav and Upadhyay 2023). Diseases such as malaria, with 249 million global cases, and dengue, with 3.9 billion people affected globally (WHO 2024), contribute to severe deaths and disabilities, and other ailments could develop once vectors disseminate across new territories. Vector-borne health issues pose an alarming danger to health in numerous regions worldwide, notably within tropical and subtropical domains such as cities in Africa, which have vector-borne diseases like malaria serving as their major public health problem, increasing at an alarming rate (Obame-Nkoghe et al. 2023). Poor-quality sewerage systems, overpopulation within city slums, and poor garbage disposal systems exacerbate the chance of epidemic outbreaks (Ogbonna et al. 2024).

Human health is significantly damaged and continues to be so because of the effects of climate change through air pollution (Shahriyari et al. 2022), heat stress (Heaviside 2020), and the increasing incidence of vector-borne illnesses (Yadav and Upadhyay 2023), which are especially acute across rapidly growing urban areas. Such regions suffer from struggles including rising population levels, confined medical facilities, inequality in socioeconomic status, and a lack of infrastructure, exposing citizens to becoming increasingly prone to health consequences brought about by changes in the climate (Smith et al. 2022; Jessel et al. 2019). Increased urbanization within nations with low-to-middle incomes generally amplifies the health threats resulting from climate change. If no action is taken against urbanization and climate change, not only does overall health decline, but it also leads to death, physical and mental illnesses, and increasingly catastrophic weather events. In order to address these issues, proper integrated urban planning for health and resilience is needed.

This chapter explores the interconnections between urbanization, climate change, and public health, emphasizing the role of urban planning in fostering climate-resilient and sustainable cities. Against this background, the main aim of this study is to analyze climate-induced health risks in urban areas and identify key adaptation and mitigation strategies that enhance urban resilience. For this purpose, a literature review was conducted, examining existing research on urbanization, climate change adaptation, mitigation strategies, and their impact on public health. The findings aim to provide urban planners and policymakers with actionable insights to enable them to develop climate-resilient cities that safeguard both the environment and public health.

Materials and Methods

This chapter draws on a wide range of literature. The chapter was conducted using a desktop research methodology that entailed the compilation and examination of pre-existing data and materials from electronic sources. This approach is increasingly favored in empirical studies due to the ease of access to a substantial array of secondary digital data (Balogun et al. 2022). Utilizing online resources allows for the accumulation of various forms of content, such as articles, reports, studies, databases, and other pertinent materials, aiding in the fulfillment of the research aims. The methodological framework of the research was structured into three iterative stages as shown in Fig. 1.1: (a) initial scoping, (b) data collection, and (c) detailed data analysis. The purpose of the review was to address the research objective and identify key interventions and adaptation strategies that enhance urban resilience and sustainability.

In the data collection stage, extensive literature and data were sourced from prominent databases such as Google Scholar, Scopus, and Web of Science, all of which are key repositories of scholarly publications. The search for literature was steered by specific keywords that included "urbanization," "climate change," "health," "sustainable cities," "resilient cities," "urban resilience," "climate change adaptation," "climate change mitigation," "resilient strategies," "governance," "urban planning," and "disaster risk reduction", among other similar subjects. These keywords helped narrow the search scope and define the study's parameters. To maintain relevance and quality, selection criteria were implemented, focusing on full-text journal articles, book chapters, and conference papers published in English, recently released, and aligned with the study's objectives. The study also included literature such as

Fig. 1.1 Diagram illustrating the desktop research method (adapted and modified from Balogun et al. 2022)

statistical data and technical reports sourced from institutional archives, websites of international development organizations, and government agencies to ensure a well-rounded understanding of the topic.

The final analysis strategy was a two-step iterative content analysis of the collected data. Resources were first organized in accordance with the search parameters, even if some items fell into more than one section. Due to a dearth of data, we iterated through each source multiple times to corroborate our hypotheses and unearth new angles. Ultimately, we blended the themes by looking at their connections and finding common ground (Vourvachis and Woodward 2015). Reviewing the available material carefully led to the discovery of several key concepts necessary for building climate-resilient and healthy cities. The content analysis method was used to categorize the primary ideas into three overarching themes, which are presented in the Results and Discussion section. The desktop research method offered an efficient and thorough strategy for collecting and evaluating pertinent data and literature for this investigation. It stands out as an economical, adaptable, and time-saving technique for data compilation and analysis in research projects (Balogun et al. 2022). Figure 1.1 depicts the workflow of the desktop research methodology utilized in this study.

Results and Discussion: Building Climate-Resilient and Healthy Cities

Urbanization and climate change are deeply interconnected, with cities facing increasing challenges related to environmental sustainability, public health, and climate resilience. As urban populations grow, the need for integrated planning approaches that address both climate adaptation and mitigation increases. This section explores key strategies for enhancing urban resilience and sustainability, focusing on the following:

Integrated Urban Planning for Health and Resilience

Integrated urban planning for health and resilience represents a framework designed for restructuring urban areas towards greater sustainability and healthier and more climate-resilient communities (UNEP 2025). The notion encourages growth in cities and simultaneously targets climate adaptation, mitigates environmental damage, and optimizes public health effects. This multifaceted plan calls on political leaders, urban planners, and the general public to move toward a comprehensive viewpoint, supporting developments in cities that are dealing with climate change concerns while promoting the quality of life of every local dweller. In the process of combining climate resilience alongside public health targets, cities could boost security, sustainability, and quality of life while simultaneously reducing the ecological harm they cause (Crane et al. 2021). Integrated urban planning strives to improve public health by minimizing the ecological causes associated with sickness and cultivating good health by means of enhanced living circumstances (Van den Bosch and Bird 2018). Following the quickly urbanizing domains, wherein population growth commonly transcends infrastructure-based limitations, integrated urban planning becomes vital for maintaining sustainability in the long run.

In addition to the above, integrated urban planning needs to include both adaptability to the shifting climate and mitigation of possible climate change repercussions (Kabisch et al. 2016). The overall components mentioned are interrelated and deserve to be coordinated with the goal of improving the resilience of city conditions. Urban planning strategies that simultaneously address climate adaptation and mitigation of environmental impacts while enhancing public health outcomes include multiple facets such as green infrastructure, low-carbon transportation, air-quality improvement, flood resilience, energy efficiency in buildings, urban heat island mitigation, and renewable energy, to name a few. In terms of green infrastructure, green spaces, parks, and urban woodlands alleviate the implications of urban heat islands, promote the beneficial utilization of excessive rainfall, and create locations for relaxation, common welfare, and psychological wellness. In addition, the building of readily available parks, neighborhood gardens, and outdoor spaces stimulates active lifestyles and boosts mental wellness through lowering anxiousness, depression, and stress whilst concurrently assisting with urban cooling and poor air purification, thereby additionally enhancing the health of the community overall (Mei and Jun 2023).

For low-carbon transportation, campaigning for alternative modes of transportation, such as cycling, public transit, and walking by means of carefully constructed transit-oriented initiatives decreases emissions from vehicles, minimizes pollution in the atmosphere, and reduces the demand for fossil fuels (Wismans et al. 2015). For air quality improvement, adopting measures to mitigate pollution in the air, notably fostering the use of cleaner modes of transport and electric automobiles, decreasing the number of emissions from industrial processes, and planting trees that filter out contaminants, may assist in avoiding the development of cardiovascular and respiratory medical conditions (Erickson and Brase 2020). Another strategy, flood resilience, involves a procedure that optimizes stormwater drainage systems and builds protection from flooding, notably through green roofs, porous pavements, and catchment ponds, thereby supporting communities in dealing with increasing water levels and more frequently occurring extreme weather conditions (Li and Burian 2023).

Energy-efficient building strategies, whether through retrofitting older structures for efficiency in energy use or creating new constructions utilizing power-efficient technological advances, consisting of solar power systems, improved insulating materials, and power-conserving equipment, could significantly lessen the carbon footprint of urban neighborhoods (El-Darwish and Gomaa 2017). Another strategy, urban heat island mitigation, entails planting trees, expanding green areas, and utilizing reflecting materials in structures, which may minimize urban heat islands, thereby diminishing the number of heat-related disorders and significantly reducing energy usage (Akbari and Kolokotsa 2016). Renewable energy is another strategy that involves the deployment of sustainable energy resources such as wind, solar power, and geothermal energies, which mitigates the release of greenhouse gases and accelerates urban progression in the direction of a less destructive energy environment (Al-Shetwi 2022).

Urban zones, despite sheltering a significant proportion of the globe's population, contribute substantially to ecological damage. A careful urban layout requires the incorporation of initiatives to lower the ecological footprint of urban areas. This calls for comprehensive, multidisciplinary approaches to reducing cities' environmental footprints, which has been observed to be attainable through sustainable resource use, circular economy, and sustainable urban mobility. Through sustainable resource use, the efficient management of energy, water, and materials contributes to urban planning by promoting practices such as landfill reduction, water conservation, and composting, thereby offsetting waste and reducing pressure on the Earth's resources (Nižetić et al. 2019). Following the circular economy structure, municipalities may embrace initiatives that emphasize resource recycling and waste reduction, thereby generating closed-loop processes that ameliorate adverse environmental impacts (Kara et al. 2022). Such a concept requires inspiring organizations to put emphasis on sustainable design, advocate life-cycle principles for supplies, and limit wastage within cities. Under the sustainable urban mobility model, electric automobile systems, interconnected mass transit systems, and ergonomically constructed pedestrian paths ultimately bring a considerable decrease in carbon footprints and pollutants in the air coming from transportation, which is a leading environmental degradation factor within cities (Biahmou 2019).

Urban Climate Adaptation and Mitigation Strategies

As urbanization increases and climate change becomes more intense, urban areas are continuously exposed to numerous hazards associated with climate change, including heatwaves, droughts, floods, elevated sea levels, and severe weather conditions (Consalo 2022). As such, strategies to adapt urban areas to climate change become necessary, and notable strategies include resilient infrastructure, renewable energy and efficient urban design, green spaces, and sustainable mobility. The concept of resilient infrastructure has become critical for towns and cities to continue functioning and rehabilitate from such issues as they contribute to climatic adaptation attempts. Because the system provides fundamental amenities as well as assisting in communities' preparedness for and recuperation from catastrophes caused by nature, resilient infrastructure has become a crucial strategy within urban areas. Resilient infrastructure further supports localities by controlling climate change's adverse consequences while fostering a sustainable future with improved public health and economic prosperity (Mehvar et al. 2021).

Resilient infrastructure signifies the development, execution, and ongoing upkeep of infrastructure, including lines of transportation, water distribution networks, energy grids, structures, and transmission systems, which are resistant to climate-related stresses and promptly bounce back from unexpected occurrences (Seager et al. 2017). By bringing together adaptability, reliability, and ecological responsibility within the infrastructure framework, communities could boost resilience against climate-related hazards and preserve the availability of crucial amenities amid catastrophic circumstances. Significant aspects of resilient infrastructure are flood security and management of stormwater systems, water distribution and management of wastewater, energy systems and clean energy facilities, urban green spaces and ecosystem-based responses, and climate-resilient transport connections (Karamouz et al. 2019; Khodadad et al. 2023; Bica and Tucan 2021; Almulhim and Cobbinah 2024). The adoption of tactics that build resilient infrastructures brings benefits such as less exposure to hazards associated with climate change, better health for the public, economic security and expansion, conservation, and responsibility toward the environment, as well as equitable social outcomes.

When metropolitan regions increase and climate change grows more intense, urban areas encounter a two-fold obstacle, which entails lowering their carbon footprint whilst addressing the mounting consequences of climate change (Abubakar and Dano 2020). Renewable energy and efficient design have been crucial instruments for creating climate-resilient urban areas, as these solutions lessen climate change by minimizing the release of greenhouse gases as well as strengthening the city' potential for adapting to growing climate problems, including extreme temperatures, rainfall, and energy scarcity (Bagheri et al. 2019). Renewable energy options such as wind power, solar power, hydroelectric power, and geothermal resources present a secure and environmentally friendly replacement for energy from fossil fuels. Making the switch to renewable energy constitutes an important aspect of urban adaptation to the climate, supporting both climate-change adaptation and mitigation and also enabling communities to cut their carbon footprint, while guaranteeing a stable and consistent energy source (Olabi and Abdelkareem 2022).

On the other hand, efficient urban design highlights maximizing the efficient utilization of resources such as land, power, and water, whilst limiting negative ecological outcomes (Ness and Xing 2017). Effective urban planning tackles the obstacles presented by climate change by improving efficiency in energy use, decreasing emission levels, and strengthening climate-related resilience. The strategy entails having buildings that are energy-efficient, green spaces in cities, adaptable transport networks, climate-resistant construction, smart infrastructure and systems for energy management, water-efficient structures, compact mixed-use urban development, and climate-responsive planning for cities. The combined use of renewable energy for environmentally conscious power and efficient urban design for optimal utilization of resources represents an integrated plan to construct climate-resilient communities that have the means to withstand the adverse consequences of climate change, helping to minimize emissions, lessen climate hazards, and boost the extent to which they can quickly adapt to the shifting environment.

Metropolitan areas have become more exposed as a result of a large density of inhabitants, dependence on unsustainable infrastructure, and inadequate green spaces. In order to reduce such dangers, green spaces and sustainable mobility options are growing as fundamental mechanisms for preparing cities for climate change issues. Such methods foster climate adaptation, elevate the health of the community, minimize carbon footprints, and transform the broad quality of life within urban regions (Vidal et al. 2020). Green spaces, comprising green roofs, parks, urban woodlands, greenhouses, and green passageways, serve as vital for the adaptation of urban centers against climate change since they deliver an abundance of societal, ecological, and financial advantages that are critical to boosting the defense of cities from climatic concerns. The strategy entails minimizing the occurrence of urban heat islands, lowering dangers from flooding and controlling stormwater for irrigation, enriching biodiversity, improving air quality, and cultivating psychological wellness and social stability (Saaroni et al. 2017).

On the other hand, sustainable mobility comprises modes of transport that are efficient in terms of energy, are inclusive, and have low-emission initiatives (Zawieska and Pieriegud 2018). Reducing the demand for personal vehicles and pushing for less polluted, healthier means of travel remains an essential plan for adapting communities to the adverse consequences of climate change. The particular approach supports the decline in emissions of greenhouse gases, reinforces resilience against extreme events, decreases pollution in the atmosphere, lessens congestion on roads, optimizes mobility, and encourages equality among people (Paradoulakis 2024). To optimize both strategies' advantages, urban areas ought to utilize a coordinated plan that matches green spaces with sustainable mobility to boost adaptability, living conditions, and environmental quality within urban settings. For instance, setting up green mobility passageways that bring green infrastructure featuring plants, landscapes, and parks together with alternative modes of transport, including biking paths and electric transit vehicles, may result in climate adaptation rewards whilst persuading people to use public transit and lowering their dependency on autos.

Urban Resilience and Health-Focused Systems

Alongside the rise in population density and the growing severity of changes in the climate, urban areas suffer from increased concerns involving risks to health, which are further amplified by conditions in the environment, inequality in socioeconomic status, and deficient infrastructure. Resilient systems, which are intended to survive, be rehabilitated from, and respond to disturbances and stressors, are fundamental in alleviating threats to health within metropolitan areas (Sharifi and Yamagata 2018). The social and economic drivers of health, as well as any direct health-improving measures undertaken at the institutional health sector level, within community settings, and at home, are all a part of the resilient health system. The six fundamental components of the resilient health system paradigm identified, as seen in Fig. 1.2, are as follows: health leadership and governance, health workforce, health information management system, health service delivery, medical products, vaccines and technologies, and health financing (Olu 2017).

The coordination of the six fundamental components of the resilient health system aids in alleviating any threats to health. Through resilient systems, protection of critical assets, better health, and maintenance of operations amidst negative and unexpected conditions are achieved. As such, to achieve health systems recovery, health disaster preparedness, emergency health response, and health disaster risk re-education, specific actions must be taken to address health vulnerabilities in urban settings. These resilient systems entail having a climate-resilient infrastructure (Vinke-De Kruijf et al. 2024), reliable health systems (Ansah et al. 2024),

Fig. 1.2 Resilient health system framework for public health disaster risk management (adapted and modified from Olu 2017)

green infrastructure (Hewitt et al. 2020), sustainable urban mobility and transportation (Nieuwenhuijsen 2020), social protection and equity (Costella et al. 2023), and climate-resilient food systems (Amoak et al. 2022). These resilient systems are further explored in the following paragraph, which aids in pointing out how each would reduce health vulnerabilities in cities and achieve resilient health systems.

Climate-resilient infrastructure, encompassing water distribution networks, transportation, housing, and electrical power lines, substantially affects the health of the community since urban areas are more able to efficiently survive extreme weather events and defend the health of their citizens when such systems are resilient to climate (Vinke-De Kruijf et al. 2024). On the other hand, a reliable health system consists of one that is capable of tackling health hazards, avoiding interruptions in the supply of healthcare, and preserving its ability to operate through situations of emergency (Ansah et al. 2024). The systems under this component alleviate health risk factors by assuring the readiness and availability of crucial medical services, particularly during disasters. Green infrastructure generates several health advantages including improved air quality and lowered health implications of smog; transformed emotional wellness and mental health, specifically in economically disadvantaged areas with constrained opportunities to enjoy nature; and being able to beneficially consume excess precipitation, hence lowering the adverse consequences of heavy downpours and preserving neighborhoods from flood-related complications for health (Hewitt et al. 2020).

Another resilient system, sustainable urban mobility and transportation, strengthens public health by minimizing pollution in the air, expanding accessibility to healthcare, and encouraging regular physical exercise (Nieuwenhuijsen 2020). In addition, resilient systems also deal with economic and social drawbacks, given that less fortunate and poor communities usually face the highest chance of facing environmental dangers and health consequences. This includes working on affordable housing, accessible health care for all, particularly during crises, and universal health coverage and initiatives involving relief during disasters (Costella et al. 2023). In another case, climate-resilient food systems have become vital for lowering food poverty and malnutrition, specifically in the aftermath of climate-induced perturbations involving extreme weather events (Amoak et al. 2022). By boosting these resilient systems, urban areas could do more to guard the health of the public, expand disaster preparedness, and guarantee the future wellness and health of all their citizens.

Strategic planning has become crucial for shielding urban areas from the damaging consequences of climate change within towns and cities; by projecting hazards related to climate and preemptively pursuing responses, strategic planning could build resilience, decrease flaws, and elevate the broader stability of communities (Wolf and Floyd 2017). It is well understood that strategic planning permits neighborhoods to acknowledge and learn about any prospective environmental hazards that could be encountered, notably when toxic air, flooding, heatwaves, and unexpected weather incidents take place. Through these cases, strategic planning assists societies in pinpointing flaws, looking at possible outcomes, and selecting steps for minimizing hazards associated with climate change (Firehock and Walker 2015).

Moreover, while climate adaptation indicates alterations to urban facilities' operations and endeavors with the goal of reducing the damaging effects of climate change, strategic planning has the role of incorporating the adaptation strategies into the complete development of urban regions, whether it be through water management systems or flood-resilient infrastructure, to name a couple of examples.

Strategic planning infuses climate resilience within every element of urban growth and development, as opposed to viewing it as a mere second thought, thus strengthening populations' adaptation to potential climate-related struggles (Kenny 2020). Moreover, strategic planning serves as vital for allowing communities to respond to existing threats from climate change while sustaining resilience over the long haul. This includes reinforcing urban systems and infrastructure in order to weather developing and unforeseen climactic conditions. Furthermore, whereas preliminary expenses for climate resilience tactics might appear costly, strategic planning for mitigation and adaptation to climate change regularly produces financial advantages over the long run. Overall, this strategy promotes citizens' quality of life while simultaneously ensuring an ecologically sound and productive urban setting for the next generation.

Conclusion

In general, the present chapter underscores the actions that are necessary to create resilient and sustainable cities whilst considering the interrelation between climate change, urbanization, and health. The research has highlighted the benefits of integrated urban planning strategies, which factor in careful consideration of both public health progress and climate-change challenges. The chapter has emphasized the adverse environmental impacts of urban areas; analyzed fundamental aspects of urban adaptation towards climate change, mitigation strategies, and resilience; and indicated the manner in which meticulous planning could help mitigate the adverse health impacts of climate change caused by urbanization. While advocating strategies for adaptation and resilient urban systems to combat global ecological problems, the present study has placed itself within the larger conversation about sustainability as well as urban health.

Despite the current study's exploration of health impacts and strategies for sustainable and resilient cities amidst growing urbanization and the outcomes of climate change, there is an important role that can be identified involving collaborative frameworks between governments, urban planners, and health experts in order to address global environmental challenges effectively (Carter et al. 2015). Considering the growing complexities of ecological issues, particularly climate change, lack of water, air contamination, and public health catastrophes, the demand for collaborative frameworks that connect crucial stakeholders such as governing bodies, city planners, and medical experts has grown urgently indispensable (Frantzeskaki et al. 2019). Such difficulties are multidimensional and interlinked, demanding an exhaustive and integrative strategy for their successful management.

Collaboration among crucial stakeholders cultivates an alignment that could accelerate the establishment of creative responses, accelerate decision-making, and push major steps on matters about climate change that have implications for city dwellers and the public's general health. Partnership across these entities facilitates organized planning and policy, encourages international cooperation and trading of information, brings together social, environmental, and health results, boosts public engagement and representation, facilitates creative strategies for sustainable cities, supports data-sharing and decision-making based on evidence, and solidifies climate equity and justice (Hughes 2015; Bodin 2017; Brink and Wamsler 2018). For instance, sustainability issues regularly harm those with disadvantaged backgrounds, such as individuals with low incomes, citizens of color, and vulnerable populations; yet nevertheless, a collaborative approach will ensure that policy regarding the environment, as well as health programs, will be drawn up with equality for all at the top of the list (Rudolph et al. 2020).

Furthermore, organized planning and policy creation must have clear, current information, as the use of collaborative systems enables the collation of information spanning several fields—notably climate information, urban infrastructure reviews, and statistics on health—to assist with decision-making processes (Corgo et al. 2024). In order to encourage the efficient transformation of the environment, it has also become necessary to encourage the public to adopt environmentally conscious behaviors and boost awareness of the health hazards corresponding to ecological challenges; hence, a process of collaboration may be useful in creating intended public relations tactics that alert urban residents regarding the relationship between public health and the sustainability of the environment (Kiss et al. 2022). Overall, actually confronting global ecological issues demands synchronized, multidimensional techniques that integrate governing bodies, health professionals, and urban planners. Through this process, collaborative frameworks permit contributors to construct incorporated, environmentally conscious approaches that jointly target climate change, urbanization, and the health of the public.

Whenever cities increase and advance, developers must incorporate health and environmental sustainability within the planning process in order to create situations that encourage the welfare of each and every citizen whilst confronting the major global problems of climate change. By integrating health and sustainability through city design, urban areas could boost their capacity for resilience and long-term sustainability, thus enhancing the quality of life for existing residents and future generations to come. Combining sustainability and health throughout the planning process is crucial for growing populations that are accessible and adaptable to climate change, public health emergencies, and limited resources. While emphasizing green infrastructure, sustainable modes of transport, climate-resilient systems, health consequence reviews, and collaborative leadership, urban areas could develop more inclusive, healthier, and more sustainable settings to benefit all residents. Governmental bodies, health professionals, city planners, and communities need to carry out these efforts cooperatively to ensure that metropolitan areas have the necessary resources for the years ahead.

References

Abubakar IR, Dano UL (2020) Sustainable urban planning strategies for mitigating climate change in Saudi Arabia. Environ Dev Sustain 22(6):5129–5152. https://doi.org/10.1007/s10668-019-00417-1

Akbari H, Kolokotsa D (2016) Three decades of urban heat islands and mitigation technologies research. Energy Build 133:834–842. https://doi.org/10.1016/j.enbuild.2016.09.067

Almulhim AI, Cobbinah PB (2023) Can rapid urbanization be sustainable? The case of Saudi Arabian cities. Habitat Int 139:102884. https://doi.org/10.1016/j.habitatint.2023.102884

Almulhim AI, Cobbinah PB (2024) Framing resilience in Saudi Arabian cities: On climate change and urban policy. Sustain Cities Soc 101:105172. https://doi.org/10.1016/j.scs.2024.105172

Al-Shetwi AQ (2022) Sustainable development of renewable energy integrated power sector: trends, environmental impacts, and recent challenges. Sci Total Environ 822:153645. https://doi.org/10.1016/j.scitotenv.2022.153645

Amoak D, Luginaah I, McBean G (2022) Climate change, food security, and health: harnessing agroecology to build climate-resilient communities. Sustainability 14(21):13954. https://doi.org/10.3390/su142113954

Ansah EW, Amoadu M, Obeng P, Sarfo JO (2024) Health systems response to climate change adaptation: a scoping review of global evidence. BMC Public Health 24(1):2015. https://doi.org/10.1186/s12889-024-19459-w

Bagheri M, Delbari SH, Pakzadmanesh M, Kennedy CA (2019) City-integrated renewable energy design for low-carbon and climate-resilient communities. Appl Energy 239:1212–1225. https://doi.org/10.1016/j.apenergy.2019.02.031

Balogun AL, Adebisi N, Abubakar IR, Dano UL, Tella A (2022) Digitalization for transformative urbanization, climate change adaptation, and sustainable farming in Africa: trend, opportunities, and challenges. J Integr Environ Sci 19(1):17–37

Biahmou A (2019) Systems engineering for sustainable mobility. Systems engineering in research and industrial practice: foundations, developments and challenges. https://doi.org/10.1007/978-3-030-33312-6_13

Bica I, Tucan L (2021) Green infrastructure and climate resilience. In: The Palgrave handbook of climate resilient societies. Springer International Publishing, Cham, pp 1–34. https://doi.org/10.1007/978-3-030-32811-5_47-1

Bodin Ö (2017) Collaborative environmental governance: achieving collective action in social-ecological systems. Science 357(6352):eaan1114. https://doi.org/10.1126/science.aan1114

Brink E, Wamsler C (2018) Collaborative governance for climate change adaptation: mapping citizen–municipality interactions. Environ Policy Gov 28(2):82–97. https://doi.org/10.1002/eet.1795

Carter JG, Cavan G, Connelly A, Guy S, Handley J, Kazmierczak A (2015) Climate change and the city: building capacity for urban adaptation. Progress Plann 95:1–66. https://doi.org/10.1016/j.progress.2013.08.001

Charlton E (2018) These are the cities with the biggest carbon footprints. World Economic Forum. https://www.weforum.org/stories/2018/07/these-are-the-cities-with-the-biggest-carbon-footprints/#:~:text=But%20a%20new%20study%20showing,University%20of%20Science%20and%20Technology

Chapman S, Watson JE, Salazar A, Thatcher M, McAlpine CA (2017) The impact of urbanization and climate change on urban temperatures: a systematic review. Landscape Ecol 32:1921–1935. https://doi.org/10.1007/s10980-017-0561-4

Clarke B, Otto F, Stuart-Smith R, Harrington L (2022) Extreme weather impacts of climate change: an attribution perspective. Environ Res Clim 1(1):012001. https://doi.org/10.1088/2752-5295/ac6e7d

Consalo K (2022) Vulnerable populations: climate change and extreme weather threats facing urban communities. Chi.-Kent J Env't Energy L 11:1. https://heinonline.org/HOL/LandingPage?handle=hein.journals/jeel11&div=5&id=&page

Corgo J, Cruz SS, Conceição P (2024) Nature-based solutions in spatial planning and policies for climate change adaptation: a literature review. Ambio 53(11):1599–1617. https://doi.org/10.1007/s13280-024-02052-1

Costella C, van Aalst M, Georgiadou Y, Slater R, Reilly R, McCord A et al (2023) Can social protection tackle emerging risks from climate change, and how? A framework and a critical review. Clim Risk Manag 40:100501. https://doi.org/10.1016/j.crm.2023.100501

Crane M, Lloyd S, Haines A, Ding D, Hutchinson E, Belesova K et al (2021) Transforming cities for sustainability: a health perspective. Environ Int 147:106366. https://doi.org/10.1016/j.envint.2020.106366

Denny E (2021) Cities are at a greater risk of heat stress, new study finds. World Economic Forum. https://www.weforum.org/stories/2021/01/cities-greater-risk-of-heat-stress-study/

Du J, Fu Q, Fang S, Wu J, He P, Quan Z (2019) Effects of rapid urbanization on vegetation cover in the metropolises of China over the last four decades. Ecol Ind 107:105458. https://doi.org/10.1016/j.ecolind.2019.105458

Dyvik EH (2024) Share of urban population worldwide in 2023, by continent. Statista. https://www.statista.com/statistics/270860/urbanization-by-continent/

El-Darwish I, Gomaa M (2017) Retrofitting strategy for building envelopes to achieve energy efficiency. Alex Eng J 56(4):579–589. https://doi.org/10.1016/j.aej.2017.05.011

Erickson LE, Brase G (2020) Reducing greenhouse gas emissions and improving air quality: two interrelated global challenges. Taylor & Francis, p 178. https://doi.org/10.1201/9781351116589

Firehock K, Walker RA (2015) Strategic green infrastructure planning: a multi-scale approach. Island Press. https://books.google.com.ph/books?hl=en&lr=&id=BV2XCgAAQBAJ&oi=fnd&pg=PR2&dq=strategic+planning+assists+societies+in+pinpointing+holes,+looking+at+possible+outcomes,+and+selecting+steps+for+minimizing+hazards+associated+with+climate+change&ots=6K6B67TSVl&sig=p6LuXn5R5tlNh6daRxfAEm9qFY8&redir_esc=y#v=onepage&q&f=false

Frantzeskaki N, McPhearson T, Collier MJ, Kendal D, Bulkeley H, Dumitru A et al (2019) Nature-based solutions for urban climate change adaptation: linking science, policy, and practice communities for evidence-based decision-making. Bioscience 69(6):455–466. https://doi.org/10.1093/biosci/biz042

Fuller R, Landrigan PJ, Balakrishnan K, Bathan G, Bose-O'Reilly S, Brauer M et al (2022) Pollution and health: a progress update. Lancet Planetary Health 6(6):e535–e547. https://doi.org/10.1016/S2542-5196(22)00090-0

Greenwalt J, Raasakka N, Alverson K (2018) Building urban resilience to address urbanization and climate change. In: Resilience. Elsevier, pp 151–164. https://doi.org/10.1016/B978-0-12-811891-7.00012-8

Health Effects Institute (2022) Air quality and health in cities: a state of global air report 2022. Health Effects Institute, Boston, MA. https://www.stateofglobalair.org/sites/default/files/documents/2022-08/2022-soga-cities-report.pdf

Heaviside C (2020) Urban heat islands and their associated impacts on health. Oxford University Press. https://doi.org/10.1093/acrefore/9780199389414.013.332

Hewitt CN, Ashworth K, MacKenzie AR (2020) Using green infrastructure to improve urban air quality (GI4AQ). Ambio 49(1):62–73. https://doi.org/10.1007/s13280-019-01164-3

Hughes S (2015) A meta-analysis of urban climate change adaptation planning in the US. Urban Clim 14:17–29. https://doi.org/10.1016/j.uclim.2015.06.003

IQAir (2025) World's most polluted cities. https://www.iqair.com/world-most-polluted-cities?srsltid=AfmBOoqpTqFWU8o8aKolXdiaoLnhFu8KSMI7B8KLieOG0kfU6i3zTYfp

IPCC (2022) Climate change 2022: impacts, adaptation, and vulnerability. In: Pörtner H-O, Roberts DC, Tignor M, Poloczanska ES, Mintenbeck K, Alegría A, Craig M, Langsdorf S, Löschke S, Möller V, Okem A, Rama B (eds) Contribution of working group II to the sixth assessment report of the intergovernmental panel on climate change. Cambridge University Press, Cambridge, UK and New York, NY, USA, pp 3056. https://doi.org/10.1017/9781009325844

Jessel S, Sawyer S, Hernández D (2019) Energy, poverty, and health in climate change: a comprehensive review of an emerging literature. Front Publ Health 7:357. https://doi.org/10.3389/fpubh.2019.00357

Jordan R (2020) Stanford researchers show how forest loss leads to spread of disease. Standford Report. https://news.stanford.edu/stories/2020/04/understanding-spread-disease-animals-human

Kabisch N, Frantzeskaki N, Pauleit S, Naumann S, Davis M, Artmann M et al (2016) Nature-based solutions to climate change mitigation and adaptation in urban areas: perspectives on indicators, knowledge gaps, barriers, and opportunities for action. Ecol Soc 21(2). https://www.jstor.org/stable/26270403

Kara S, Hauschild M, Sutherland J, McAloone T (2022) Closed-loop systems to circular economy: a pathway to environmental sustainability? CIRP Ann 71(2):505–528. https://doi.org/10.1016/j.cirp.2022.05.008

Karamouz M, Taheri M, Khalili P, Chen X (2019) Building infrastructure resilience in coastal flood risk management. J Water Resour Plan Manag 145(4):04019004. https://doi.org/10.1061/(ASCE)WR.1943-5452.0001043

Kaur R, Pandey P (2021) Air pollution, climate change, and human health in Indian cities: a brief review. Front Sustain Cities 3:705131. https://doi.org/10.3389/frsc.2021.705131

Kenny M (2020) Planning in practice for resilience and climate change in extreme and extreme-ing urban environments, Doctoral dissertation, University of Warwick. http://webcat.warwick.ac.uk/record=b3678561

Khalili S, Rantanen E, Bogdanov D, Breyer C (2019) Global transportation demand development with impacts on the energy demand and greenhouse gas emissions in a climate-constrained world. Energies 12(20):3870. https://doi.org/10.3390/en12203870

Khodadad M, Aguilar-Barajas I, Khan AZ (2023) Green infrastructure for urban flood resilience: a review of recent literature on bibliometrics, methodologies, and typologies. Water 15(3):523. https://doi.org/10.3390/w15030523

Kiss B, Sekulova F, Hörschelmann K, Salk CF, Takahashi W, Wamsler C (2022) Citizen participation in the governance of nature-based solutions. Environ Policy Gov 32(3):247–272. https://doi.org/10.1002/eet.1987

Li J, Burian SJ (2023) Evaluating real-time control of stormwater drainage network and green stormwater infrastructure for enhancing flooding resilience under future rainfall projections. Resour Conserv Recycl 198:107123. https://doi.org/10.1016/j.resconrec.2023.107123

Mehvar S, Wijnberg K, Borsje B, Kerle N, Schraagen JM, Vinke-de Kruijf J et al (2021) Towards resilient vital infrastructure systems–challenges, opportunities, and future research agenda. Nat Hazards Earth Syst Sci 21(5):1383–1407. https://doi.org/10.5194/nhess-21-1383-2021

Mei L, Jun W (2023) Urban environment, green spaces, and mental health: an interdisciplinary investigation. J Human Appl Sci Res 6(4):14–35. https://journals.sagescience.org/index.php/JHASR/article/view/85

Mohajerani A, Bakaric J, Jeffrey-Bailey T (2017) The urban heat island effect, its causes, and mitigation, with reference to the thermal properties of asphalt concrete. J Environ Manage 197:522–538. https://doi.org/10.1016/j.jenvman.2017.03.095

Ness DA, Xing K (2017) Toward a resource-efficient built environment: a literature review and conceptual model. J Ind Ecol 21(3):572–592

Nieuwenhuijsen MJ (2020) Urban and transport planning pathways to carbon neutral, liveable and healthy cities; a review of the current evidence. Environ Int 140:105661. https://doi.org/10.1016/j.envint.2020.105661

Nižetić S, Djilali N, Papadopoulos A, Rodrigues JJ (2019) Smart technologies for promotion of energy efficiency, utilization of sustainable resources and waste management. J Clean Prod 231:565–591. https://doi.org/10.1016/j.jclepro.2019.04.397

Obame-Nkoghe J, Makanga BK, Zongo SB, Koumba AA, Komba P, Longo-Pendy NM et al (2023) Urban green spaces and vector-borne disease risk in Africa: The case of an unclean forested

park in libreville (Gabon, Central Africa). Int J Environ Res Public Health 20(10):5774. https://doi.org/10.3390/ijerph20105774
Ogbonna SN, Ochie CN, Aniwada EC (2024) Urban slum housing quality, and its public health implications in Nigeria: a case of urban slum residents in Enugu metropolis, South East, Nigeria. BMC Publ Health 24(1):3231. https://doi.org/10.1186/s12889-024-20764-7
Olabi AG, Abdelkareem MA (2022) Renewable energy and climate change. Renew Sustain Energy Rev 158:112111. https://doi.org/10.1016/j.rser.2022.112111
Olu O (2017) Resilient health system as conceptual framework for strengthening public health disaster risk management: an African viewpoint. Front Publ Health 5:263. https://doi.org/10.3389/fpubh.2017.00263
Paradoulakis A (2024) Sustainable mobility, climate change and geopolitics. https://doi.org/10.26267/unipi_dione/4093
Pasquini L, van Aardenne L, Godsmark CN, Lee J, Jack C (2020) Emerging climate change-related public health challenges in Africa: a case study of the heat-health vulnerability of informal settlement residents in Dar es Salaam, Tanzania. Sci Total Environ 747:141355. https://doi.org/10.1016/j.scitotenv.2020.141355
Pereira C, Flores-Colen I, Mendes MP (2024) Guidelines to reduce the effects of urban heat in a changing climate: green infrastructures and design measures. Sustain Dev 32(1):57–83. https://doi.org/10.1002/sd.2646
Rudolph L, Maizlish N, North S, Dervin K (2020) A public health learning collaborative on climate change for urban health departments, 2016–2018. Public Health Rep 135(2):189–201. https://doi.org/10.1177/0033354920902468
Saaroni H, Amorim JH, Hiemstra JA (2017) Urban green infrastructure as a tool for mitigating the urban heat island-a review of research methods. In: Book of abstracts green infrastructure: nature based solutions for sustainable and resilient cities, p 88. https://www.greeninurbs.com/wp-content/uploads/2017/04/book_of_abstract.pdf#page=81
Saxena V (2025) Water quality, air pollution, and climate change: investigating the environmental impacts of industrialization and urbanization. Water Air Soil Pollut 236(2):1–40. https://doi.org/10.1007/s11270-024-07702-4
Seager TP, Clark SS, Eisenberg DA, Thomas JE, Hinrichs MM, Kofron R et al (2017) Redesigning resilient infrastructure research. In: Resilience and risk: methods and application in environment, cyber and social domains. Springer, Netherlands, pp 81–119. https://doi.org/10.1007/978-94-024-1123-2_3
Shahriyari HA, Nikmanesh Y, Jalali S, Tahery N, Zhiani Fard A, Hatamzadeh N et al (2022) Air pollution and human health risks: mechanisms and clinical manifestations of cardiovascular and respiratory diseases. Toxin Reviews 41(2):606–617. https://doi.org/10.1080/15569543.2021.1887261
Sharifi A, Yamagata Y (2018) Resilience-oriented urban planning. Resilience-oriented urban planning: theoretical and empirical insights, 3a–27. https://doi.org/10.1007/978-3-319-75798-8_1
Smith GS, Anjum E, Francis C, Deanes L, Acey C (2022) Climate change, environmental disasters, and health inequities: the underlying role of structural inequalities. Curr Environ Health Rep 9(1):80–89. https://doi.org/10.1007/s40572-022-00336-w
UNEP (2025) Integrated urban planning. UN environment programme. https://www.unep.org/topics/cities/integrated-planning/integrated-urban-planning#:~:text=Integrated%20planning%20is%20a%20powerful,%2C%20housing%2C%20health%20and%20biodiversity
Van den Bosch M, Bird W (eds) (2018) Oxford textbook of nature and public health: the role of nature in improving the health of a population. Oxford University Press. https://books.google.com.ph/books?hl=en&lr=&id=vr9EDwAAQBAJ&oi=fnd&pg=PP1&dq=).+Integrated+urban+planning+strives+at+improving+public+health+through+minimizing+the+ecological+causes+associated+with+sickness+and+cultivating+good+health+by+means+of+enhanced+living+circumstances&ots=wYJMuniwlH&sig=sEb2fiFipul-bsQoh_2PATsTlgQ&redir_esc=y#v=onepage&q&f=false

Vourvachis P, Woodward T (2015) Content analysis in social and environmental reporting research: trends and challenges. J Appl Acc Res 16(2):166–195

Vidal DG, Barros N, Maia RL (2020) Public and green spaces in the context of sustainable development. In: Sustainable cities and communities. Springer International Publishing, Cham, pp 479–487. https://doi.org/10.1007/978-3-319-95717-3_79

Vinke-De Kruijf J, Heim LaFrombois ME, Warbroek B, Morris JC, Kuks SM (2024) Climate-resilient water infrastructure: a call to action. J Crit Infrastruct Policy. https://doi.org/10.1002/jci3.12017

Wang X, Sun D, Yao T (2016) Climate change and global cycling of persistent organic pollutants: a critical review. Sci China Earth Sci 59:1899–1911. https://doi.org/10.1007/s11430-016-5073-0

World Health Organization (WHO) (2024) Vector-borne diseases. https://www.who.int/news-room/fact-sheets/detail/vector-borne-diseases

Wismans J, Grahn M, Denbratt I (2015) Low-carbon transport–health and climate benefits. In: Background paper for the intergovernmental ninth regional environmentally sustainable transport (EST) forum in Asia, pp 17–20. https://publications.lib.chalmers.se/records/fulltext/233671/233671.pdf

Wolf C, Floyd SW (2017) Strategic planning research: toward a theory-driven agenda. J Manag 43(6):1754–1788. https://doi.org/10.1177/0149206313478185

Yadav N, Upadhyay RK (2023) Global effect of climate change on seasonal cycles, vector population and rising challenges of communicable diseases: a review. J Atmos Sci Res 6(1):21–59. https://doi.org/10.30564/jasr.v6i1.5165

Yoro KO, Daramola MO (2020) CO_2 emission sources, greenhouse gases, and the global warming effect. In: Advances in carbon capture. Woodhead Publishing, pp 3–28. https://doi.org/10.1016/B978-0-12-819657-1.00001-3

Zawieska J, Pieriegud J (2018) Smart city as a tool for sustainable mobility and transport decarbonisation. Transp Policy 63:39–50. https://doi.org/10.1016/j.tranpol.2017.11.004

Zhao Q, Guo Y, Ye T, Gasparrini A, Tong S, Overcenco A et al (2021) Global, regional, and national burden of mortality associated with non-optimal ambient temperatures from 2000 to 2019: a three-stage modelling study. Lancet Planet Health 5(7):e415–e425. https://doi.org/10.1016/S2542-5196(21)00081-4

Chapter 2
Expert Perspectives on Climate Change and Public Health in Port Cities: A Case Study of Mumbai

Abeer Sanyour and Sadashiv Chaturvedi

Abstract Port cities like Mumbai are particularly vulnerable to the escalating impacts of climate change, which include rising sea levels, more frequent cyclones, and intensified flooding. These environmental challenges exacerbate public health risks such as heat stress, air pollution, and the spread of waterborne diseases, especially in low-lying and densely populated urban areas. This study aims primarily to identify the most pressing issues related to climate change and its impact on public health by deriving criteria for dealing with climate change and examining expert evaluations from urban planners, public health professionals, and environmental scientists regarding the health-related impacts of climate change, in Mumbai, a major coastal city with critical socio-economic and environmental significance. Moreover, link the expert's assessments to their field of experience. Using an importance-performance analysis framework, this research assesses experts' perspectives on key Climate Change indicators, revealing (1) Significant discrepancies between the perceived importance and actual performance of current climate adaptation strategies, with "Water recycling" emerging as the most pressing issue needing improvement. (2) Experts exhibited varied views on vulnerable regions, with urban planners highlighting infrastructure and spatial vulnerabilities in coastal or flood-prone areas. Public health professionals emphasized the need for enhanced disease prevention and heat stress mitigation across the city or within neighborhoods near the port. (3) The research underscores the importance of formulating integrated, interdisciplinary policies that prioritize climate resilience and public health within the urban planning processes of port cities. The study provides critical insights into how Mumbai and similarly situated port cities can better address the health risks posed by climate change. The findings are intended to guide policymakers in developing targeted strategies that bolster public health systems and infrastructure resilience, promoting sustainable urban development in a changing climate.

A. Sanyour (✉)
Department of Architecture, Planning and Design, IIT BHU, Varanasi, India
e-mail: abeersanyour.rs.apd23@itbhu.ac.in

S. Chaturvedi
School of Hydrology and Water Resources, Nanjing University of Information Science and Technology, Nanjing (NUIST), Nanjing, China

© The Author(s), under exclusive license to Springer Nature Switzerland AG 2025

A. I. Almulhim et al. (eds.), *Urbanization, Climate Change, and Health*, Climate Change Management, https://doi.org/10.1007/978-3-031-99132-5_2

Keywords Importance-performance analysis · Climate change · Public health · Port cities · Sustainable urban development · Mumbai

Introduction

Climate change is an unparalleled global challenge of the twenty-first century, with profound and far-reaching consequences for ecosystems, economies, and human societies (Masson-Delmotte et al. 2021). Among its numerous impacts, the effects on public health are significant, especially in urban areas where dense populations, industrial activity, and environmental vulnerabilities converge to create unique challenges (Watts et al. 2018). The rapid pace of urbanization further amplifies these challenges by straining already limited infrastructure, leaving populations vulnerable to climate-related health crises such as heat stress, vector-borne diseases, and respiratory illnesses caused by declining air quality. Climate change exacerbates these risks, posing significant threats to the achievement of the Sustainable Development Goals (SDGs), particularly those related to poverty reduction (SDG 1), health and well-being (SDG 3), and sustainable Cities and communities (SDG 11) (FAO 2020; UNDP 2020).

Port cities' vulnerability to climate change's health impacts stems from their unique environmental, social, and economic combination. Coastal regions such as Mumbai, Chennai, and Kolkata face heightened risks of extreme weather events, rising sea levels, and flooding, which have severe implications for public health. These risks are intensified by inadequate urban infrastructure, including poor drainage systems and insufficient healthcare facilities, particularly in informal settlements that house marginalized communities. Such conditions create fertile ground for spreading vector-borne diseases like dengue and malaria, and waterborne illnesses such as cholera and typhoid (World Health Organization 2018). Moreover, the concentration of industrial activity and transportation in port cities contributes to elevated levels of air pollution, further exacerbating respiratory illnesses during heat waves.

In India, the health impacts of climate change are becoming increasingly pronounced. Rising global temperatures, a hallmark of climate change, have contributed to a more favorable environment for vector-borne diseases, with dengue cases surging by over 40% in the past decade, driven by extended monsoon seasons and warmer conditions that enhance mosquito breeding (Ministry of Health and Family Welfare (MoHFW) 2019). The 2019 Kerala floods, exacerbated by climate-induced extreme rainfall, exposed the heightened vulnerability of coastal and low-lying areas, leading to widespread public health emergencies, including outbreaks of waterborne diseases and displacement-related health issues (Ifejube et al. 2024). Moreover, intensifying heatwaves, increasingly linked to anthropogenic climate change, have led to hundreds of deaths annually in urban centers such as Delhi and Ahmedabad, particularly affecting marginalized and elderly populations who lack adequate shelter or access to cooling technologies (de Bont et al. 2024). All these challenges come under the global issue of climate change.

Despite these challenges, the intersection of climate change, public health, and urban planning in port cities remains underexplored, particularly in the context of developing countries like India. This study addresses this gap and employs Importance-Performance Analysis (IPA) as a case study to investigate the relationship between the perceived importance and actual performance of climate change adaptation measures in addressing public health challenges in Mumbai Port City. By identifying criteria for dealing with climate change, this study aims to capture expert perspectives with the IPA method utilized to identify the most pressing issues related to climate change, evaluate discrepancies between the importance and performance of the adaptation strategies discussed in this paper while also analyzing the correlation between the experts' field characteristics (e.g., experience) and their assessments. The findings focus on Mumbai, India, and aim to generalize insights for comparable coastal cities globally. The study attempts to contribute to the growing body of knowledge on the intricate relationship between climate change and public health in urban coastal regions. Despite the critical importance of climate change adaptation measures, current research in this domain remains insufficient compared to other public health areas. Moreover, there is a notable lack of expert-driven assessments evaluating the importance and performance of these adaptation strategies, particularly in coastal and port cities, which face immense pressures on their built environments. This study addresses this research gap by integrating expert input to assess and enhance climate resilience in these vulnerable urban areas and proposing integrated strategies to enhance climate resilience.

Materials and Methods

The methodology was divided into several sections to achieve the research objectives, illustrated in Fig. 2.1.

Significant Challenges of Climate Change in Coastal and Port Cities

Climate change presents significant and multifaceted challenges to global public health as a direct and indirect driver of health risks. Rising temperatures and extreme weather events, such as heat waves, floods, and hurricanes, exacerbate health issues, including heat-related illnesses, respiratory disorders, and injuries. Furthermore, climate change facilitates the spread of vector-borne diseases, such as malaria and dengue, by altering the habitats and lifecycles of disease-carrying organisms such as mosquitoes (Haines et al. 2006).

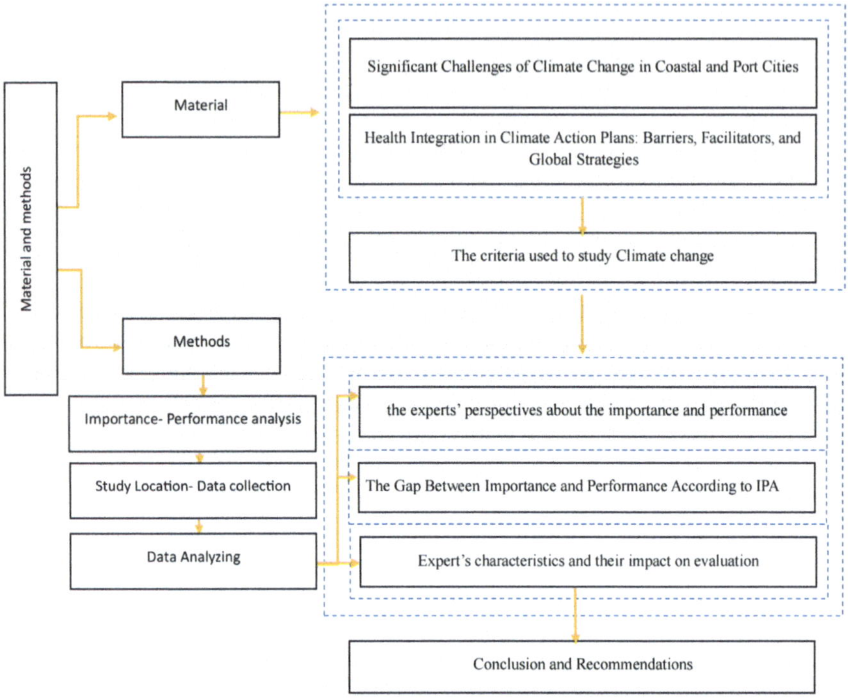

Fig. 2.1 Methodology adopted for selection of studies. Source by Author

Indirectly, climate change affects food security, water quality, and mental health, with marginalized populations bearing the most significant burden due to existing vulnerabilities and limited adaptive capacities (McMichael & World Health Organization 2003). Walimah and others identified the significant health risks associated with climate change, including vector-borne diseases, heat-related illnesses, respiratory disorders, and mental health issues (Walimah et al. 2024). Patil discussed the direct effects of heatwaves on vulnerable populations, particularly the elderly and children in urban areas of developing countries. In these areas, the urban heat island effect often exacerbates air pollution (Patil and Deepa 2007).

Coastal and port cities, with their dense populations, unique geography, and economic activities, face heightened public health risks from climate change. These risks include flooding, air pollution, heat stress, and waterborne diseases. Addressing these challenges necessitates the integration of climate resilience with health system strengthening. Between 1985 and 2018, floods caused by rainfall accounted for 70% of climate-related health impacts (Taji et al. 2024). While sea level rise is generally considered a long-term chronic phenomenon, flooding is often a short-term event driven by weather and geography. In some cases, the two intersect, amplifying the risks and leading to fatalities in coastal areas. Table 2.1 illustrates the contribution of climate change to public health impacts, with a particular focus on port cities.

Table 2.1 Health impacts of climate change

Climate change	Impact on public health	Reference
The effect of recurrent floods and soil saturation with water	– Spread of waterborne diseases (cholera and diarrhea) – Contamination of drinking water sources	Shahid (2010)
The effect of water on diseases	– Diseases can survive in a moist environment	
Rising river levels	Disruption of sewage systems and the spread of diseases	
Sea level rise	– Public health impacts of floods: disruption of sewage systems and thus the spread of diseases – Increased salinity in freshwater sources and, consequently, its impact on the quality of drinking water – Diarrhea and skin diseases – Vector-borne diseases and creating a fertile environment for the reproduction of mosquitoes, malaria, and dengue fever – Destruction of agricultural land and threat to food security	
Rainfall-induced floods	– Hypertension, 12% of all the reports – Wounds 10.8% minor trauma	Taji et al. (2024)
Floods	– Waterborne disease – Vector-borne diseases – Malaria (floods create stagnant water) – Cholera (the contamination of drinking water) – Scabies and taeniasis (poor sanitation conditions) – Rhodesian sleeping – Sickness	Suhr and Steinert (2022)
Rising sea levels	– Disease vectors such as mosquitoes – Affecting drinking water quality	Porru et al. (2024)
Tropical cyclone frequency	– Intestinal infectious diseases – Sepsis, especially in developing societies – Carbon monoxide poisoning from generator use is notable	Mitchell et al. (2014)

In many developing countries, drinking water is sourced from wells. In Bangladesh, for example, 97% of the population faces health issues due to rising sea levels that release arsenic (a carcinogen) from sediments into wells (Frisbie et al. 2024). Similarly, the Dammam metropolitan area faces significant urban sustainability risks due to rising sea levels and increased flooding, leading to heat islands, groundwater depletion, and exposure to disease epidemics (Dano et al. 2023). Additionally, studies have found a linear relationship between the intensity of tropical cyclones and the increased risk of hospitalization for infectious diseases (Huang et al. 2024).

India, one of the largest coastal nations, is experiencing significant public health challenges from climate change, particularly in its coastal cities and ports. The high population density in these regions intensifies the strain on health systems and infrastructure. In response, India is implementing adaptive strategies, such as strengthening early warning systems and enhancing public health services. Notable examples of extreme weather events include the June 2005 heatwave in Bhubaneswar, Odisha, which reached temperatures of 46.3 °C, and severe droughts in Cherrapunji, a region known for its rainfall. These events, along with cyclones, floods, and rising temperatures, underscore the tangible impacts of global climate change (Patil and Deepa 2007). In July 2005, Mumbai experienced unprecedented flooding, marking one of its most severe weather events. Over 1,000 mm of rainfall within 24 h overwhelmed the city's drainage and sewage systems, submerging significant portions of the city and affecting more than 20% of its area. The disaster caused extensive loss of life and property, highlighting the vulnerability of urban infrastructure to extreme climate events (Ranger et al. 2011). Yadav et al. observed that the Earth's surface temperature rises at 0.3 °C annually. Additionally, rainfall during the monsoon period (June to September) has increased, with 55% of urban areas in Mumbai now vulnerable to flooding (Yadav et al. 2024).

Health Integration in Climate Action Plans: Barriers, Facilitators, and Global Strategies

Studies have emphasized the principle of human rights, adopting a rights-based approach by involving affected individuals and communities in decision-making processes for strategies to mitigate the effects of climate change (Meier et al. 2022). A survey was conducted among members of twelve global health professional organizations to examine the health impacts of climate change. The data, analyzed using SPSS, revealed significant adverse health effects. Notably, heat-related illnesses, vector-borne diseases, and water- and food-borne illnesses were identified as primary concerns, with these categories accounting for 36–47% of reported impacts (Kotcher et al. 2021).

Additionally, a questionnaire-based study employed the Normalization Process Theory to explore barriers and facilitators for incorporating health impacts into climate action plans. This study targeted individuals working in local authorities, utilizing open-ended responses and applying a weighted ranking method to prioritize findings (Dodd et al. 2023).

In response to the climate change crisis and the global mobilization of researchers and nations, various multifaceted strategies and adaptive measures have emerged to mitigate its effects on public health. Bangladesh has integrated climate change into its national development strategy (Shahid 2010). As part of these efforts, the Bangladesh Climate Change Strategy and Action Plan (BCCSAP), implemented from 2009 to 2018, was developed as a long-term strategy alongside the National Adaptation Programme of Action (NAPA), which ensures access to safe water supplies in climate-sensitive areas vulnerable to sea level rise (Hadi 2019). In Pakistan, the focus has been on assessing the resilience of municipal water supply systems, particularly in flood-affected regions. This approach emphasizes community participation and relies on interviews with local water managers to identify gaps and enhance the resilience of essential services (Tufail et al. 2025).

Similarly, Japan has incorporated climate change into its national agenda through the Climate Change Adaptation Act of 2018, which outlines the development of a National Adaptation Plan (NAP) and Local Climate Change Adaptation Plans (LCCAPs). These plans aim to reduce health impacts and minimize exposure to climate change risks by fostering collaboration across various sectors, including healthcare, urban planning, and environmental management (Kim et al. 2023). China has implemented several policies and initiatives to safeguard public health in the context of climate change and its associated health impacts. Notable among these is the "Synergy Implementation Plan for Reducing Pollution and Carbon Emissions," which prioritizes improving urban air quality to mitigate respiratory and cardiovascular diseases. Additionally, the country has focused on reducing risks linked to sea level rise, including floods, erosion, storms, and saltwater intrusion. An estimated 42 million people were identified as being at risk of floods associated with sea level rise by 2023, highlighting the critical need for climate information services to provide relevant and actionable data (Zhang et al. 2023). Australia has also developed a comprehensive approach through its National Health and Climate Strategy, which integrates health considerations into climate change policies. The 2023–24 national budget allocated specific funding to support climate adaptation and mitigation efforts, reinforcing the country's commitment to addressing the intersection of health and climate change (Walter et al. 2024).

Importance-Performance Analysis (IPA)

The Importance-Performance Analysis (IPA) model, introduced by Martilla and James in 1977, evaluates the significance and performance of various attributes from the perspective of consumers or experts. This framework simultaneously assesses the importance of specific factors and their perceived performance, enabling the prioritization of improvements (Martilla and James 1977). The IPA framework employs a

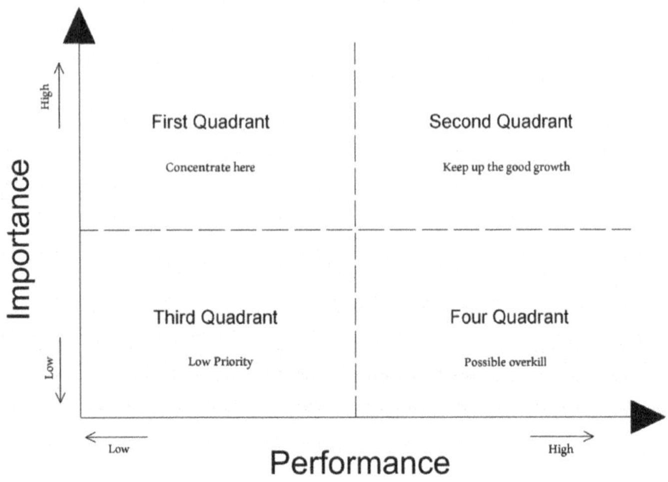

Fig. 2.2 Importance-performance analysis grid (redrawn and adapted from Martilla and James 1977)

grid divided into four quadrants (Fig. 2.2), plotting attributes based on their importance and performance. The positioning of the grid's cross-point (where the axes intersect) can vary depending on the approach used. Martilla and James emphasized that grid lines are subject to judgment, as the IPA concerns relative rather than absolute measures of importance.

- Scale-Centered Approach: The cross-point is centrally on the scale, such as four on a 7-point scale.
- Data-Centered Approach: The cross-point is determined by the mean or median values of the data, relying on calculated mean importance and performance scores.

Typically, importance and performance are measured using 5- or 7-point scales (Bacon 2003). Researchers can identify gaps between users' expectations and their actual experiences by evaluating performance. This is essential for uncovering areas needing improvement and aligning strategies with user needs (Munim et al. 2023). IPA is widely regarded as a data-driven decision-making method that effectively determines development priorities.

Study location:

The city of Mumbai, India, was selected as the case study to examine the phenomenon of climate change and its associated diseases on a global scale. Mumbai's size, population density, historical significance, and functional characteristics closely resemble those of many global port cities expected to face the impacts of climate change. Like other coastal cities in India, Mumbai is projected to be severely affected by rising sea levels by 2050, potentially losing approximately 50% of its land area (Subramanian et al. 2023).

Site Selection and Rationale

This study focuses on two sites within Mumbai (Fig. 2.3): Mazagaon on the eastern coast and Worli on the western coast. These locations were selected based on their high population density and consideration of both historical and modern factors. India ranks seventh among countries most affected by extreme weather events linked to climate change (Hussain et al. 2024). The selected sites are coastal areas particularly vulnerable to the effects of high tides and rising sea levels, making them ideal for addressing global issues faced by ports and coastal cities.

Mazagaon, one of Mumbai's oldest neighborhoods, represents historical importance and modern challenges. Situated on the eastern coast near the Arabian Sea, Mazagaon is characterized by a low elevation above sea level (Khatri et al. 2023). Its proximity to work centers and ports has contributed to the growth of slum settlements in nearby areas. The eastern coastline, including Mazagaon, is highly susceptible to erosion due to natural and human factors.

- Natural Factors: Increased wave energy, rising sea levels, and frequent hurricanes exacerbate coastal erosion.
- Human Factors: The construction of port docks and other infrastructural developments intensifies these challenges (Herlekar et al. 2023).

Worli, strategically located on the western coast of Mumbai, has evolved from a traditional fishing village into a modern urban area with significant commercial

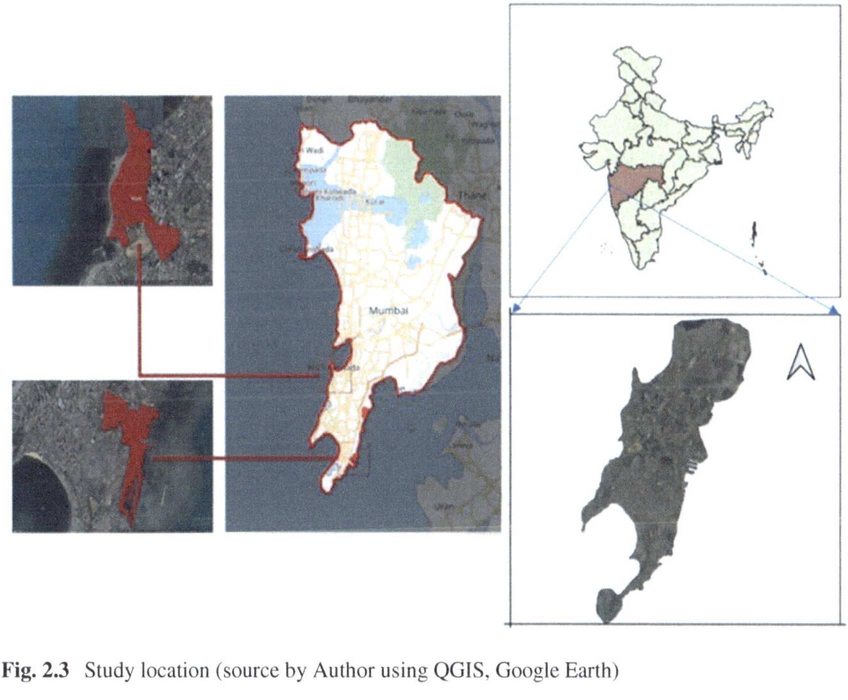

Fig. 2.3 Study location (source by Author using QGIS, Google Earth)

spaces. It forms part of the Lower Parel-Worli business district and embodies urban development's dual challenges and opportunities in a global port city.

According to expert opinion, Worli is highly vulnerable to climate change due to:

- High Sensitivity: Dependency on fishing and low adaptive capacity.
- Context Sensitivity: Vulnerability to rising sea levels and changing weather patterns.
- Health Concerns: Increased cases of malaria and dengue fever among the fishing community in Worli (Senapati and Gupta 2017).

Worli is a microcosm of global challenges in urban coastal areas, illustrating the intricate relationship between climate change, urban development, and public health.

Data Collection: This study used a structured questionnaire to gather expert opinions on climate adaptation measures in Mumbai. The questionnaire comprised four sections. The first section collected socio-demographic data and the experts' professional experiences. The second and third sections focus on assessing the importance and performance of climate adaptation measures in addressing public health challenges in Mumbai. A five-point Likert scale was employed for this purpose (1 = "Strongly Disagree," 2 = "Disagree," 3 = "Neutral," 4 = "Agree," 5 = "Strongly Agree") (Gai et al. 2023). The fourth section invited open-ended responses, allowing participants to propose strategies or measures for adapting to climate change, specifically focusing on Climate Change challenges in Mumbai as a coastal city. The findings aim to inform a similar approach in other global coastal cities.

The criteria used in the questionnaire were developed based on an extensive review of relevant studies addressing the impacts of climate change on coastal cities. These criteria, presented in Table 2.3, were further refined to align with the context of coastal and port cities. Insights from this review also guided revisions to the questionnaire text to enhance its relevance and applicability.

Data was collected through face-to-face interviews in October 2024, with participants selected based on availability and expertise. At the beginning of each interview, the purpose of the questionnaire was explained to participants, as outlined in the introductory section. Questions raised by participants were addressed to ensure clarity and understanding before they proceeded to complete the questionnaire.

This structured approach ensured that the questionnaire captured both quantitative assessments and qualitative insights, providing a comprehensive understanding of climate adaptation measures and their implications for public health in Mumbai and other coastal cities worldwide.

Data Analysis:

Approximately 30 experts were invited to participate in the study. However, after accounting for attendance and willingness to complete the questionnaire, responses were received from 20 participants. The reliability of the data was assessed using Cronbach's alpha coefficient, which was calculated in Excel to ensure the consistency and validity of the responses. Descriptive statistics were employed to analyze the

Table 2.2 Socio-demographic of experts

Variables	Category	Number/percentage
Gender	Male	70% (percentage)
	Female	30% (percentage)
Level of education	Bachelor	4 (number)
	Master	8 (number)
	Doctoral	8 (number)
The average OF years' experience	–	4.5 (number)
Occupation	Urban planners	6
	University academics (lecturers)	3
	Environmental specialists	3
	Health specialists (doctors)	5
	Disaster management specialists	3

socio-demographic characteristics of the participants, including education levels, professional experience, and areas of expertise.

The participants were categorized into five professional groups: urban planners, university academics (lecturers), health specialists (doctors), environmental specialists, and disaster management specialists. The average professional experience among respondents was 4.5 years, with male experts constituting 70% of the sample. A summary of the participants' socio-demographic characteristics is presented in Table 2.2.

An Analysis of Variance (ANOVA) (Ntumi 2021) It was conducted to explore statistical relationships and detect significant effects. The importance-performance Analysis (IPA) framework was also used to evaluate the factors and criteria outlined in Table 2.3. In the IPA analysis, performance scores were plotted on the X-axis, and importance scores on the Y-axis. The mean value method was applied to determine the grid's cross-point, dividing the data into four quadrants (Nuryasman and Nuringsih 2020). This approach facilitated prioritizing improvement areas and aligning results with strategies tailored to stakeholders' perspectives.

Results and Discussion: Aligning Experts' Perspectives with Climate Change

Identify Pressing Climate-Health Issues in Mumbai by IPA in the two sites (Mazagaon and Worli):

The data presented in Table 2.4 highlights that experts place significant importance on infrastructure services that mitigate the spread of pollution associated with climate change, contributing to the prevalence of various diseases. Specifically, the factor

Table 2.3 The criteria used in the questionnaire were developed based on literature studies

Strategy	Factor/criteria	Sub criteria	Reference
Dealing with climate change	Adapting to sea level rise	Sustainable drainage systems	Vigar (2009)
		Water recycling (rainwater percentage recovered)	Girard (2011)
	Pollution/air, water, and noise	Roads dedicated to smooth mobility or reducing emissions	Medeiros and van der Zwet (2020)
		The quality of water	Giovinazzi and Moretti (2009)
		The quality of the air	Ogara et al. (2023)
	Urban environment	Open spaces created or rehabilitated in urban areas	Medeiros and van der Zwet (2020)
		Increase afforestation and reforestation	United Nations (2015)
		Effectively regulate harvesting	

"roads dedicated to smooth mobility or reducing emissions" received an importance rating of 3.9. This aligns with prior studies emphasizing the substantial number of diseases transmitted through contaminated water in Mazagaon (Deshpande et al. 2009).

In Mazagaon, the experts assigned the highest importance values in water quality, sustainable drainage systems, and water recycling (measured as the percentage of rainwater recovered).

For the second site, Worli, experts attributed high importance to similar factors, including water quality, water recycling (rainwater percentage recovered), and air quality. Notably, the performance value given to the air quality in Mazagaon was lower than Worley, and this is consistent with studies that attributed air pollution at the Mazagaon site to the proximity of the shipyard, where shipping activity takes place (Korhale et al. 2023).

The importance of sustainable drainage systems was rated as 4.35 for the first site (Mazagaon) and 3.8 for the second site (Worli), corresponding performance gaps of 1.55 and 1.15. Although experts considered it an essential factor for both sites, its performance was notably weak at the Mazagaon site, leading to dissatisfaction among the experts. Water recycling (measured by the percentage of rainwater recovered) was identified as one of the highest-gap criteria for both sites, indicating a significant difference between the perceived importance and the current performance. The open spaces created or rehabilitated factor showed moderate importance and performance, with the gap between the two being relatively small, suggesting a balanced evaluation

Table 2.4 The results of performance and importance ratings

Criteria for climate change and public health		Mean of importance		Mean of performance		Gap	
		Mazagaon	Worli	Mazagaon	Worli	Mazagaon	Worli
A	Sustainable drainage systems	4.35	3.8	2.75	2.65	1.55	1.15
B	Water recycling (rainwater percentage recovered)	4.3	4.05	2.4	2.5	1.85	1.55
C	Roads dedicated to smooth mobility or reducing emissions	3.85	3.9	2.5	3.1	1.35	0.8
D	The quality of water	3.9	4	2.8	2.65	0.75	1.35
E	The quality of the air	3.4	3.95	2.35	2.6	1.1	1.35
F	Open spaces created or rehabilitated in urban areas	3.3	3.7	2.55	2.7	0.75	1
G	Increase afforestation and reforestation	3.25	3.65	2.45	2.9	0.75	0.75
H	Effectively regulate harvesting	2.8	3.35	2.4	2.55	0.3	0.8

from the experts. Effective regulation of harvesting was considered a low priority at the first site, but was deemed relevant at Worli, reflecting context-specific concerns and priorities (Brihanmumbai Municipal Corporation 2022).

Figure 2.4 presents expert opinions for both sites, categorized into four quadrants based on the importance and performance of various factors.

First Quadrant: "Concentrate here": This quadrant includes factors B (water recycling, i.e., rainwater percentage recovered) and C (roads dedicated to smooth mobility or reducing emissions), considered critical but low performance. These factors require immediate attention and improvement.

Second Quadrant: "Keep Up the Good Work": This quadrant includes factors A (sustainable drainage systems) and D (quality of water), which experts see positively,

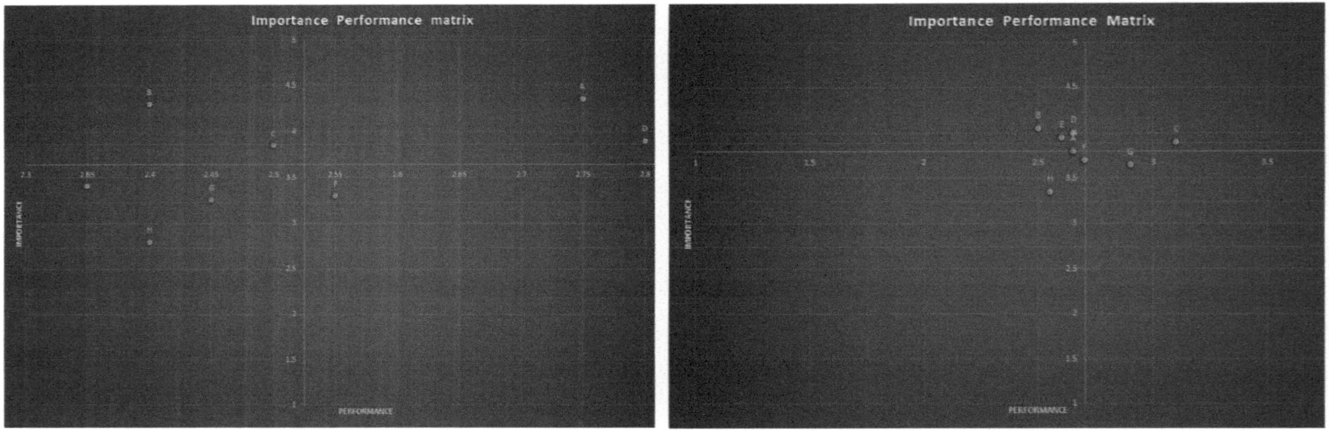

Fig. 2.4 IPA scatter plots for eight criteria for climate change and public health sub-items for all respondents. The right plot shows the results for Worli site, while the left plot shows the results for Mazagaon site (source by Author using IPA matrix)

as their performance aligns with their importance. Experts are satisfied with the current performance of these factors. Experts at both sites agreed on the importance of factor B, while for the second site (Worli), factors E, D, and A also fell into this quadrant, indicating that experts were satisfied with these. Factor C showed satisfactory performance at the second site (Worli).

Third Quadrant: "Low Priority": For both sites, factor H (effectively regulating harvesting) was classified in the "low priority" quadrant, with low importance and low performance, indicating minimal focus from the experts.

Fourth Quadrant: "Possible Overkill": Only factor F (open spaces created or rehabilitated) appeared in this quadrant for the first site. Experts did not deem this factor highly important despite its poor performance. In the second site (Worli), this factor moved slightly closer to the third quadrant, indicating some improvement in perception. Factor G (increase afforestation and reforestation) was also added to the third quadrant in the second site, suggesting that, although important, its performance still needs attention.

The Gap Between Importance and Performance According to IPA:

The Importance-Performance Analysis (IPA) results reveal notable discrepancies between the evaluated criteria and factors' perceived importance and actual performance.

These gaps highlight areas requiring strategic focus and improvement.

Key Findings:

1. Sustainable Drainage Systems: This factor was rated as highly important, reflecting its critical role in managing urban floods and promoting water sustainability. However, its poor performance on the first site underscores the challenges in its implementation and the urgent need for improvement.
2. Organized Harvesting: This criterion was assigned low priority, particularly at Mazagaon, as experts' interests focused more on other pressing issues (e.g., urban water management or transportation initiatives).
3. Significant gaps in water recycling and air quality criteria were observed, signaling an urgent need for strategic interventions. Conversely, the gap between transportation methods and emissions reduction was relatively small, owing to targeted urban transport initiatives implemented at Worli (Chawla et al. 2025).

Expert Characteristics and Their Impact on Evaluation Results:

This study not only identifies the experts' preferences regarding the importance and performance values of factors selected to evaluate climate change adaptation strategies in Mumbai but also seeks to understand the specific influence of the occupation of experts as a key social characteristic. The aim is to assess how their professional backgrounds shape their perspectives on climate adaptation.

Multiple linear regression analysis was conducted to assess the influence of expert characteristics on the evaluation of climate adaptation factors. This analysis aimed to examine the relationship between a set of independent variables, precisely, education level, occupation, and gender, and the dependent variable, which in this research refers to the experts' views on the importance or performance of climate adaptation factors.

The model's fit was assessed using R^2 values, which indicate how well the regression model explains the variance in the data (Ozili 2023). The results showed no significant associations between the social and functional characteristics of the experts and their perspectives on the importance and performance of the eight climate adaptation criteria for the coastal city. Specifically, the R^2 values were 0.15 for Mazagaon (the first Site) and 0.18 for Worli (the second Site), suggesting a low explanatory power of the model.

A correlation was found between expert characteristics and the importance of the "sustainable drainage systems" factor. However, no correlation was observed between expert characteristics and the importance of the "water quality" factor. Additionally, expert characteristics were linked to performance and satisfaction ratings for the following factors: sustainable drainage systems and "effectively regulated harvesting."

Regarding *p*-values, which assess the statistical significance of expert characteristics in the model, occupation was found to have a notable impact on expert assessments at the first location, with a *p*-value of approximately 0.78. In contrast, at the second site (Worli), education level significantly influenced the importance of factors related to climate adaptation strategies—Table 2.5.

The expert groups were divided into four categories based on occupation. A notable difference emerged upon comparing the average importance values for the factors selected by each group. Environmental experts assigned the highest average importance values to Factor D (Water Quality) (Fig. 2.5). In contrast, disaster management experts gave the highest values to Factors E (The quality of the air) and B (Water recycling (rainwater percentage recovered)). Factor H (effectively regulates harvesting) consistently gave the lowest average importance value across all expert categories.

Regarding performance assessments, health experts rated Factor G (increase afforestation and reforestation) as having the highest average importance values. However, there were no significant differences in the average importance values chosen by the experts from the second location (Worli) (Fig. 2.6).

Table 2.5 The results of ANOVA

	Variable	Importance		Performance	
		P-value	*F*	*P*-value	*F*
Mazagaon	Gender	0.57	0.22	0.43	1.07
	Education	0.96		0.46	
	Occupation	0.78		0.26	
Worli	Gender	0.26	1.16	0.13	1.30
	Education	0.96		0.41	
	Occupation	0.17		0.47	

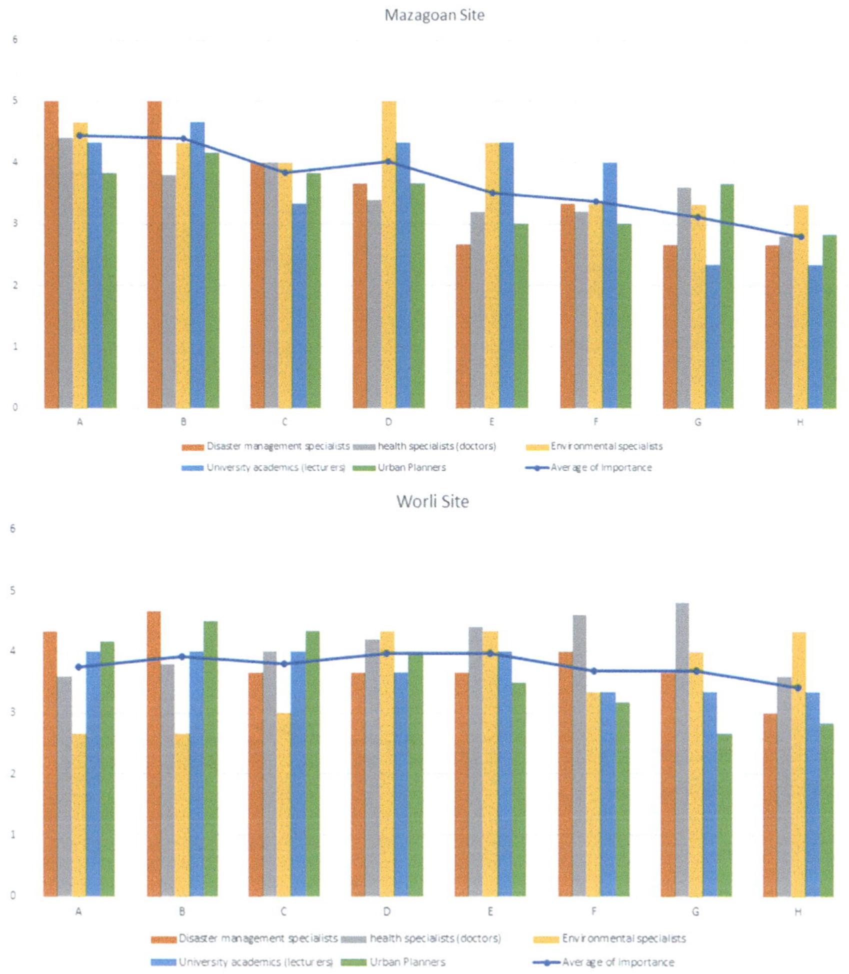

Fig. 2.5 Importance values for eight criteria for climate change and public health

Despite the observed variations in the importance ratings, all expert groups demonstrated some differences in the performance ratings of the eight factors. Urban planning experts rated the performance of Factor D (The quality of water) as very poor. Additionally, all experts agreed that the performance of Factor G (increase afforestation and reforestation) was below average, indicating that implementing air quality measures is perceived as insufficient.

The lack of significant differences in experts' choices in this study may be attributed to the homogeneity of the sample. All experts share a common understanding of the risks and effects of climate change on public health. Furthermore, the minor variations in preferences across the groups reflect the experts' tendency to prioritize aspects of climate adaptation that are most closely aligned with their

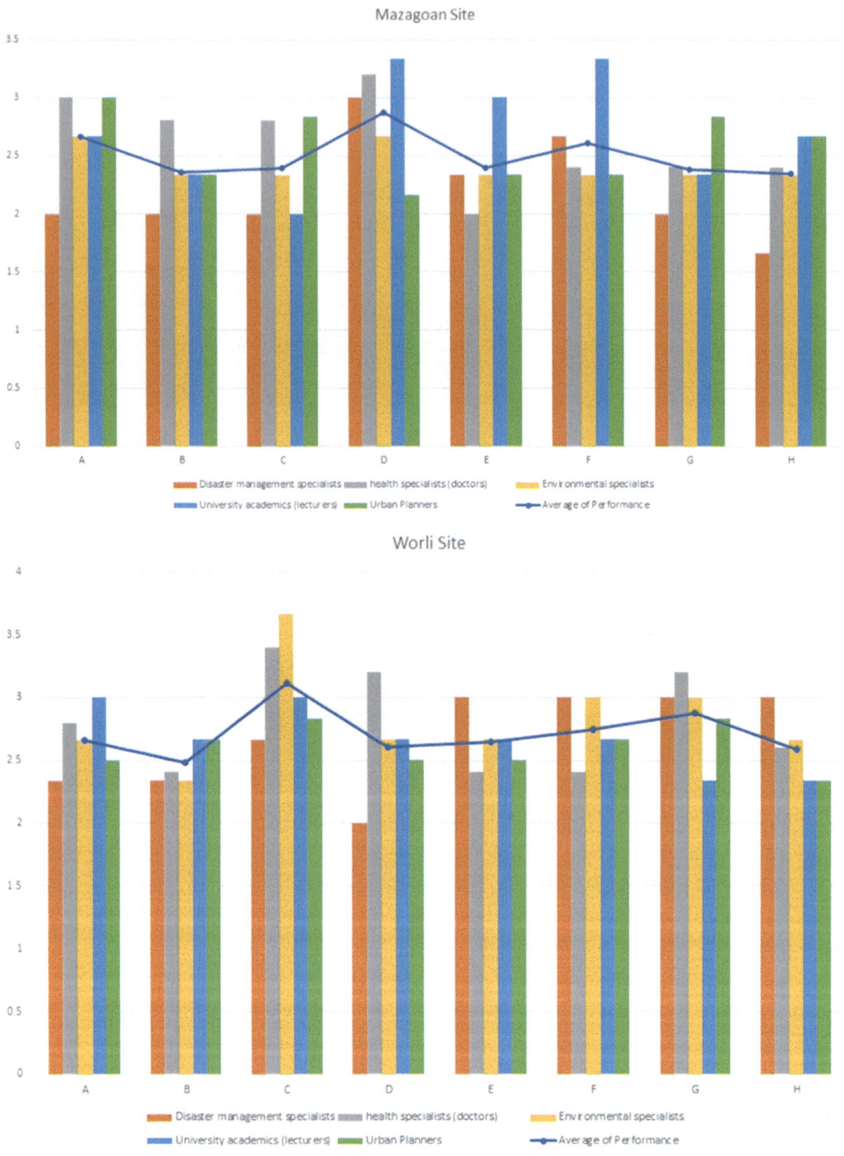

Fig. 2.6 Performance values for eight criteria for climate change and public health

specific professional expertise and experience. In order to enhance the reliability and relevance of the questionnaire results, priority is given to selecting experts whose field of work directly aligns with the study's thematic areas. This approach ensures that the insights gathered are contextually valid and professionally informed.

Recommendations by Experts: The experts provided additional insights on factors and criteria not included in the questionnaire but deemed essential for addressing challenges in coastal and port cities:

1. Renewable Energy Adoption: Experts emphasized the importance of adopting renewable energy solutions, aligning with India's ongoing efforts to implement such measures in both newly developed and existing cities (Sawhney 2021).
2. Early Warning Systems: Developing advanced early warning systems was recommended to mitigate the severity and impact of floods. This aligns with the need for resilient infrastructure (Dash 2024)
3. Healthcare Infrastructure: Health experts stressed the importance of distributing and diversifying healthcare facilities to address diseases associated with climate change. They also recommended placing these facilities at safe distances from residential areas, particularly those at higher risk of climate-related impacts (Cheng and Sha 2024).
4. Stakeholder Collaboration: Experts agreed on the critical role of enhanced stakeholder cooperation in driving sustainable initiatives. This would ensure the success of projects and contribute to the protection and resilience of coastal cities.

By addressing these gaps and incorporating the recommendations, urban planners and policymakers can strengthen the adaptation strategies necessary for ensuring sustainable development and climate resilience in coastal and port cities.

Conclusion

This paper examines the perspectives of experts from diverse professional backgrounds, including urban planners, academics, health specialists, environmental experts, and disaster management professionals, on climate change impacts and risk mitigation strategies in Mumbai. Although climate change is widely studied, particularly in recent years, this study employed the Importance-Performance Analysis (IPA) approach. Addressing our three key objectives, the study:

(1) Identified pressing climate-health priorities through Importance-Performance Analysis (IPA), revealing critical gaps in water recycling and sustainable mobility infrastructure, particularly the disconnect between their high importance and low implementation performance.
(2) Evaluated location-specific disparities in adaptation needs: where Mazagaon's heritage-urban hybrid context prioritized drainage systems, Worli's commercial port activities demanded focused attention on water recycling. These findings validate the need for place-based adaptation frameworks.
(3) Assessed professional bias through cross-disciplinary analysis, finding that while urban planners emphasized infrastructure resilience and public health experts prioritized disease prevention (e.g., heat stress mitigation), the IPA

results showed remarkable consensus on fundamental priorities across all sectors.

The study contributes to climate-resilient urban planning by demonstrating how IPA can bridge disciplinary divides in coastal cities. By aligning the findings with SDG 11 (sustainable cities) and SDG 13 (climate action), this study provides a replicable framework for other high-density port cities facing similar climate health challenges. Future research should expand stakeholder inclusion to validate these expert-driven priorities with community perspectives.

Limitations and Future Directions: A key limitation of this study was its focus on a select group of experts, which can be addressed in future research by incorporating random samples from the general population and experts. This would provide a more comprehensive understanding of varying perspectives. Additionally, future studies could focus on specific aspects of climate change impacts, delving deeper into these areas to lay the groundwork for broader research that includes professionals from various fields.

Acknowledgements ChatGPT was used to edit and improve language, solely to enhance the clarity, coherence, and formatting of the text. Specifically, AI-assisted language processing was employed to refine grammar, improve readability, and ensure consistency in academic writing style. However, all ideas, analyses, interpretations, and conclusions presented in this chapter are entirely own of the authors and based on their original research.

The author Abeer Sanyour acknowledges the Indian Institute of Technology (Banaras Hindu University), Varanasi, India, and Indian Council for Cultural Relations (ICCR), India, for providing a supporting.

References

Bacon DR (2003) A comparison of approaches to importance-performance analysis. Int J Mark Res 45(1):1–15. https://doi.org/10.1177/147078530304500101

Brihanmumbai Municipal Corporation (2022) Home. Mumbai Climate Action Plan. https://mcap.mcgm.gov.in/

Chawla R, Joshi A, Khaleel L (2025) Project management analysis: case of Mumbai metro project as an alternate mode of transportation. In: Varma A, Chand Sharma V, Tarsi E (eds) Proceedings of the 2nd international conference on trends in architecture and construction. Springer Nature, pp 349–366. https://doi.org/10.1007/978-981-97-4988-1_21

Cheng Q, Sha S (2024) Last defense in climate change: assessing healthcare inequities in response to compound environmental risk in a megacity in Northern China. Sustain Cities Soc 115:105886. https://doi.org/10.1016/j.scs.2024.105886

Dano UL, Abubakar IR, AlShihri FS, Ahmed SMS, Alrawaf TI, Alshammari MS (2023) A multi-criteria assessment of climate change impacts on urban sustainability in Dammam Metropolitan Area, Saudi Arabia. Ain Shams Eng J 14(9):102062. https://doi.org/10.1016/j.asej.2022.102062

Dash B (2024) Gap between impact-based and impact forecast and warning: implications for people-centric early warning system (EWS) in India. Nat Hazards Res 4(1):110–117. https://doi.org/10.1016/j.nhres.2023.09.005

de Bont J, Nori-Sarma A, Stafoggia M, Banerjee T, Ingole V, Jaganathan S, Mandal S, Rajiva A, Krishna B, Kloog I, Lane K, Mall RK, Tiwari A, Wei Y, Wellenius GA, Prabhakaran D, Schwartz

J, Prabhakaran P, Ljungman P (2024) Impact of heatwaves on all-cause mortality in India: a comprehensive multi-city study. Environ Int 184:108461. https://doi.org/10.1016/j.envint.2024.108461

Deshpande A, Bhendigeri S, Shirsekar T, Dhaware D, Khandekar RN (2009) Analysis of heavy metals in marine fish from Mumbai Docks. Environ Monit Assess 159(1):493–500. https://doi.org/10.1007/s10661-008-0645-3

Dodd S, Butterfield S, Davies J, Furbo MK, Morris A, Brown H (2023) Assessing the barriers and facilitators of climate action planning in local governments: a two-round survey of expert opinion. BMC Publ Health 23(1). Scopus. https://doi.org/10.1186/s12889-023-16853-8

FAO, 2020; UNDP, 2020 (n.d.) The state of food security and nutrition in the World 2020. FAO, IFAD, UNICEF, WFP and WHO. https://doi.org/10.4060/ca9692en

Frisbie SH, Mitchell EJ, Molla AR (2024) Sea level rise from climate change is expected to increase the release of arsenic into Bangladesh's drinking well water by reduction and by the salt effect. PLoS ONE 19(1 JANUARY). Scopus. https://doi.org/10.1371/journal.pone.0295172

Gai S, Fu J, Rong X, Dai L (2023) Importance–performance analysis and improvement of an urban park's cultural ecosystem services based on users' perspectives: a Beijing case study. J Asian Archit Build Eng 22(2):726–739. https://doi.org/10.1080/13467581.2022.2049800

Giovinazzi O, Moretti M (2009) Port cities and urban waterfront: transformations and opportunities. TeMA J Land Use Mobility Environ. https://doi.org/10.6092/1970-9870/123

Girard LF (2011) Multidimensional evaluation processes to manage creative, resilient and sustainable city. Aestimum 123–139. https://doi.org/10.13128/Aestimum-10464

Hadi T (2019) An analysis of water policies and strategies of Bangladesh in the context of climate change. Asia-Pacific J Rural Develop 29(1):111–123. https://doi.org/10.1177/1018529119860958

Haines A, Kovats RS, Campbell-Lendrum D, Corvalan C (2006) Climate change and human health: impacts, vulnerability and public health. Publ Health 120(7):585–596. https://doi.org/10.1016/j.puhe.2006.01.002

Herlekar MA, Kamble PB, Gawali PB, Hanamgond PT, Aher SP (2023) Quantitative assessment of shoreline changes along the tropical West Coast, Maharashtra, India: a remote sensing and GIS approach. J Earth Syst Sci 132(1):31. https://doi.org/10.1007/s12040-023-02047-8

Huang W, Vogt T, Park J, Yang Z, Ritchie EA, Xu R, Zhang Y, Hales S, Yu W, Hundessa S, Otto C, Yu P, Liu Y, Ju K, Lavigne E, Ye T, Wen B, Wu Y, Kliengchuay W (2024) Risks of infectious disease hospitalisations in the aftermath of tropical cyclones: a multi-country time-series study. Lancet Planet Health 8(9):e629–e639. https://doi.org/10.1016/S2542-5196(24)00158-X

Hussain S, Hussain E, Saxena P, Sharma A, Thathola P, Sonwani S (2024) Navigating the impact of climate change in India: a perspective on climate action (SDG13) and sustainable cities and communities (SDG11). Front Sustain Cities. https://doi.org/10.3389/frsc.2023.1308684

Ifejube OJ, Kuriakose SL, Anish TS, van Westen C, Blanford JI (2024) Analysing the outbreaks of leptospirosis after floods in Kerala, India. Int J Health Geog 23(1):11. https://doi.org/10.1186/s12942-024-00372-9

Khatri S, Kokane P, Kumar V, Pawar S (2023) Prediction of waterlogged zones under heavy rainfall conditions using machine learning and GIS tools: a case study of Mumbai. GeoJournal 88(1):277–291. https://doi.org/10.1007/s10708-022-10731-3

Kim Y, Oka K, Kawazu EC, Ng CFS, Seposo X, Ueda K, Hashizume M, Honda Y (2023) Enhancing health resilience in Japan in a changing climate. Lancet Reg Health Western Pacific 40. https://doi.org/10.1016/j.lanwpc.2023.100970

Korhale N, Anand V, Panicker A, Beig G (2023) Measurements of surface ozone and its precursors in different microenvironments of coastal Indian metropolis of Mumbai. Int J Environ Sci Technol 20(2):2141–2158. https://doi.org/10.1007/s13762-022-03910-9

Kotcher J, Maibach E, Miller J, Campbell E, Alqodmani L, Maiero M, Wyns A (2021) Views of health professionals on climate change and health: a multinational survey study. Lancet Planet Health 5(5):e316–e323. https://doi.org/10.1016/S2542-5196(21)00053-X

Martilla JA, James JC (1977) Importance-performance analysis. J Mark 41(1):77–79. https://doi.org/10.1177/002224297704100112

Masson-Delmotte V, Zhai P, Pirani A, Connors SL, Péan C, Berger S, Caud N, Chen Y, Goldfarb L, Gomis MI, Huang M, Leitzell K, Lonnoy E, Matthews JBR, Maycock TK, Waterfield T, Yelekçi Ö, Yu R, Zhou B (eds) (2021) Climate change 2021: the physical science basis. Contribution of working group I to the sixth assessment report of the intergovernmental panel on climate change. Cambridge University Press. https://doi.org/10.1017/9781009157896

McMichael AJ, World Health Organization (eds) (2003) Climate change and human health: risks and responses. World Health Organization

Medeiros E, van der Zwet A (2020) Sustainable and integrated urban planning and governance in metropolitan and medium-sized cities. Sustainability 12(15), Article 15. https://doi.org/10.3390/su12155976

Meier BM, Bustreo F, Gostin LO (2022) Climate change, public health and human rights. Int J Environ Res Publ Health 19(21), Article 21. https://doi.org/10.3390/ijerph192113744

Ministry of Health and Family Welfare (MoHFW) (2019) Home: National Center for Vector Borne Diseases Control (NCVBDC). https://ncvbdc.mohfw.gov.in/

Mitchell RD, Aitken P, Franklin RC (2014) The physical health impacts of tropical cyclones. Ann Australas College Trop Med 15:2–8

Munim ZH, Maditati DR, Kummer S, Schramm H-J (2023) Cross-country comparison of operant resources in logistics outsourcing relationships. Int J Emerg Mark, ahead-of-print(ahead-of-print). https://doi.org/10.1108/IJOEM-11-2021-1753

Ntumi S (2021) Reporting and interpreting one-way analysis of variance (ANOVA) using a data-driven example: a practical guide for social science researchers. J Res Educ Sci (JRES) XII(14):38–47

Nuryasman MN, Nuringsih K (2020) Using IPA model for ensuring sustainable development in the rural tourism destination, 67–74. https://doi.org/10.2991/assehr.k.201209.008

Ogara DAE, Morishita J, Davies PJ, Mbui M, Gamoyo M, Njoroge N, Chann IO, Igesa BS, Ochieng SO (2023) An indicator-based approach to assess sustainability of port-cities and marine management in the Global South. Front Mar Sci. https://doi.org/10.3389/fmars.2023.1052128

Ozili PK (2023) The acceptable R-square in empirical modelling for social science research. In: Social research methodology and publishing results: a guide to non-native English speakers. IGI Global Scientific Publishing, pp 134–143. https://doi.org/10.4018/978-1-6684-6859-3.ch009

Patil RR, Deepa TM (2007) Climate change: the challenges for public health preparedness and response—an Indian case study. Ind J Occup Environ Med 11(3):113. https://doi.org/10.4103/0019-5278.38460

Porru MC, Arras C, Biddau R, Cidu R, Lobina F, Podda F, Wanty R, Da Pelo S (2024) Assessing recharge sources and seawater intrusion in coastal groundwater: a hydrogeological and multi-isotopic approach. Water (Switzerland) 16(8). Scopus. https://doi.org/10.3390/w16081106

Ranger N, Hallegatte S, Bhattacharya S, Bachu M, Priya S, Dhore K, Rafique F, Mathur P, Naville N, Henriet F, Herweijer C, Pohit S, Corfee-Morlot J (2011) An assessment of the potential impact of climate change on flood risk in Mumbai. Clim Change 104(1):139–167. https://doi.org/10.1007/s10584-010-9979-2

Sawhney A (2021) Striving towards a circular economy: climate policy and renewable energy in India. Clean Technol Environ Policy 23(2):491–499. https://doi.org/10.1007/s10098-020-01935-7

Senapati S, Gupta V (2017) Socio-economic vulnerability due to climate change: deriving indicators for fishing communities in Mumbai. Marine Policy 76:90–97. Scopus. https://doi.org/10.1016/j.marpol.2016.11.023

Shahid S (2010) Probable impacts of climate change on public health in Bangladesh. Asia Pacific J Publ Health 22(3):310–319. https://doi.org/10.1177/1010539509335499

Subramanian A, Mosur Nagarajan A, Vinod S, Chakraborty S, Sivagami K, Theodore T, Shalini Sathyanarayanan S, Tamizhdurai P, Mangesh L, V. (2023) Long-term impacts of climate change on coastal and transitional eco-systems in India: an overview of its current status, future projections, solutions, and policies. RSC Adv 13(18):12204–12228. https://doi.org/10.1039/D2RA07448F

Suhr F, Steinert JI (2022) Epidemiology of floods in sub-Saharan Africa: a systematic review of health outcomes. BMC Publ Health 22(1):268. https://doi.org/10.1186/s12889-022-12584-4

Taji A, Yumiya Y, Chimed-Ochir O, Fukunaga A, Tsurugi Y, Kiwaki K, Akahoshi K, Toyokuni Y, Chishima K, Mimura S, Wakai A, Kondo H, Koido Y, Kubo T (2024) Medical needs during the Kumamoto heavy rain 2020: analysis from emergency medical teams' responses. BMC Emerg Med 24(1). Scopus. https://doi.org/10.1186/s12873-024-01009-7

Tufail M, Nasir MJ, Rahman A, Kakakhel SI, Tariq A (2025) A geospatial assessment of the resilience of municipal water supply to flooding in Nowshera District, Pakistan. HydroResearch 8:89–98. https://doi.org/10.1016/j.hydres.2024.09.009

United Nations (2015) Goal 15: life on land. The Global Goals. Retrieved 2 Jan 2025, from https://globalgoals.org/goals/15-life-on-land/

Vigar G (2009) Towards an integrated spatial planning? Eur Plan Stud 17(11):1571–1590. https://doi.org/10.1080/09654310903226499

Walimah E, Andayani HF, Sonjati E, Siagian IO, Silapurna EL (2024) Mitigating the impact of climate change on public health: strategies for resilience. Glob Int J Innov Res 2(9). https://doi.org/10.59613/global.v2i9.308

Walter TG, Bricknell LK, Preston RG, Crawford EGC (2024) Climate change adaptation methods for public health prevention in Australia: an integrative review. Curr Environ Health Rep 11(1):71–87. https://doi.org/10.1007/s40572-023-00422-7

Watts N, Amann M, Arnell N, Ayeb-Karlsson S, Belesova K, Berry H, Bouley T, Boykoff M, Byass P, Cai W, Campbell-Lendrum D, Chambers J, Daly M, Dasandi N, Davies M, Depoux A, Dominguez-Salas P, Drummond P, Ebi KL (2018) The 2018 report of the Lancet Countdown on health and climate change: shaping the health of nations for centuries to come. Lancet (London, England) 392(10163):2479–2514. https://doi.org/10.1016/S0140-6736(18)32594-7

World Health Organization (2018) COP24 special report: health and climate change. https://www.who.int/publications/i/item/9789241514972

Yadav N, Wu J, Banerjee A, Pathak S, Garg RD, Yao S (2024) Climate uncertainty and vulnerability of urban flooding associated with regional risk using multi-criteria analysis in Mumbai, India. Environ Res 244:117962. https://doi.org/10.1016/j.envres.2023.117962

Zhang S, Zhang C, Cai W, Bai Y, Callaghan M, Chang N, Chen B, Chen H, Cheng L, Dai H, Dai X, Fan W, Fang X, Gao T, Geng Y, Guan D, Hu Y, Hua J, Huang C (2023) The 2023 China report of the lancet countdown on health and climate change: taking stock for a thriving future. Lancet Publ Health 8(12):e978–e995. https://doi.org/10.1016/S2468-2667(23)00245-1

Chapter 3
Climate Change and Mental Health Resilience in Northern Coast of Java's Island, Indonesia Coastal Communities: Challenges and Adaptations

Hari Harjanto Setiawan, Rizal Akbar Aldyan, Badrun Susantyo, Nyi Raden Irmayani, Habibullah, Mery Ganti, Abdullah, Luthfi Setiadiwibawa, Euis Setiawati, Ineu Maryani, Suhandy Siswoyo, Djaenudin, Mahdi Ibrahim Tanjung, Karno, Ruslan Halifu, and Mohamad Miftah

Abstract Climate change significantly impacts the mental well-being and psychological resilience of people living along the north coast of Java, Indonesia. This study seeks to explore the effects of climate change on mental health, evaluate

H. H. Setiawan · R. A. Aldyan (✉) · B. Susantyo · N. R. Irmayani · Habibullah · M. Ganti · Djaenudin · R. Halifu
National Research and Innovation Agency (BRIN), Jakarta, Indonesia
e-mail: riza014@brin.go.id

Abdullah
Telkom University, Bandung, Indonesia

L. Setiadiwibawa
Winaya Mukti University, Bandung, Indonesia

E. Setiawati
Religious Training Center Bandung, Indonesia's Ministry of Religious Affairs (MORA), Bandung, Indonesia

I. Maryani
Guidance and Counseling Study Program, Faculty of Educational Sciences, Indonesia University of Education, Bandung, Indonesia

S. Siswoyo
Faculty of Engineering and Industrial Education, Indonesian University of Education, Bandung, Indonesia

M. I. Tanjung
Ministry of Public Works, Bandung, Indonesia

Karno
Institute of Home Affairs Government, Jakarta, Indonesia

© The Author(s), under exclusive license to Springer Nature Switzerland AG 2025

A. I. Almulhim et al. (eds.), *Urbanization, Climate Change, and Health*, Climate Change Management, https://doi.org/10.1007/978-3-031-99132-5_3

community-driven adaptation strategies, and examine the effectiveness of governmental initiatives in strengthening psychological resilience within coastal populations. The research utilized quantitative and qualitative methods, blending survey data with in-depth interviews. This mixed-methods investigation was carried out in five regions identified as high-risk: DKI Jakarta, Tangerang Regency, Cirebon Regency, Demak Regency, and Pasuruan Regency. The survey involved 518 respondents to measure anxiety, stress, and depression using the GAD-7 and PHQ-9, as well as risk perception and social support. In-depth interviews explored daily emotional impacts and perceptions of support from the government. The results reveal that coastal communities face increased levels of anxiety, stress, and depression as a direct consequence of environmental disruptions, including rising sea levels and recurrent flooding. Economic hardship and limited access to mental health services exacerbate these effects. Community adaptation strategies include improving family ties and social solidarity and strengthening positive mindsets. Governments are crucial in tackling climate change by providing mental health support and enhancing community awareness of disaster preparedness. The study concluded that climate change causes significant psychological distress in coastal communities on the North Coast of Java. Mental health support should be integrated into climate adaptation policies and community-based approaches to improve community mental resilience. This study emphasizes the importance of a holistic approach incorporating mental health aspects into climate adaptation policy planning in coastal areas.

Keywords Climate change · Mental health · Coastal communities · Mental resilience · Climate adaptation

Introduction

Coastal populations experience significant mental health impacts from climate change. The lives of individuals in these areas are being profoundly impacted by rising sea levels and other climate-related challenges, resulting in heightened levels of stress and anxiety (Mousoulidou et al. 2024). Research highlights that individuals living in coastal areas face heightened susceptibility to mental health challenges, including anxiety, depression, and post-traumatic stress disorder, often stemming from environmental disruptions and climate change-related impacts (Gifford and Gifford 2016). These effects extend beyond individual well-being, influencing coastal communities' social and economic structures. Addressing these mental health challenges is crucial for fostering resilience and supporting communities in adapting to climate change's persistent impacts.

Coastal communities frequently face profound mental health issues, such as trauma and psychological distress, directly linked to the impact of extreme weather

M. Miftah
Central Java Regional Research and Inovation Agency (BRIDA), Semarang, Indonesia

events intensified by climate change (Hoog and Verboon 2019). Disasters like flooding frequently cause devastating loss of life, widespread destruction of homes, disruption to livelihoods, and profound societal upheaval, affecting entire communities. These incidents can result in severe outcomes, frequently triggering post-traumatic stress disorder along with a variety of other mental health challenges. Over time, the persistent impacts of climate change—such as rising sea levels, coastal erosion, and the escalating frequency and severity of tidal flooding—are anticipated to affect mental health and well-being significantly (Mousoulidou et al. 2024).

The persistent challenges and far-reaching impacts of climate change profoundly influence the mental well-being of coastal communities. The uncertainty about the future, the anguish triggered by environmental degradation, and the growing challenges of adapting to shifting conditions intensify mental health issues, including depression, anxiety, and other psychological difficulties. Additionally, economic instability and the disruption of livelihoods caused by climate change can exacerbate existing mental health issues while also triggering new psychological challenges. Along the northern coast of Java, unpredictable weather patterns and rising sea levels have triggered environmental changes that significantly impact the livelihoods of coastal residents, especially those dependent on fisheries (Mulyasari et al. 2019; Osman et al. 2022). The disruption of livelihoods in coastal communities often results in feelings of helplessness and heightened anxiety, driven by the stress of losing income and food security (Adityawitari et al. 2020). Damage to coastal ecosystems, such as mangroves, due to climate change not only impacts physical health through increased exposure to environmental hazards but also undermines community well-being and social cohesion (Septiarani and Yesiana 2020).

These psychological changes often lead to mental health challenges, such as depression and anxiety, which have especially severe impacts on vulnerable groups, including the elderly and those with pre-existing mental health conditions (Reyes et al. 2021; Corvalán et al. 2022). The mental health impacts of climate change are especially pronounced in communities with limited adaptive capacity (Butler et al. 2015; Solecki and Friedman 2021). On the northern coast of Java, limited resources and inadequate support systems further hinder effective adaptation, resulting in heightened psychological distress (Stilita 2024). These findings highlight the critical need for holistic strategies incorporating mental health support into climate adaptation efforts within community planning and policymaking frameworks (Corvalán et al. 2022).

This study aims to explore the profound impacts of climate change on the mental health of coastal communities in northern Java, Indonesia. Specifically, it examines how rising sea levels, frequent flooding, and economic challenges contribute to anxiety, stress, and depression among these populations. Furthermore, this study investigates community adaptation strategies, such as fostering social solidarity and strengthening family ties, alongside government interventions like mental health services and disaster preparedness programs. This research also aims to analyze the role of social and institutional networks in mitigating mental health risks associated with climate change. By examining how community-led initiatives and governmental policies influence resilience, this study seeks to identify effective approaches for integrating mental health considerations into broader adaptation strategies. Ultimately, the study underscores the importance of embedding mental health support into climate adaptation policies to bolster the resilience and well-being of these vulnerable communities.

The Rationale of the Study

Climate Change in Coastal Communities

Communities along the northern coast of Java face severe repercussions from climate change, with rising sea levels, intensified flooding, and ecological degradation presenting significant and ongoing threats. This region's geographical features and socio-economic reliance on marine resources make it especially susceptible to the adverse effects of climate change.

Climate change has led to accelerating sea-level rise, posing an increasingly pressing challenge. Studies indicate that sea levels along Java's northern coast are increasing by 6–10 mm annually (Nashrrullah et al. 2014). This trend intensifies the risk of tidal flooding, which disrupts the livelihoods of coastal communities reliant on fisheries and agriculture (Purnomo et al. 2023). The frequency and severity of tidal floods have grown, particularly during tidal cycles, causing significant economic losses, especially in salt farming and other coastal industries (Nirwansyah and Braun 2019; Wijayanti 2023). These challenges are further exacerbated by land subsidence, which significantly heightens the susceptibility of coastal areas to flooding (Sarah et al. 2021; Andari 2023).

In addition, climate change has profound ecological impacts. Rising sea levels and pollution resulting from increased urbanization and industrial activities threaten coastal ecosystems, including mangroves and fisheries (Handiani et al. 2022). Damage to these ecosystems impacts biodiversity and weakens the natural barriers that protect coastal communities from tidal waves and flooding (Marfai 2014).

Communities along the North coast of Java are highly dependent on marine capture fisheries, which are directly affected by changing climate conditions, including shifts in seasonal rainfall and temperature patterns (Mulyasari et al. 2019). The vulnerability of coastal communities is further intensified by their limited adaptive capacity, driven by socio-economic challenges such as poverty and restricted access to vital resources (Kartika et al. 2019). This highlights the critical necessity for comprehensive adaptation strategies that effectively tackle climate change's immediate and long-term challenges (Adityawitari et al. 2020).

North Java coastal communities face multiple threats from climate change, characterized by rising sea levels, increased flooding, and ecological degradation. Tackling these challenges demands a cohesive strategy that combines efficient disaster risk management, active community participation, and sustainable development practices to strengthen resilience against the continuous impacts of climate change.

Mental Health

The concept of mental health encompasses a multifaceted and comprehensive understanding that incorporates one's emotional state, psychological condition, and social functioning. The concept of mental health encompasses more than just the absence of psychological disorders. It includes positive attributes such as the ability to bounce back from adversity, handle stress effectively, and employ robust strategies for dealing with life's challenges. The World Health Organization (WHO) highlights that mental health is fundamental to overall health and well-being, influencing how individuals think, feel, and behave in their daily lives. It is characterized not only by the absence of mental disorders but also by the presence of qualities that enhance overall well-being (McAneney et al. 2015; Davoren et al. 2013).

Research indicates that various factors influence mental health, including individual traits, community environments, and social determinants. Socioeconomic status, social support, and community involvement significantly influence mental health outcomes (Lente et al. 2011). The stigma around mental health problems often leads to disparities in access and utilization of care, particularly among marginalized populations (Heboyan et al. 2021). The interplay of these factors highlights the intricate nature of mental health. It reinforces the need for a comprehensive approach that addresses personal struggles and systemic obstacles to accessing care.

Mental health is a dynamic and essential component of overall well-being, encompassing more than just the absence of mental disorders. Multiple factors influence mental health, including individual characteristics and broader social determinants. Fostering mental health necessitates a holistic approach that considers individual requirements and the structural obstacles that hinder access to care. This approach ultimately fosters an environment that supports overall well-being.

Community Resilience and Social Adaptation Strategies

Building community resilience and implementing social adaptation strategies are essential for tackling the challenges brought about by environmental changes, social upheavals, and economic transitions. The interaction of social capital, coping strategies, and community engagement significantly influences how communities adapt and thrive when faced with adversity.

Social capital plays a pivotal role in strengthening community resilience. It encompasses networks, relationships, and shared norms that enable collective action for the benefit of all. Social structures are crucial in empowering communities to design and execute effective adaptation strategies, especially in responding to the challenges posed by climate change. A strong foundation of social capital greatly enhances a community's adaptive capacity, particularly in those most vulnerable to climate-related risks (Hagedoorn et al. 2019).

Adaptive coping strategies are essential to fostering community resilience. Leveraging social networks can act as a critical mechanism for managing stress, with the success of this approach mainly depending on how well it aligns with the specific challenges and stressors experienced by individuals and communities (Wegmann et al. 2023).

Community resilience is deeply intertwined with adaptability and the ability to learn and evolve. Resilience thinking is a vital theoretical framework that underscores the importance of adaptability and the capacity for transformation within social-ecological systems (Folke et al. 2010). To craft effective strategies for addressing the challenges of climate change, it is crucial to acknowledge and prioritize the social dimensions of resilience (Villasante et al. 2022).

Community resilience and social adaptation strategies are closely interlinked through social capital mechanisms and coping strategies. Evidence shows that communities with strong social networks are better positioned to adopt effective adaptation measures, cope with stressors, and ultimately increase their resilience to environmental and social challenges.

Material and Methods

This study employs a mixed-methods approach, combining qualitative and quantitative techniques to deliver a comprehensive analysis of climate change's effects on the mental health of coastal communities. The quantitative phase employed structured surveys to collect data on levels of anxiety, stress, and depression, as well as perceptions of climate change risks, availability of social support, and access to mental health services (Hayes and Poland 2018). The qualitative phase included in-depth interviews and focus group discussions that explored climate change's emotional and everyday impacts, mental adaptation strategies, and perceptions of support from government agencies and local organizations (Hayes et al. 2020). By integrating these methods, the study successfully uncovered patterns of adaptation and resilience within coastal communities impacted by climate change (Fongnzossie et al. 2017).

This study was carried out in five high-risk areas along the northern coast of Java: DKI Jakarta, Tangerang Regency, Cirebon Regency, Demak Regency, and Pasuruan Regency. These regions are especially susceptible to the impacts of climate change, with rising sea levels and intensified flooding serving as prominent examples.

The quantitative approach involved a sample of 518 respondents, with 104 randomly selected from each study site. The data were gathered through a structured questionnaire aimed at evaluating levels of anxiety, stress, and depression, alongside perceptions of climate change risks, the availability of social support, and access to mental health services. Mental health was measured using the General Anxiety Disorder-7 (GAD-7) and Patient Health Questionnaire-9 (PHQ-9), while risk perception and social support were assessed using Likert scales. The analysis employed descriptive statistics to examine the distribution of mental health conditions across different locations. Inferential statistical methods were employed to explore the correlation between the impacts of climate change and mental health outcomes (Hayes et al. 2019). Correlation analyses were also performed to explore the connections between factors such as climate change risk perception, social support, and mental health outcomes (Obradovich et al. 2018).

During the qualitative phase, 5–10 participants from each location who had experienced significant mental health effects were selected for interviews, resulting in 25–50 respondents. Through in-depth interviews and focus group discussions (FGDs), the study explored climate change's daily and emotional impacts, mental adaptation strategies, and perceptions of government and local support. Thematic analysis was conducted to uncover key themes related to psychological impacts, adaptive strategies, and social support systems.

The integration of quantitative and qualitative data occurs at the interpretation stage (Putra et al. 2019). Quantitative data highlights the link between climate change and mental health, while qualitative data offers more profound insights into personal experiences and the broader social context (Massazza et al. 2022). This integration helps identify patterns of adaptation and resilience among coastal communities facing climate change. Ethically, the study ensured participants' consent, confidentiality of identity, and respect for their privacy.

Result

The Impact of Climate Change on Mental Health

Stress, Anxiety and Depression Levels in Society

Coastal communities along Java's northern shoreline are experiencing escalating stress, anxiety, and depression driven by the intensifying effects of climate change. The region's vulnerability to environmental threats, such as rising sea levels, extreme weather events, and the degradation of marine ecosystems, has significantly intensified the escalating mental health challenges. These issues pose severe threats to coastal communities' livelihoods and overall well-being.

The survey results indicate significant mental health challenges among coastal communities affected by climate change. Based on responses from 518 participants across the five study locations, the mean Generalized Anxiety Disorder-7 (GAD-7) score was 10.4 ($SD = 4.6$), indicating moderate anxiety levels. The mean Patient Health Questionnaire-9 (PHQ-9) score was 11.1 ($SD = 5.2$), suggesting moderate depressive symptoms among respondents. The prevalence of mental health conditions showed that 38.2% of respondents experienced mild anxiety (GAD-7: 5–9), 42.5% had moderate anxiety (GAD-7: 10–14), and 19.3% suffered from severe anxiety (GAD-7: 15+). Similarly, 35.7% of participants reported mild depression (PHQ-9: 5–9), 44.8% had moderate depression (PHQ-9: 10–14), and 19.5% experienced severe depression (PHQ-9: 15+). A one-way ANOVA test showed statistically significant differences in anxiety and depression levels across study locations ($F = 5.37$, $p < 0.001$), with the highest mean anxiety and depression scores reported in Demak Regency and Cirebon Regency, where climate-related disasters are most frequent.

Table 3.1 presents the mean GAD-7 (Generalized Anxiety Disorder-7) and PHQ-9 (Patient Health Questionnaire-9) scores across five study locations along the northern coast of Java. The table provides insights into the prevalence of anxiety and depression among coastal communities affected by climate change.

- **Highest Levels of Psychological Distress**:

  Demak Regency recorded the highest mean GAD-7 (12.2) and PHQ-9 (12.8) scores, indicating severe psychological distress, likely due to frequent tidal flooding and economic instability.
  Cirebon Regency also exhibited high anxiety (11.5) and depression (12.1) scores, showing a similar trend.

- **Mild, Moderate, and Severe Cases**:

  Moderate anxiety and depression were the most common categories across all regions, with around 42–44% of respondents experiencing mild anxiety and 44–47% experiencing mild depression.
  Severe anxiety and depression were more prevalent in Pasuruan and DKI Jakarta, suggesting localized stressors affecting mental well-being.

- **Regional Variations**:

  Pasuruan Regency had the lowest mean GAD-7 (9.5) and PHQ-9 (9.8) scores, suggesting a relatively lower mental health burden than other regions.

Table 3.1 Presents the mean GAD-7 and PHQ-9 scores across different study locations, highlighting variations in anxiety and depression levels

Region	Mean gad-7 score	Mean PHQ-9 score	Mild anxiety (%)	Moderate anxiety (%)	Severe anxiety (%)	Mild depression (%)	Moderate depression (%)	Severe depression (%)
DKI Jakarta	9.8	10.3	35.4	42.1	22.5	33.9	45.2	20.9
Tangerang Regency	10.1	10.7	36.7	41.8	21.5	34.5	44.1	21.4
Cirebon Regency	11.5	12.1	39.2	43.5	17.3	36.2	46.5	17.3
Demak Regency	12.2	12.8	40.5	44.0	15.5	37.0	47.8	15.2
Pasuruan Regency	9.5	9.8	34.8	41.2	24.0	33.0	43.9	23.1

Source Research data

While DKI Jakarta recorded lower anxiety scores, the percentage of severe depression cases (20.9%) was still notable.

This table highlights the significant mental health burden experienced by coastal communities, emphasizing the need for targeted interventions, particularly in high-risk regions like Demak and Cirebon.

Figure 3.1 shows the distribution of mean Generalized Anxiety Disorder-7 (GAD-7) and Patient Health Questionnaire-9 (PHQ-9) scores in five coastal areas of northern Java: DKI Jakarta, Tangerang Regency, Cirebon Regency, Demak Regency, and Pasuruan Regency. This graph shows the variation in levels of anxiety and depression experienced by coastal communities affected by climate change.

The analysis results show that Demak Regency recorded the highest scores for anxiety (GAD-7: 12.2) and depression (PHQ-9: 12.8), reflecting high levels of psychological distress due to recurring tidal flooding and economic instability. Cirebon Regency also showed high levels of anxiety and depression, with scores of 11.5 and 12.1, respectively. In contrast, Pasuruan Regency had the lowest GAD-7 and PHQ-9 scores (9.5 and 9.8), indicating milder psychological impacts than other areas.

Overall, the patterns observed in Fig. 3.1 suggest that levels of anxiety and depression tend to be higher in areas with more frequent climate disasters and lower economic resilience. These results highlight the importance of targeted policy interventions to improve the mental resilience of coastal communities in the face of climate change impacts.

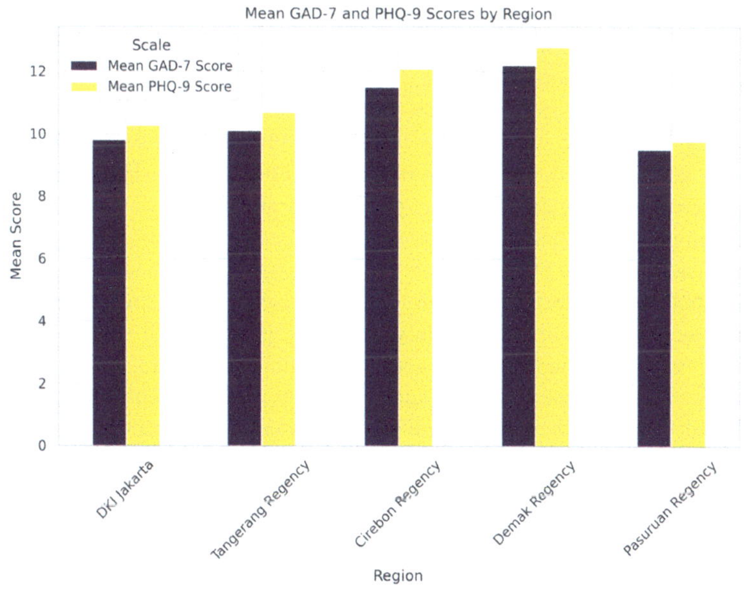

Fig. 3.1 Mean GAD-7 and PHQ-9 scores by region. *Source* Research Data

Kabir's (2023) study highlights that climate change has increased environmental stress and exacerbated mental health problems among coastal communities in Bangladesh. Likewise, the northern coast of Java, one of Indonesia's most climate-sensitive regions, is grappling with severe challenges posed by rising sea levels and changing ocean currents. These changes pose serious threats to local fisheries, a primary source of livelihood for coastal communities (Septiarani and Yesiana 2020; Mulyasari et al. 2019).

Economic instability, combined with the persistent fear of losing homes and critical community infrastructure due to climate change impacts, dramatically amplifies feelings of helplessness and vulnerability. This overwhelming sense of vulnerability affects individuals on a personal level and extends to entire communities, giving rise to widespread despair. People find it difficult to envision a stable future as the combined pressures of financial instability and environmental threats continue to mount, creating an environment where optimism is replaced by despair and uncertainty dominates daily life (Saltzman et al. 2023).

The mental health effects of climate change extend far beyond the immediate environmental challenges. They include profound and long-lasting psychological effects. Studies indicate that more vulnerable groups, particularly women, experience heightened levels of anxiety and distress associated with climate change-related issues (Searle and Gow 2010). Increased awareness and media coverage of climate-related threats contribute to widespread fear and worry about the future, exacerbating public mental health issues (Searle and Gow 2010; Hayes et al. 2019).

Stress induced by environmental changes can result in severe mental health issues, as demonstrated by research linking concerns about climate change to heightened levels of distress among affected populations (Searle and Gow 2010; Rivara and Fihn 2019). The mental health consequences of climate change often manifest as post-traumatic stress disorder (PTSD), intense grief, and survivor's guilt, especially in the aftermath of catastrophic events (Hayes et al. 2019). These impacts are also compounded by chronic stress arising from environmental change, which can lead to ongoing anxiety and depression among affected populations (Hayes and Poland 2018). Severe psychological effects are particularly prevalent in developing regions, where communities often lack the necessary resources to adapt effectively to climate change impacts (Shanthi et al. 2017).

The mental health effects of climate change are deeply interconnected with socio-economic factors. Coastal communities frequently grapple with intersecting challenges such as poverty and restricted access to mental health care, which can intensify feelings of helplessness and despair (Shanthi et al. 2017). This emphasizes the critical need for comprehensive mental health services and carefully targeted intervention strategies. Mental health professionals are vital in addressing these challenges through community engagement and awareness-raising initiatives (Shoib 2023).

The concept of "ecological grief" has emerged as a profound psychological response among coastal communities experiencing the loss of their natural environment due to climate change (Cunsolo et al. 2020). This form of grief stems from a deep sense of loss tied to the degradation of ecosystems that are integral to both cultural and economic identities. As climate change continues to reshape these landscapes, the mental health impacts are expected to intensify, resulting in heightened levels of anxiety and depression (Gebhardt 2023).

Coastal communities along Java's northern coastline are facing profound and multifaceted challenges, including heightened stress, anxiety, and depression, as a result of the impacts of climate change. The interaction between environmental stressors, socio-economic vulnerabilities, and mental health challenges highlights the critical need for a holistic approach to effectively address the psychological impacts of climate change.

Psychological Trauma Due to Disasters

The psychological trauma experienced by coastal residents along Java's northern coast due to climate change is a complex issue, encompassing mental health challenges, displacement, and the loss of cultural identity. As environmental threats like rising sea levels and intensified flooding grow more severe, the psychological impact on these communities becomes increasingly evident.

Qualitative findings from in-depth interviews ($n = 30$) and focus group discussions ($n = 5$) further support the quantitative results. Participants frequently described experiences of displacement due to tidal flooding, economic insecurity, and loss of cultural identity, which contributed to prolonged stress and trauma. A thematic analysis identified displacement-related trauma as a major concern, cited by 73% of interviewees, while financial insecurity was mentioned by 86% of participants as a significant stressor. Additionally, 62% of respondents reported intergenerational transmission of climate-related anxiety, highlighting the long-term psychological effects of environmental stressors.

Java's northern coast is highly susceptible to climate change's impacts, including tidal flooding and coastal erosion, which profoundly affect the mental well-being of local communities. Tidal flooding disrupts daily life and fosters helplessness and anxiety among those affected (Edifianto 2023).

Maharani's research highlights a strong connection between displacement caused by climate-related disasters and the onset of mental health issues, including anxiety and depression, among coastal populations in high-risk areas (Maharani 2024). These communities often perceive their experiences with climate change through the lens of personal and collective trauma, which increases their sense of vulnerability and fear (Peñalba et al. 2021).

The psychological toll of climate change is further exacerbated by the loss of a sense of belonging as residents grapple with environmental degradation and the ever-present threat of displacement. The emotional toll of severed ties to home and community often manifests as profound sadness and a sense of loss, further affecting mental well-being (McMichael and Powell 2021). This emotional distress is frequently referred to as "climate trauma," encapsulating the psychological burden of living in an environment undergoing significant change (Woodbury 2019).

A critical dimension of this issue lies in the connection between environmental changes and the health impacts of climate changes. Coastal regions face growing exposure to waterborne diseases and other health hazards due to climate change, significantly amplifying psychological stress. This stress not only undermines individual mental health but also weakens coastal communities' social cohesion and resilience (Maharani 2024).

The psychological trauma experienced by coastal communities along the northern coast of Java as a result of climate change is a critical concern, encompassing mental health issues, forced displacement, and the erosion of cultural identity. The complex interaction between environmental degradation, health risks, and the emotional toll of losing one's home underscores the critical need for comprehensive mental health support and resilience-building strategies to assist coastal communities in adapting to the challenges posed by climate change.

Resilience and Adaptation Strategies for Coastal Communities on the North Coast of Java

Coastal communities facing climate-related stressors have developed various coping mechanisms to mitigate psychological distress, with social and familial support playing a critical role in mental resilience. Strengthening family ties was identified as the most crucial factor, with 81% of respondents emphasizing its importance in maintaining psychological stability amid environmental uncertainties. Additionally, 67% of participants acknowledged the role of social support networks, including religious and traditional support groups, in fostering emotional well-being and mitigating the adverse psychological impacts of climate change. These findings underscore the significance of informal social structures in sustaining mental health, particularly in regions where formal mental health interventions remain limited. However, despite the reliance on communal and familial support, only 39% of respondents reported accessing government mental health services following climate-induced disasters, citing issues such as limited availability, inadequate outreach, and logistical barriers. The low utilization of institutional mental health support suggests a gap in the integration of formal psychological interventions with existing community-based coping strategies, highlighting the need for targeted policy responses that enhance service accessibility while leveraging traditional support systems.

A logistic regression analysis further reinforced the protective role of social cohesion in mental health outcomes, which demonstrated a significant association between strong social networks and reduced psychological distress. Individuals with robust social support systems exhibited 58% lower odds of developing severe anxiety (OR = 0.42, 95% CI: 0.31–0.57, $p < 0.001$) and 49% lower odds of experiencing severe depression (OR = 0.51, 95% CI: 0.36–0.69, $p < 0.001$), suggesting that the presence of reliable social connections serves as a crucial buffer against the mental health repercussions of climate-related adversities. These findings highlight the necessity of incorporating social cohesion strategies into mental health policies for coastal communities, ensuring that interventions address clinical aspects of psychological well-being and strengthen existing communal resilience mechanisms. Future research should explore ways to integrate community-driven support structures with formal mental health frameworks, facilitating a more comprehensive and culturally responsive approach to mental health care in climate-vulnerable regions.

Positive Thinking

The psychological effects of climate change are profound, often manifesting as ecological anxiety—a persistent fear of impending environmental disaster (Gifford and Gifford 2016). While it is crucial to acknowledge these distressing emotions, it is equally important to highlight the role of positive thinking in alleviating such despair. Coastal communities along Java's northern coast exhibit extraordinary resilience and a steadfast sense of optimism when confronting the challenges posed by climate change. In the face of rising sea levels and extreme weather events, these communities maintain a positive outlook and remain steadfast in adapting to changing conditions.

Embracing a positive mindset can strengthen cognitive abilities, enhance problem-solving skills, and nurture vital social relationships, all of which are essential for tackling the challenges brought about by climate change. By cultivating a mindset oriented towards hope, empowerment, and concerted action, both individuals and communities can develop the resilience necessary to effectively deal with this global crisis (Fritze et al. 2008). By expanding the narrative beyond acknowledging ecological anxiety, we can harness the power of positive thinking to support our response to climate change. Cultivating a mindset emphasizing hope, empowerment, and collective action enables individuals and communities to develop resilience and deal with global crises effectively.

Positive thinking can improve cognitive performance, enhance problem-solving skills, and foster stronger social connections, all of which are critical to effectively addressing the challenges of climate change (Wilson and Orlove 2021). Adopting this constructive perspective can empower us to take meaningful steps toward a more sustainable and resilient future. A positive mindset can inspire proactive engagement in sustainable practices, drive advocacy for climate-conscious policies, and foster collaboration to develop transformative solutions to combat the challenges of climate change.

Taking Wisdom

Coastal communities along the North Coast of Java face numerous challenges from climate change, affecting their social, economic, and environmental well-being. In this context, exploring how these communities can draw lessons from past disasters and adapt to ongoing environmental changes is essential.

Coastal communities along the North Coast of Java have gained valuable insights from climate change-related disasters, leading to effective adaptation strategies such as diversifying livelihoods and enhancing capacities in natural resource management. Communities also learned lessons from climate change disasters by raising awareness about the importance of environmental conservation.

For instance, mangrove planting has proven to be an effective strategy for preventing coastal erosion and safeguarding coastal ecosystems (Risa et al. 2021; Manek et al. 2023). Coastal communities along the North Coast of Java can learn from climate change-related disasters by adopting practical adaptation strategies and promoting awareness of the importance of environmental conservation.

Increasing Family Ties

Climate change poses significant challenges for coastal communities along the North Coast of Java. These communities need to strengthen family bonds and resilience to address these challenges. As the smallest social unit, families are critical in enhancing human resource quality and fostering social resilience. Research shows that lower-educated and underprivileged families are more vulnerable to the effects of climate change (Defina 2023). Improving family welfare in coastal areas is crucial, enabling communities to better adapt to environmental changes.

One approach that can be taken is through effective family communication. Good communication can increase family resilience by strengthening relationships between family members and fostering a supportive environment (Prawita and Jayanti 2023). Engaging in activities such as group discussions and introspection can help families share experiences and develop strategies to overcome challenges related to climate change. In addition, integrating local wisdom and spiritual values into daily life can strengthen family resilience (Nafisah 2023). This is particularly important in coastal areas, where local wisdom often contains relevant information for sustainable natural resource management.

Women's empowerment is important in improving family resilience in coastal communities. Women are often the primary managers of household resources and play an important role in decision-making regarding family finances (Bibin et al. 2021). Empowerment initiatives that involve women, such as skills training and access to resources, can increase family income and strengthen their economic resilience (Ahmadi 2023). Women's involvement in community activities can strengthen important social networks critical in dealing with crises, including climate change (Susanti 2022). Through a holistic and integrated approach, families' involvement in addressing climate change is expected to strengthen so that they can adapt and survive in increasingly challenging conditions.

Social Solidarity

Java's coastal communities on the North Coast face significant challenges due to climate change, necessitating the emphasis on social solidarity as an important element of mental resilience social solidarity, characterized by unity and mutual support within communities, is critical to strengthening the adaptive capacity of coastal communities along the north coast of Java to cope with climate change-induced stressors. Drawing from relevant literature, this response emphasizes the pivotal role of social solidarity in fostering these communities' mental resilience.

Social resilience is essential for understanding how communities navigate external pressures, such as those brought on by climate change. Favor emphasizes that social resilience reflects a community's capacity to adapt and manage environmental changes, particularly among fishing communities whose livelihoods are directly impacted by climate variability (Favour 2020). This concept is echoed by Deb and Haque, who emphasize the importance of local institutions and community networks in mediating responses to climate change, thereby enhancing collective coping mechanisms (Deb and Haque 2017). Integrating local knowledge and community involvement in adaptation strategies fosters a sense of ownership and responsibility, essential for building resilience (Hiwasaki et al. 2014).

Participatory approaches involving community members in decision-making are crucial for strengthening social solidarity. Combining top-down and bottom-up adaptation planning enhances effectiveness by fostering stakeholder engagement and incorporating diverse knowledge systems (Butler et al. 2015). This participatory framework empowers local communities and strengthens social bonds as individuals collaborate to achieve common goals. Participatory approach efforts can significantly improve mental resilience by fostering a supportive environment where community members feel valued and connected.

The importance of education and capacity building in the promotion of social solidarity must be taken into account. Havea emphasizes that vocational training focused on climate change and health provides individuals with the critical skills needed to adapt to the challenges posed by climate change effectively (Havea et al. 2018). By improving individual competencies, communities can collectively design adaptive strategies that are culturally relevant and context-specific. This educational approach contributes to a more informed society capable of engaging in proactive action, strengthening social cohesion and resilience.

Managing natural resources and the environment to foster social solidarity is also important. For example, mangrove forest conservation has enhanced community resilience by providing ecological benefits and promoting social cohesion among community members (Gevaña et al. 2021). The presence of healthy ecosystems not only supports livelihoods but also serves as a common interest that unites community members in conservation efforts, thereby strengthening social bonds.

Social solidarity serves as a fundamental pillar of mental resilience for coastal communities along Java's northern coast as they navigate the challenges posed by climate change. By fostering community engagement, promoting education, and prioritizing sustainable natural resource management, coastal communities along the northern coast of Java can increase their adaptive capacity and build resilience to the impacts of climate change.

Correlation Between Climate Change Perceptions and Mental Health Outcomes

Pearson correlation analysis examined the relationships between climate change risk perception, social support, and mental health outcomes. The results indicate that climate change risk perception was significantly correlated with both anxiety ($r = 0.48$, $p < 0.001$) and depression ($r = 0.53$, $p < 0.001$). This suggests that individuals who perceive a higher risk from climate change tend to experience greater psychological distress, particularly in the form of anxiety and depressive symptoms. The heightened awareness of environmental threats, uncertainties about the future, and concerns over personal and community well-being may contribute to this increased mental burden. These findings align with existing research that links environmental stressors to adverse mental health outcomes, reinforcing the need for targeted interventions to address climate-related psychological distress.

Conversely, social support was found to have a significant negative correlation with anxiety ($r = -0.39$, $p < 0.001$) and depression ($r = -0.41$, $p < 0.001$). This indicates that individuals with stronger social networks and a sense of community support tend to experience lower levels of anxiety and depressive symptoms. Social support may serve as a crucial protective factor, helping individuals cope with the psychological stress associated with climate change by providing emotional reassurance, practical assistance, and a sense of belonging. These findings highlight the importance of fostering social connections and community resilience to mitigate the mental health impacts of climate change, emphasizing the need for policies and programs that strengthen social cohesion and support systems in vulnerable communities.

Discussion

The Role of Government in Improving Community Mental Resilience

The Indonesian government plays a pivotal role in fostering the mental resilience of coastal communities along Java's northern coast as they confront the challenges posed by climate change. This multifaceted role includes policy development, community engagement, and incorporating mental health services into climate adaptation strategies.

The government's dedication to mental health is evident through coordinated efforts across various ministries. In 2011, the country established a program involving seven ministries to address mental illness, which aligns with World Health Organization (WHO) recommendations for a collaborative approach to mental health care (Kadar et al. 2019). Collaboration between ministries is essential for tackling the mental health challenges faced by coastal communities impacted by climate change-induced disasters. This approach allows for a more comprehensive response integrating economic support and mental health services (Shoib 2023).

The Indonesian government acknowledges the critical role of localized planning and active community engagement in driving practical climate change adaptation efforts. Strong local governance is crucial for executing climate adaptation strategies tailored to the unique vulnerabilities of coastal communities (Measham et al. 2011). Instruments like the Coastal Community Hazard Protection Resilience Index (CCHPR Index) are essential for evaluating adaptation strategies. They highlight the vital importance of local leadership and proactive community involvement in fostering resilience (Ewing and Grifman 2018). This approach fosters a sense of responsibility among community members and enhances their emotional resilience by involving them in decision-making processes that directly impact their lives (Hayes and Poland 2018).

Integrating mental health considerations into climate change policies is crucial to the government's responsibilities. Research shows that emotional resilience can be strengthened by improving access to mental health care, which should be a priority after extreme weather events (Hayes et al. 2018; Nicholas et al. 2020). The Indonesian government can enhance these initiatives by expanding access to mental health services and incorporating mental health professionals into emergency preparedness plans (Nicholas et al. 2020; Trombley et al. 2017). A proactive approach like this can help alleviate the psychological impacts of climate change, such as anxiety and depression, which are common in communities facing environmental stressors (Ramadan and Ataallah 2021; Trombley et al. 2017).

Furthermore, the Indonesian government can enhance community resilience by promoting educational programs that provide individuals with coping strategies to address the psychological effects of climate change (Njeru et al. 2022). These programs can be implemented in collaboration with health departments and nongovernmental organizations (NGOs), offering training and resources tailored to the unique needs of coastal communities (Nikmah 2023). By fostering a culture of preparedness and resilience, the government can empower communities to adapt to the physical impacts of climate change while effectively addressing the accompanying mental health challenges.

The Indonesian government is responsible for enhancing the mental well-being of coastal communities in North Java by implementing a range of strategic initiatives. This requires fostering inter-ministerial collaboration, strengthening local governance, integrating mental health services into climate adaptation policies, and implementing public education initiatives. Such initiatives are essential for equipping communities to confront the challenges posed by climate change, thereby enhancing their overall resilience.

Conclusions

This study underscores the profound impact of climate change on the mental health and resilience of coastal communities along Java's northern coast. The findings reveal a strong correlation between climate-induced stressors—rising sea levels, frequent flooding, and environmental degradation—and increased levels of anxiety, depression, and psychological distress. Economic instability and the fear of displacement further exacerbate these challenges. However, social adaptation strategies, such as strengthening family ties, fostering social solidarity, and cultivating positive mindsets, have emerged as crucial mechanisms for maintaining mental well-being and enhancing community resilience.

A key takeaway from this research is the critical role of social and institutional networks in mitigating climate-related mental health risks. While community-based coping mechanisms provide essential psychological support, the findings highlight gaps in government interventions, particularly in the accessibility and integration of mental health services within climate adaptation policies. Strengthening institutional support through targeted policies, localized mental health initiatives, and economic resilience programs is necessary for fostering long-term community well-being.

Future research should explore the long-term psychological consequences of climate change-induced displacement and its intergenerational effects. Additionally, more studies are needed on the effectiveness of government-led mental health interventions and how they can be integrated with community-driven strategies. There is also a growing need to develop culturally sensitive mental health programs tailored to the specific socio-economic and environmental contexts of coastal populations.

To enhance resilience, policymakers and stakeholders should prioritize:

1. **Integrating Mental Health into Climate Adaptation Strategies**—Embedding psychological well-being into climate resilience frameworks.
2. **Strengthening Community-Based Mental Health Programs**—Leveraging local social structures to provide accessible mental health support.
3. **Enhancing Access to Mental Health Services**—Expanding outreach through mobile clinics, telehealth, and community-based care.
4. **Economic and Livelihood Diversification**—Supporting alternative income sources to reduce economic stress and strengthen adaptive capacity.
5. **Leveraging Traditional Ecological Knowledge**—Promoting conservation practices such as mangrove restoration to enhance environmental and psychological resilience.

By addressing these priorities, climate adaptation efforts can move beyond physical and economic responses to include holistic strategies that safeguard mental health, ensuring the sustainability and well-being of vulnerable coastal communities in the face of the ongoing climate crisis.

Acknowledgment of AI Assistance In the preparation of this book chapter, AI-assisted tools were utilized to enhance efficiency in language refinement, literature search optimization, and structural coherence. AI was employed to generate preliminary drafts, conduct grammar and plagiarism checks, and streamline citation management. However, all intellectual inputs, analytical interpretations, and conclusions were formulated by the authors. The AI tools served as an auxiliary mechanism to improve clarity and coherence while ensuring adherence to academic writing standards. The final manuscript underwent rigorous manual review and editing by the authors to maintain scholarly integrity and originality.

References

Adityawitari F, Saputra D, Sutarto A (2020) Preliminary assessment for coastal climate adaptation and resilience in kepetingan hamlet, sawohan village, sidoarjo regency. In: International conference on sustainability science and management: advanced technology in environmental research (CORECT-IJJSS 2019). E3s Web of Conferences 153, 03004

Ahmadi A (2023) Pemanfaatan limbah sabut kelapa sebagai upaya pemberdayaan perempuan pesisir. Resona Jurnal Ilmiah Pengabdian Masyarakat 7(1):133

Andari L (2023) Identification of sea level rise and land subsidence based on sentinel 1 data in the coastal city of pekalongan, central java, indonesia. Jurnal Kelautan Tropis 26(2):329–339

Bibin M, Nirmasari D, Suhendra S (2021) Peran perempuan nelayan dalam meningkatkan perekonomian keluarga di kelurahan ponjalae kota palopo. Jurnal Sains Dan Teknologi Perikanan 1(2):36–45

Butler J, Wise R, Skewes T, Bohensky E, Peterson N, Suadnya W, Rochester W (2015) Integrating top-down and bottom-up adaptation planning to build adaptive capacity: a structured learning approach. Coast Manag 43(4):346–364

Corvalán C, Gray B, Prats E, Sena A, Hanna F, Campbell-Lendrum, D (2022) Mental health and the global climate crisis. Epidemiol Psychiatric Sci 3:e86, 1–10

Cunsolo A, Harper S, Minor K, Hayes K, Williams K, Howard C (2020) Ecological grief and anxiety: the start of a healthy response to climate change? Lancet Planet Health 4(7):e261–e263

Davoren M, Fitzgerald E, Shiely F, Perry I (2013) Positive mental health and well-being among a third level student population. Plos One 8(8):e74921 1–8

Deb A, Haque C (2017) Multi-dimensional coping and adaptation strategies of small-scale fishing communities of Bangladesh to climate change induced stressors. Int J Clim Change Strat Manag 9(4):446–468

Defina D (2023) Upaya meningkatkan kesejahteraan keluarga dan kualitas pengasuhan pada keluarga nelayan di pesisir jawa barat. Policy Brief Pertanian Kelautan Dan Biosains Tropika 5(4):822–828

Edifianto G (2023) Water level monitoring system simulation using flutter framework in pekalongan city. Emerg Inf Sci Technol 2(2):46–56

Ewing L, Grifman P (2018) Resilient communities: a vulnerabiltiy assessment is only step one. Coast Eng Proceed 36:77

Favour V (2020) Climate change and coping strategies of Ibeno and Onna coastal fishing communities in Niger Delta, Nigeria. GSC Biol Pharm Sci 10(3):045–052

Folke C, Carpenter S, Walker B, Scheffer M, Chapin T, Rockström J (2010) Resilience thinking: integrating resilience, adaptability and transformability. Ecol Soc 15(4):1–9

Fongnzossie EF, Jean SD, Victor K, Claudia M (2017) Assessing climate change vulnerability and local adaptation strategies in adjacent communities of the Kribi Campo coastal ecosystems, South Cameroon. Elsevier BV 24:1037–1051

Fritze JG, Blashki G, Burke S, Wiseman J (2008) Hope, despair and transformation: climate change and the promotion of mental health and wellbeing. BioMed Central 2(1):1–10

Gebhardt N (2023) The relationship of climate change awareness and psychopathology in persons with pre-existing mental health diagnoses. Front Psych 14(1):1–11

Gevaña D, Garcia J, Ruzol C, Malabayabas F, Grefalda L, O'Brien E, Camacho L (2021) Climate change resiliency through mangrove conservation: the case of alitas farmers of infanta, philippines. Book Chapter Springer Nature Link pp 195–214
Gifford E, Gifford R (2016) The largely unacknowledged impact of climate change on mental health. Taylor & Francis 72(5):292–297
Hagedoorn L, Brander L, Beukering P, Dijkstra H, Franco C, Hughes L, Segal B (2019) Community-based adaptation to climate change in small island developing states: an analysis of the role of social capital. Clim Dev 11(8):723–734
Handiani D, Heriati A, Suciaty F (2022) Coastal vulnerability assessment along the north java coastlines-Indonesia. Jurnal Segara 18(1):1–12
Havea P, Siga A, Rabuatoka T, Tamani A, Devi P, Senikula R, Combes H (2018) Using vocational education to provide development solutions in the pacific: an emphasis on climate change and health. Biorxiv Prepr Serv Biol
Hayes K, Poland B (2018) Addressing mental health in a changing climate: incorporating mental health indicators into climate change and health vulnerability and adaptation assessments. Int J Environ Res Public Health 15(9):1–19
Hayes K, Blashki G, Wiseman J, Burke S, Reifels L (2018) Climate change and mental health: risks, impacts and priority actions. Int J Ment Heal Syst 12(1):1–12
Hayes K, Berry P, Ebi K (2019) Factors influencing the mental health consequences of climate change in Canada. Int J Environ Res Public Health 16(9):1583
Hayes K, Poland B, Cole DC, Agic B (2020) Psychosocial adaptation to climate change in High River, Alberta: implications for policy and practice. Springer Science+Business Media 111(6) 880–889
Heboyan V, Douglas M, McGregor B, Benevides T (2021) Impact of mental health insurance legislation on mental health treatment in a longitudinal sample of adolescents. Med Care 59(10):939–946
Hiwasaki L, Luna E, Syamsidik S, Shaw R (2014) Process for integrating local and indigenous knowledge with science for hydro-meteorological disaster risk reduction and climate change adaptation in coastal and small island communities. Int J Disaster Risk Reduction 10:15–27
Hoog ND, Verboon P (2019) Is the news making us unhappy? The influence of daily news exposure on emotional states. Br J Psychol 111(2):157–173
Kabir S (2023) Psychological health ramifications of sea-level rise in Bangladesh coastal populations. Research Square. Preprints
Kadar K, Andriani A, Tandi D (2019) Exploring mental health nursing practice in Indonesian rural area. J Public Ment Health 19(2):119–126
Kartika F, Helmi M, Amirudin A (2019) Meta-analysis of community's adaptation pattern with tidal flood in Pekalongan city, central Java, Indonesia. In: The 4th international conference on energy, environment, epidemiology and information system (ICENIS 2019). E3s Web of Conferences 125 09001
Lente E, Barry M, Molcho M, Morgan K, Watson D, Harrington J, McGee H (2011) Measuring population mental health and social well-being. Int J Public Health 57(2):421–430
Maharani A (2024) Depression among people who live in coastal hazard areas in Indonesia: evidence from a population-based national survey. Research Square. Preprints
Manek A, Hasan M, Sukmawati S (2023) Penanaman mangrove untuk pencegahan abrasi di pantai sekitar pelabuhan atapupu desa jenilu kecamatan kakuluk mesak kabupaten belu. Kelimutu J Community Serv 3(1):1–9
Marfai M (2014) Impact of sea level rise to coastal ecology: a case study on the northern part of Java Island, Indonesia. Quaestiones Geographicae 33(1):107–114
Massazza A, Teyton A, Charlson FJ, Benmarhnia T, Augustinavicius J (2022) Quantitative methods for climate change and mental health research: current trends and future directions. Elsevier BV 6(7):e613–e627

McAneney H, Tully M, Hunter R, Kouvonen A, Veal P, Stevenson M, Kee F (2015) Individual factors and perceived community characteristics in relation to mental health and mental well-being. BMC Public Health 15(1):1–13
McMichael C, Powell T (2021) Planned relocation and health: a case study from Fiji. Int J Environ Res Public Health 18(8):1–17
Measham T, Preston B, Smith T, Brooke C, Gorddard R, Withycombe G, Morrison C (2011) Adapting to climate change through local municipal planning: barriers and challenges. Mitig Adapt Strat Glob Change 16(8):889–909
Mousoulidou M, Taxitari L, Christodoulou A (2024) Social media news headlines and their influence on well-being: emotional states, emotion regulation, and resilience. Multidiscipl Dig Publish Inst 14(6):1647–1665
Mulyasari G, Waluyati L, Suryantini A, Prawito P (2019) Climate change represented by seasonal rainfall, temperature, and wind speed trends in north coastal of Java, Indonesia: implications on marine capture fisheries. Int J Adv Sci Eng Inf Technol 9(6):2067–2073
Nafisah D (2023) Ketahanan keluarga di desa karangsalam kecamatan kedungbanteng kabupaten banyumas berbasis kearifan lokal dan spiritual. Jurnal Cakrawala Ilmiah 2(11):4469–4488
Nashrrullah S, Pasaribu J, Hazarika M, Samarakoon L (2014) Study on flood inundation in Pekalongan, Central Java. Int J Remote Sens Earth Sci (IJRESES) 10(2):1–9
Nicholas P, Breakey S, White B, Brown M, Fanuele J, Starodu R, Ros A (2020) Mental health impacts of climate change: perspectives for the ed clinician. J Emerg Nurs 46(5):590–599
Nikmah L (2023) Analysis of the level of community resilience to tidal flood disaster in karangagung village, palang district, tuban regency, east java. J Mar Resour Coast Manag 4(2):7–10
Nirwansyah A, Braun B (2019) Mapping impact of tidal flooding on solar salt farming in northern java using a hydrodynamic model. ISPRS Int J Geo Inf 8(10):1–22
Njeru M, Arasa J, Musau J, Kihara M (2022) The effects of climate change on the mental health of smallholder crop farmers in Embu and Meru counties of kenya. Afr J Clim Chang Resour Sustain 1(1):1–12
Obradovich N, Migliorini R, Paulus MP, Rahwan I (2018) Empirical evidence of mental health risks posed by climate change. Nat Acad Sci 115(43):10953–10958
Osman N, Saleh E, Ismail N, Sentian J, Sidek M (2022) Climate change impacts to coastal communities in Kudat, Sabah, Malaysia. Malays J Soc Sci Human (MJSSH) 7(9):e001774 1–8
Peñalba E, David A, Mabanta M, Samaniego C, Ellamil S (2021) Climate change adaptation: the case of coastal communities in the Philippines. J Geogr Inst 71(2):115–133
Prawita E, Jayanti A (2023) Penguatan ketahanan keluarga melalui komunikasi efektif di desa guwosari bantul. Jurnal Pengabdian Masyarakat 4(1):71–78
Purnomo S, Widiyanto W, Suroso S (2023) Landfilling as individual adaptation for coastal flooding (rob) countermeasures in north central java. In: 3rd international conference on disaster management. IOP conference series earth and environmental science, vol 1173(1), p 012057
Putra GA, Koestoer RH, Lestari I (2019) Psycho-social performance towards understanding local adaptation of coastal flood in cilincing community, North Jakarta, Indonesia. In: The first international conference on environmental geography and geography education (ICEGE), vol 243, pp 012005. IOP Publishing
Ramadan A, Ataallah A (2021) Are climate change and mental health correlated?. Gen Psychiatry 34(6):e100648 1–12
Reyes M, Carmen B, Luminarias M, Mangulabnan S, Ogunbode C (2021) An investigation into the relationship between climate change anxiety and mental health among gen z Filipinos. Curr Psychol 42(9):7448–7456
Risa N, Alamsyah R, Tenriwaruwaty A, Wahyuni A, Liswahyuni A (2021) Peran masyarakat desa pasimarannu kabupaten sinjai untuk melindungi pantai dari ancaman abrasi dengan melakukan penanaman mangrove. Tarjih Agribus Dev J 1(2):32–40
Rivara F, Fihn S (2019) Climate change and health. Jama Netw Open 2(9):e1912502 1–11

Saltzman L, Hansel T, Ferreira R (2023) Climate change and health equity: expanding our reach using technology. Plos Climate 2(7):e0000242 1–4
Sarah D, Soebowo E, Satriyo N (2021) Review of the land subsidence hazard in Pekalongan Delta, Central Java: insights from the subsurface. Rudarsko-Geološko-Naftni Zbornik 36(4):163–176
Searle K, Gow K (2010) Do concerns about climate change lead to distress? Int J Clim Change Strat Manag 2(4):362–379
Septiarani, Yesiana R (2020) Affecting factors on community based mangrove replantation programs in Semarang coastal area. Ind J Plann Develop 5(2):87–94
Shanthi B, Mahalakshimi P, Chandrasekaran V (2017) Assessment of challenges faced by the coastal women due to the impact of climatic change in selected coastal districts of Tamil Nadu, India. Ind J Fish 64:153–159
Shoib S (2023) Prevention of suicides associated with global warming: perspectives from early career psychiatrists. Front Psych 14(1):1–7
Solecki W, Friedman E (2021) At the water's edge: coastal settlement, transformative adaptation, and well-being in an era of dynamic climate risk. Annu Rev Public Health 42(1):211–232
Stilita G (2024) Keeping sane in a changing climate: assessing psychologists' preparedness, exposure to climate-health impacts, willingness to act on climate change, and barriers to effective action. Int J Environ Res Public Health 21(2):1–14
Susanti E (2022) Peran perempuan pesisir terhadap ketahanan keluarga pasca bencana alam gempa bumi di desa maliaya, kabupaten majene tahun 2021. Jurnal Ketahanan Nasional 27(3):417
Trombley J, Chalupka S, Anderko L (2017) Climate change and mental health. Ajn Am J Nurs 117(4):44–52
Villasante S, Macho G, Silva M, Lopes P, Pita P, Simón A, Calvo N (2022) Resilience and social adaptation to climate change impacts in small-scale fisheries. Front Mar Sci 9:1–18
Wegmann E, Schiebener J, Brand M (2023) Social-networks use as adaptive or maladaptive strategy for coping with stress. Sci Rep 13(1):1–9
Wijayanti E (2023) Extreme coastal flood inundation mapping based on sentinel 1 using google earth engine. In: ICST UGM 2023—The 4th geoscience and environmental management symposium, vol 468 04002. E3s Web of Conferences
Wilson AJ, Orlove B (2021) Climate urgency: evidence of its effects on decision making in the laboratory and the field. Elsevier BV 51:65–76
Woodbury Z (2019) Climate trauma: toward a new taxonomy of trauma. Ecopsychology 11(1):1–8

Chapter 4
A Study of Small and Medium-Sized Towns of Eastern Uttar Pradesh in the Context of Climate Change Implications, Environment and Sustainability Challenges

Ankit Kumar Gupta, Kapil Kumar Gavsker, and Vimal Kumar Lahari

Abstract The different regions of the world are grappling with disaster adaptation and prevention as climate change fuels extreme weather occurrences exacerbates natural catastrophes and human health implications. Cities and towns are being included in adaptation efforts and planning process. Western large cities have made the most practical advances in climate adaptation so far, with small and medium-sized communities trailing behind. The public and communities of small and medium-sized towns face institutional and political impediments to advancing adaptation programs. This is important to their long-term viability given their limited capabilities and resources. Regional implications of climate change are inevitable and spatially variable. Climate change presents considerable sustainability difficulties for small and medium-sized towns in the Eastern region of Uttar Pradesh. The towns, which are critical to regional growth, are being affected by rising air temperatures, erratic rainfalls, water related problems and more frequent extreme weather events. Climate change is seen as it exacerbates existing vulnerabilities such as inadequate infrastructure, weak governance, lack of urban design and threatening socio-economic sustainability. The social and economic resilience is also crucial in dealing with natural catastrophes. The present research paper analyses geo-climate conditions of select two small and medium-sized urban centers (Tanda and Padrauna towns) of eastern region of Uttar Pradesh, their social organization, levels of basic amenities, and public health infrastructure. In order to respond to impacts associated with climate change, the study highlights the urgent need for sustainable measures such as integrated urban design,

A. K. Gupta (✉) · K. K. Gavsker
Department of Geography, Institute of Science, Banaras Hindu University, Varanasi, India
e-mail: ankitgupta@bhu.ac.in

K. K. Gavsker
e-mail: kkgavsker@bhu.ac.in

V. K. Lahari
Department of Sociology, Faculty of Social Sciences, Banaras Hindu University, Varanasi, India
e-mail: Vimalk.lahari1@bhu.ac.in

© The Author(s), under exclusive license to Springer Nature Switzerland AG 2025

A. I. Almulhim et al. (eds.), *Urbanization, Climate Change, and Health*, Climate Change Management, https://doi.org/10.1007/978-3-031-99132-5_4

improved urban resource management, urban governance and the use of geospatial technologies. Study further explores avenues to enhance sustainable development practices and urban resilience. Study assumes that small and medium-sized towns be governed and managed in such a way so as to raise their quality of life, livability and physical infrastructure, amenities to promote climate-resilient urban society, sustainable growth and urban livability.

Keywords Climate change · Eastern region · Geo-climate · Small and medium size towns · Social organization · Sustainability

Introduction

For the last several decades, global warming and its related observed and predicted climate changes in the twenty first century are pressing issues. Climate change, being a complicated worldwide issue, has profound effects on ecological, environmental, socio-political and socio-economic systems (Adger et al. 2005; Leal Filho et al. 2021; Feliciano et al. 2022). The phenomenal hallmarks include rising sea levels, melting ice sheets, sudden changes in weather patterns, rising global temperatures, etc. and such have a huge impact on both the natural and human systems (Battisti and Naylor 2009; Lipczynska-Kochany 2018; Michel et al. 2021). In the context of climate change, the issue of climate change is much more pressing, and further actions are needed as fast and in a coordinated way to at least weaken its catastrophic effects (Leppänen et al. 2014). However, because of the inherent uncertainties in knowing how climate will actually change its dynamics (Ayers et al. 2017; Jurgilevich et al. 2017; Izaguirre et al. 2021), new frameworks of municipal and national policies are required to pounder upon the uncertain effects of the global climate change phenomenon.

Cities are sensitive to climate change because their populations are dense, they include crucial infrastructure and they have complicated governance processes. Cities are the gathering places of people, goods and economic activity and therefore are highly vulnerable to the effects of meteorological hazards and other climate related risk. These governance systems of cities are fragmented with the different departments that oversee urban planning, water resources, energy, housing, emergency services, at different levels of government (Seto et al. 2014). Urban growth that is common in many developing countries, such as South Asian nations, is informal urban expansion, entails exacerbating vulnerabilities and cities unprepared for the hazards brought on climate change. It is generally accepted that the urban poor bear the burden of health hazards brought on by climate change, and that climate change is predicted to increase health risks, especially for those living in the poorest areas of developing country cities (Chaudhry 2023). The social and environmental determinants of health (SEDH) have a mediating role in both the direct health effects of climate change like an increase in the prevalence of heat strokes from heat waves, and the indirect health effects like under nutrition from rising food prices. Therefore, in addressing climate change impacts in cities and urban places require integrated development approaches, urban management and pursuing urban sustainability.

Urbanization is defined as the spread of urban areas as a result of concentrated human and economic activity, serves two major functions. Firstly, it is an important driver of economic growth, propelling industrialization and raising living standards (Ochoa et al. 2018). Secondly, rapid and unplanned urbanization exacerbates environmental degradation, contributes to greenhouse gas emissions, and exposes urban inhabitants to climate-related dangers. The interaction between urbanization, industrial structure adjustment and urban land-use efficiency has a substantial impact on the sustainability of urban growth (Das et al. 2024). Cities provide potential for economic expansion and innovation, but they also hold concentrated vulnerabilities. The emerging issues include effectiveness of governance institutions and infrastructure provisions in relation to rising urban population growth.

Urbanization and climate change has cumulative effects on the urban public health. Rising global temperatures, increased frequency of extreme weather events, and urban heat island effects all pose serious threats to human health, particularly in densely populated cities (Schuurmans 2021; Weisheimer and Palmer 2005). Poor air quality, caused by industrial emissions and traffic pollution, worsens respiratory ailments and other health issues (Yadav et al. 2015). Furthermore, climate change has aided the development of vector-borne diseases, with rising temperatures and shifting precipitation patterns providing ideal circumstances for disease vectors (Lipczynska-Kochany 2018). Floods, droughts, and heat waves disproportionately affect vulnerable people, especially in developing nations with poor healthcare access and adaptation capacity (Ayers et al. 2017). With the rise of urban agglomerations and changing rural landscape through small and medium towns, the significance of urban places in achieving sustainability has been recognized world over. The significance of advancing urban sustainability was brought to light by the 1987 Brundtland Report and the 1992 Rio Declaration on Environment and Development (Brundtland 1987; UN 1992). The importance of cities in climate change adaptation and mitigation was highlighted in the Intergovernmental Panel on Climate Change's (IPCC) 2007 Fourth Assessment Report. This emphasis was further developed in the Fifth Assessment Report (Seto et al. 2014). In order to meet the problems presented by climate change, these frameworks emphasize the vital necessity of sustainable urban planning and the incorporation of climate issues into urban governance.

The vulnerability of developing countries to climate change impacts is widely acknowledged. These countries are especially vulnerable to climate-related threats due to their fragile ecosystems, economic vulnerabilities, and poor adaptive capacity (Panda 2009). Rapid urbanization in underdeveloped nations frequently coincides with inadequate infrastructure, unplanned settlements, and high population concentrations, worsening their vulnerability to catastrophic weather events (Ahmed et al. 2022). Changes in land use, caused by urbanization, industrialization, and agricultural development, exacerbate climate change impacts. Changes in river patterns, erosion, desertification, and deforestation disturb ecosystems and increase climate vulnerability (Chakraborty 2009; Sreenivasulu and Bhaskar 2010). These changes have an impact on the hydrological cycle, causing water scarcity, changing precipitation patterns, and increased flooding risks.

India is one of the prime illustrations of the complicated interaction between urbanization and climate change and its regional implications. As one of the world's fastest-growing economies, India has experienced rapid urbanization and industrialization in recent decades. However, this expansion has come at a significant environmental cost, with India currently ranking among the top contributors to global carbon dioxide and greenhouse gas emissions (Pathak 2023). Between 1970 and 2021, the country witnessed 573 climate-related disasters that killed approximately 138,000 people (IMD Annual Report 2021). These catastrophes, which include extreme heat waves, lengthy droughts, and intense rainfall all that causes deadly floods, highlight the critical need for regional climate action.

In recent years, India has had its highest temperatures and longest dry spells, with major rice-producing states such Uttar Pradesh, Bihar, Jharkhand, and West Bengal enduring consecutive drought episodes (Amrit et al. 2021). Simultaneously, the country has seen regular and severe flooding, including the Uttarakhand floods of 2013, the Kashmir floods of 2014, and the Kerala floods of 2018 (Akram et al. 2023). These occurrences show the rising frequency and severity of climate-related dangers in India, which are being driven by the combined effects of climate change and urbanization. India's status as the world's most populated nation, with 1.4 billion people, exacerbates its climatic issues. The expanding demand for resources and energy, along with inadequate infrastructure and governance mechanisms, has resulted in a larger carbon footprint and increased vulnerability to climate threats (Salunke 2022).

Urban Scene of Eastern Uttar Pradesh

Uttar Pradesh is India's most populous state. It has different regions affected by various event including metrological and hydrological of nature which may get influenced by the climate change and its regional ramifications. The state planning division of Uttar Pradesh divided the state into four economic regions as shown in Table 4.1. The Western part of the state is highly urbanized compared to other region of the state. The Central part of Uttar Pradesh is second highest urbanized region of the state and the Bundelkhand is third in the level of urbanization. The Eastern Uttar Pradesh is having second highest number of urban centers, but it is the least urbanized of the four economic regions, with a sizable urban population concentrated in big agglomerations like Allahabad, Gorakhpur, and Varanasi (Gupta and Gavsker 2024). Eastern Uttar Pradesh has the second highest number of urban centers. These cities and urban places demonstrate urban domination and predominance, putting tremendous strain on their resources and infrastructure. Surrounding these major cities are numerous small and medium towns (SMTs), which represent the urban future of the region.

However, these towns face significant challenges due to climate change, including inadequate infrastructure, limited access to basic services, and vulnerabilities to extreme weather events. Surrounding these big cities are numerous small and medium

Table 4.1 Regional pattern of urbanization in Uttar Pradesh during 2001–2011

Economic regions	Percent of the urban population		No of urban centers		Change in level of urbanization (%)
	2001	2011	2001	2011	
Western U.P	28.25	31.38	343	426	11.07
Central U.P	25.15	25.73	107	125	2.3
Bundelkhand	22.39	22.67	55	59	1.25
Eastern U.P	11.74	12.19	203	305	3.83

Source Using District Census Handbooks, Census of India 2001 and 2011

towns (SMTs), which symbolize the region's future urban development. However, climate change poses substantial problems to small towns, such as weak infrastructure, restricted access to essential services, and vulnerability to extreme weather events. In this context, small and medium-sized towns (SMTs) in eastern Uttar Pradesh are critical to determining the region's urban and environmental future. While large urban centres face resource restrictions and overcrowding, SMTs provide opportunity for balanced regional growth and climate resilience. However, small towns confront a number of obstacles, including limited public health infrastructure, insufficient basic amenities, and socioeconomic inequality (Ochoa et al. 2018; Panda et al. 2017). For the present study, the selected urban centers are termed as small and medium city here and there in the analysis offered in this chapter. Climate change exacerbates these challenges, with rising temperatures, irregular rainfall patterns, and an increase in the frequency of extreme weather events posing serious threats to urban populations' well-being. This study tries to address the following objectives: (i) To study urban population trends of the selected towns; (ii) To analyze social and economic profile of the towns; (iii) To examine land use land cover and environmental condition of the towns; and (iv) To explore the health and environmental challenges faced by towns and outlining sustainability measures.

As region continues to urbanize and industrialize, addressing the challenges posed by climate change will require coordinated efforts at local, regional, and national levels. Urban sustainability must be prioritized in policymaking, with a focus on enhancing infrastructure, improving governance systems, and building adaptive capacities. Public health systems must be strengthened to address the risks posed by climate change, particularly in vulnerable urban areas. The experiences of two towns i.e. Tanda and Padrauna in Eastern Uttar Pradesh highlight commonality and differences and the importance of tailored solutions that consider local contexts and priorities. By fostering sustainable urban development process and building climate resilience, provinces and regions can navigate the complex interplay between urbanization and climate change and environmental effects in order to ensure a better future for its urban living.

Selection of Towns

Eastern Uttar Pradesh is divided into eight divisions for administrative purposes. This division comprises of districts. These division are Azamgarh division (Azamgarh, Ballia and Mau disticts), Basti division (Basti, Sant Kabir Nagar and Siddtharnagar districts), Devipatan division (Baharaich, Balrampur, Shravasti and Gonda districts), Gorakhpur division (Deoria, Gorakhpur, Kushinagar and Maharajganj districts), Mirzapur division (Mirzapur, Sonbhadra and Sant Ravidas Nagar districts), Varanasi division (Chanduali, Ghazipur, Jaunpur and Varanasi districts), Prayagraj division (Fatehpur, Kaushambi, Pratapgarh and Prayagraj districts) and Ayodhya division (Ambedkar Nagar, Amethi, Ayodhya, Barabanki and Sultanpur districts). Fatehpur and Barabanki districts do not fall within the eastern region of Uttar Pradesh. Therefore, these districts are part of the Prayagraj and Ayodhya Divisions, respectively, have been excluded from this study. To identify suitable towns for analysis, we ranked towns within these divisions based on their population size, excluding Class I cities with populations exceeding 100,000. One town from each division ranked highest by population size and meeting the study's criteria, was selected. Following the classifications outlined by Fielmua (2021) and Srinivas (n.d.), towns with populations between 5000 and 50,000 are categorized as small towns, while those with populations above 50,000 but less than 100,000 are categorized as medium towns. In this context, the study identified five medium towns and three small towns (as shown in Table 4.1). For detailed analysis, we selected one medium town (Tanda, classified as a Nagar Palika Parishad or NPP) and one small town (Padrauna, also an NPP), both of which have the highest population sizes within their respective categories as shown in Fig. 4.1 and Table 4.2.

This study focuses on two small and medium-sized towns in eastern Uttar Pradesh: Tanda (population: 95,516) and Padrauna (population: 49,723). These towns serve as useful case studies for understanding the interplay of geo-climatic variables, social organization, and urban governance in the context of climate change. The study's goal is to find ways for improving climate resilience and encouraging sustainable urban development by examining the distribution of geoclimatic conditions, levels of basic amenities, and public health infrastructure in these communities.

Study Area

Padrauna (49,723 persons) is a city and the administrative headquarters of Kushinagar district in Uttar Pradesh, India (Fig. 4.2). Geographically, it is located at 26°53′54″N latitude and 83°58′46″E longitude. The town enjoys connectivity to several key cities via National Highway No. 28B and State Highway 64. Padrauna's economy is predominantly driven by agriculture, followed by the services sector, with minimal contribution from the industrial sector. The town is home to a majority Hindu population, along with significant communities of Muslims, Sikhs, and smaller groups of

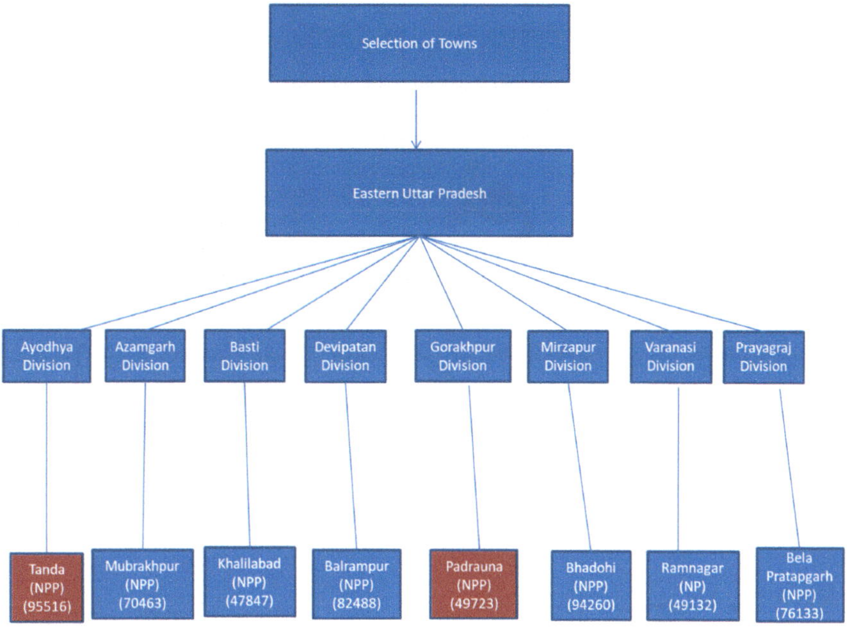

Fig. 4.1 Description of the selection of towns. *Selected towns are highlighted

Table 4.2 Description of selected towns

Division	District	Town	Population in 2011	Class
Azamgarh	Azamgarh	Mubarakpur (NPP)	70,463	II
Basti	Sant Kabir Nagar	Khalilabad (NPP)	47,847	III
Devipatan	Balrampur	Balrampur (NPP)	82,488	II
Gorakhpur	**Kushinagar**	**Padrauna (NPP)**	**49,723**	**III**
Mirzapur	Sant Ravidas Nagar	Bhadohi (NPP)	94,620	II
Varanasi	Varanasi	Ramnagar (NP)	49,132	III
Prayagraj	Pratapgarh	Bela Pratapgarh (NPP)	76,133	II
Ayodhya	**Ambedkarnagar**	**Tanda (NPP)**	**95,516**	**II**

Source Census of India, 2011

Buddhists and Christians. An irrigation canal also passes through the town, adding to its agrarian prominence.

Tanda (95,516 persons) is a town located in the Ambedkar Nagar district of Uttar Pradesh, India, situated along the banks of the Ghaghara River (Fig. 4.2). Tanda has a rich historical legacy, dating back to 1714 A.D., when Mughal Emperor Farrukhsiyar tasked Raja Syed Mohammad Hayat with building a hamlet on the banks of the Ghaghara River. Tanda, once a barren plain, grew into a thriving estate that stretched from the borders of the old Awadh State to those of Bihar. It is positioned at 26°55′N

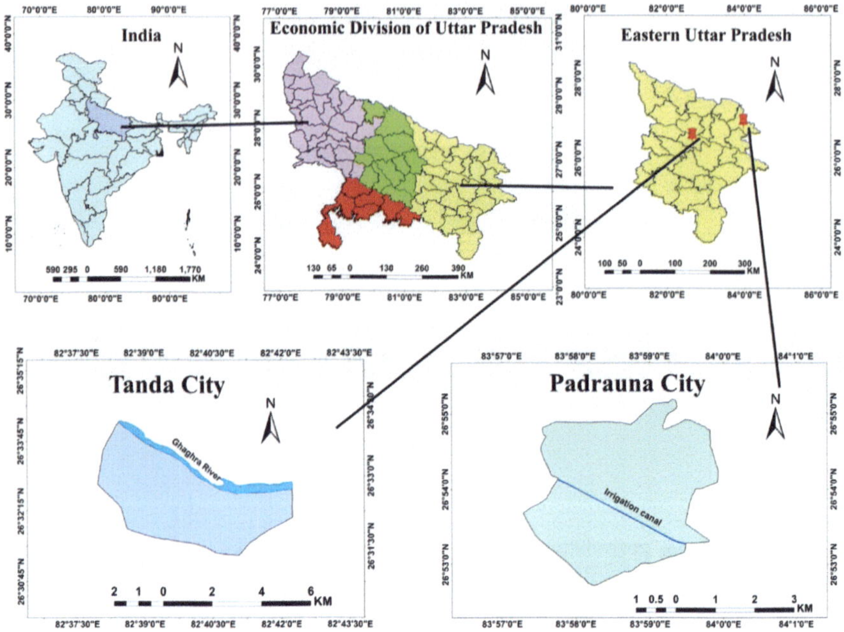

Fig. 4.2 Locational setting of Tanda and Padrauna City. *Source* Location map prepared based on State Planning Board, Uttar Pradesh

latitude and 82°65′E longitude. As part of the Ayodhya division, Tanda serves as a tehsil and is classified as a Nagar Palika Parishad. The town lies approximately 20 km (12 miles) northeast of its district headquarters, Akbarpur. Tanda has a rich industrial heritage, being widely recognized for its production of Tanda Terricot fabric, which is manufactured on power looms. Historically, the town was also known for handloom weaving, a tradition that has contributed to its reputation as an industrial hub.

Methodology

The current study is primarily based on secondary of data sources. This is obtained from government publications, such as the District Census Handbook and the Primary Census Abstract (PCA), for data on demographic and social characteristics such as population size, decadal urban growth rate, educational and health infrastructure from 1901 to 2011. On the other hand, data on land use and land cover change over the previous three decades have been collected and obtained from the United States Geological Survey (USGS) website. Accordingly, Landsat 7 data from April 4, 2001, Landsat 5 data from March 25, 2011, and Landsat 8 data from March 4, 2021 have been used (Table 4.3). All of these images have undergone the necessary atmospheric correction, geometric registration, supervised image classification (using the maximum technique), and accuracy assessment. The accuracy of each class of the three classified images (2001, 2011, and 2021) was tested using a stratified random sample approach, and an accuracy assessment was performed for each data error matrix for classification (Fig. 4.3).

Projection of Population

To project the population for 2021, geometric projection method is applied using the formula

$$P2021 = P2011(1 + K/100)^{n},$$

where, P2021 is the projected population for 2021, and P2011 is population for 2011, n is the number of decades since last census and $\sqrt[4]{K1 \times K2 \times K3 \times K4}$ and K1, K2,

Table 4.3 Description of the satellite image data

Year	Date	Cloud cover	Satellite name	Spatial resolution
2001	22-Apr-01	< 1	Landsat 7	30 Meter
2011	25-Mar-11	< 1	Landsat 5	30 Meter
2021	4-Mar-21	< 2	Landsat 8	30 Meter

*Using these images for Analysis

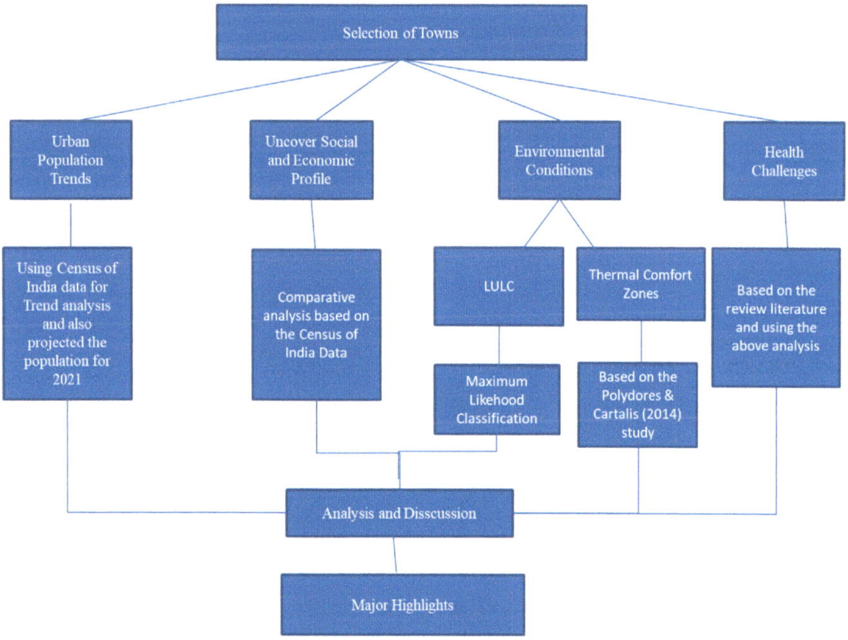

Fig. 4.3 Methodology flow chart

K3 and K4 are population growth rate for 1981, 1991, 2001 and 2011 respectively (Gupta and Gavsker 2024).

Retrieval of Monthly Thermal Discomfort Patterns

Discomfort indices are widely employed due to their simplicity, but calculating comfort indices such as PMV using the Rayman, ENVI-MET, or other models necessitates parameterization (Mohan et al. 2013; Roelofsen 2016; Shastry et al. 2016). Furthermore, thermal indices are straightforward to calculate; for example, the ET requires just the outdoor temperature to calculate indoor thermal comfort, whereas the DI (°C) requires only the temperature (°C) and humidity (%). For this reason, the DI was employed in this study to analyse outdoor thermal discomfort and was computed using mean air temperature (Ta [°C]) and mean relative humidity (RH [%]) retrieved from NASA's power website for monthly data for the years 2001, 2011, and 2021 for both cities. Polydoros and Cartalis (2014) provided the equation for calculating DI based on air temperature and relative humidity, as well as the discomfort classification criteria (Table 4.4).

$$DI(°C) = Ta - .55(1 - .01RH)(Ta - 14.5)$$

Table 4.4 Classification of the thermal comfort according to DI values

Class number	DI (°C)	Discomfort conditions
1	DI < 21	No discomfort
2	21 ≤ DI < 24	Less than 50% feels discomfort
3	24 ≤ Di < 27	More than 50% feels discomfort
4	27 ≤ DI < 29	Most of the population feels discomfort
5	29 ≤ DI < 32	Everyone feels severe stress
6	DI ≥ 32	State of medical emergency

Source Polydoros and Cartalis (2014)

Results and Analysis

Social Organization of the Towns

The Scheduled Castes (SC), known historically as the lowest rung of the social ladder of Indian traditional society. They have had to overcome sorts of exclusion from traditional social structure. They have remained systemically discriminated and socially excluded by dominant society for long and they have been a sufferer of graded inequality. Through extensive efforts to rectify these imbalances, the social and economic conditions of SC communities are still widely debated. In post-independence period in India this group has been at the center of affirmative actions of the state and certain safeguards they are entitled to.

Two important urban centres of area in eastern Uttar Pradesh allow us a glimpse into the social and religious structure of the region. The population composition shows a very significant dominance of the general category including the upper castes and backward communities. According to the census 2011, more than 92% of population in Tanda and Padrauna rate themselves as belonging to general category. In both the towns, about 7% of the population is scheduled Caste and tribal population is negligible, i.e., less than 1%.

This makes the social structure of these towns more complicated as it adds religious identity to the mix. Census 2011 also shows a high degree of religious diversity in Tanda and Padrauna. Approximately 72% of the population of Tanda regards itself as Muslim. Second largest religious group in Tanda is Hindus. Hence, this demographic pattern is relative contrary to Padrauna, where Hindus are the dominant demographic at around 70%, while Muslims are 29% of the population.

These towns had distinct religious and caste compositions, which in turn mirror the more general socio-economic and cultural dynamics of eastern Uttar Pradesh. Tanda's Muslim majority was a historical-and culturist influence while Padrauna's Hindu majority implies a different trajectory of settlement and community formation. Even though these towns are different, both towns share a common trend of Scheduled Castes being a minority in both towns, constituting marginality in urban settings.

Demographic and Social Pattern

Padrauna and Tanda cities' demographic trends have been exemplified in the Fig. 4.4a, b. The population of the Padrauna increased from 7031 in 1901 to 49,723 in 2011, resulting into a growth of 607.59% during the period. Just like Padrauna, population in Tanda increased from 19,853 in 1901 up to 95,515 in the year 2011, drastically by 381.11%. Both cities remained stagnant from population levels between 1901 and 1921, because diseases were the norm. Since the 1951–1991 period, the decadal growth rates for both cities were particularly high, but no pattern was discerned. Since 1991 the decadal growth rate has slightly declining over the last two decades.

Tanda and Padrauna have projected populations of 59,668 and 116,667 as based on historical trends. These projections suggest that Tanda will classify as a Class I city (population greater than 100,000) while Padrauna is probably to be a Class II city (population between 50,000 to 99,999) as shown in Fig. 4.4a. The growth is a sign that these towns are continuing to be urbanized and becoming more important demographically, as shown in Fig. 4.4b.

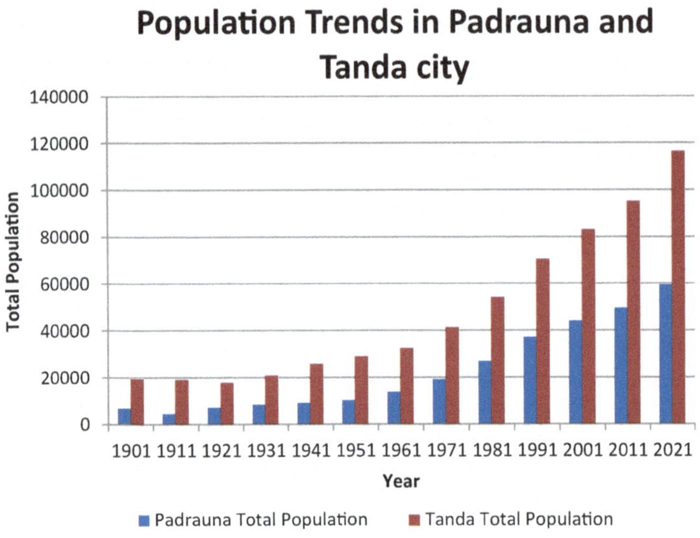

(i)

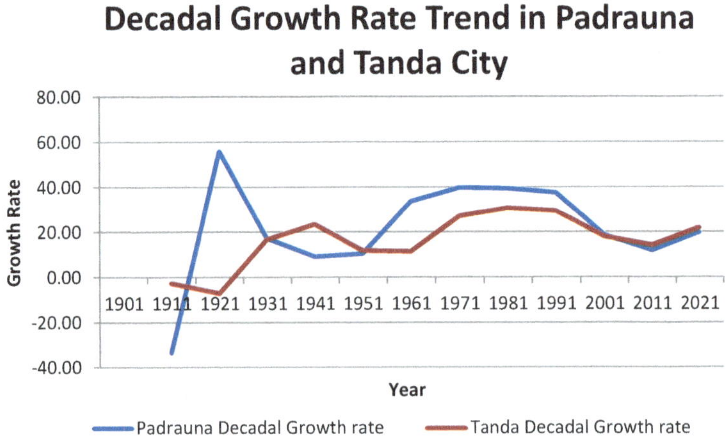

(ii)

Fig. 4.4 **a, b** Demographic trends of population and decadal growth rate of Padrauna and Tanda City. *Source* District Census Handbook, Census of India 2011 and Calculated by authors

In 2011 census, there is much similar pattern in the literacy status and educational infrastructure in Tanda and Padrauna cities. The literacy rate is 77.6% and schools (Primary, Middle, College, and Senior Secondary) are very less available per 10,000 populations and no college is available during studied time period in Tanda. Similar to Padrauna, the literacy rate here is 78.98% and the number of schools per 10,000 population is also very little. In fact, Padrauna has one college which provides one small advantage in higher education access, but lacks Tanda's dual colleges.

Data for the 2011 Census is used to produce the comparison between the medical infrastructure and the key manufactured commodities in Tanda and Padrauna towns. With 7 hospitals per 10,000 population in Tanda, such health care accessibility is evident. The town is famous for its vast production of clothes and agricultural instruments giving it an industrial touch. However, the Tanda has relatively better medical infrastructure as it has 12 beds per 10,000 population which means better healthcare availability in Padrauna. Padrauna is an agrarian and a household industry based economy so it is economically known to make jaggery, sugar and utensils.

The data shows that Padrauna barely has more advantage over Tanda in terms of healthcare infrastructure. Nevertheless, both towns have only small amount of medical facilities per capita and there is a need for better medical services. They also have a contrast in industrial and agricultural dependence in their economic activities.

Land Use and Land Cover

Changes in land use and land cover (LU/LC) are well acknowledged as one of the main factors causing changes in the environment in spatial and temporal scales (Turner et al. 1993). It is essential to monitor these changes to study urban growth and development pattern (Mundhe and Jaybhaye 2014). Environmental degradation is caused by rapid population growth, urbanization and industrialization that place great burden on land resources. Urban expansion leads to decreased landscape diversity, increasing dominance of urban areas, and a more continuous, homogeneous, and clustered urban landscape (Dewan et al. 2012). Land use pattern in any urban area is directly related with population growth and diversity of economic activity. One such land use change phenomenon is the result of various causes, of which the most prevalent are in developing countries where urban activities are spreading more extensively and population is growing rapidly. There are no exceptions in the two towns studied. The fast growing population, increasing urban functions, has heavily taxed land resource and the environment and urban services. Over the past decades, such changes have led to significant shifts in land use within the towns and its fringe areas. The land use and land cover (LULC) data have been classified into five major groups namely; water bodies, built up area, vegetation, agriculture and barren land (Table 4.5) to examine the way of the nature and level of these spatio-temporal land use changes.

The LULC changes in Padrauna and Tanda cities were observed between 2001 and 2021. The built up area of the city increased substantially at Padrauna in the

Table 4.5 Description of land use land cover type

Classes	Description of land use land cover type
Built-up	Residential, commercial, industrial and transportation facilities areas
Barren land	Parks garden, playground
Water bodies	River, lake, streams, wetlands, fishing ponds, canal
Vegetation	Open forest, dense forest. shrubs and bushes, trees, planation
Agriculture	Cropland, orchards, pastures

Source Based on the general scheme of the first level classification of land use and land cover

beginning, causing massive vegetation cover and agricultural land to be replaced by it (Fig. 4.5). Rapid urbanization is reflected in a 40.27% of the total area built up area which equals 19.33% in the year 2001, as shown in Table 4.6. Vegetation cover also decreased from 25.67% in 2001 to 17.57% in 2021 as well as agricultural land fell from 38.67 to 28.27% during the same period. From this time, LULC categories were transformed as urban functions were becoming highly dominant in the city. From 2001 to 2011, the most noticeable change has been conversion of agricultural land to built-up areas of 106.4 ha. Similarly, 42.76 ha of vegetated area had been turned into built up areas, which shows that urban expansion was still present, as shown in Table 4.7. From 2011 to 2021, this pattern persisted as concomitant agricultural land and vegetation were converted to built up spaces for national population and urban activities growth. In this sense, some changes in vegetation also registered some changes in agriculture, marking that there is a dynamic interplay between these categories, especially in the city's fringe area. From 2001 to 2011, about 26.83 ha of vegetation were converted to agricultural land, whereas this amount rose to 61.03 ha in the period from 2011 to 2021, as shown in Table 4.7.

This indicates that perhaps agricultural activities are slightly expanding among some who live in the peripheral areas where many residents are on the farm as their main source of livelihood. Besides, 8.44 ha of agricultural land transformed into barren during the period of 2011—2021. The fact that this has changed might suggest there is increasing pressure on such agriculture land on account of degraded soil or other environmental issues.

On the contrary, 8.01 ha of barren land were converted to built up areas during the same period as they are part of the ongoing urbanization efforts, as shown in Table 4.7. An important shift was that agricultural land was turned into vegetation: 19.17 ha in 2001–2011 and 33.64 ha in 2011–2021. It could be attributed to afforestation or replanting in some regions that were perhaps undertaken with intentions of environmental restoration or greenery in urban landscape.

Overall, the LULC changes in Padrauna city can highly be described by a complex pattern of rapid urbanization, decrease in vegetation cover, and dynamic change among major land use patterns such as agriculture and others. The city's development bears the fruit in increase of built up areas, but also suffers the side effects, the incipient loss of agricultural land and vegetation. The changes in these areas reflect

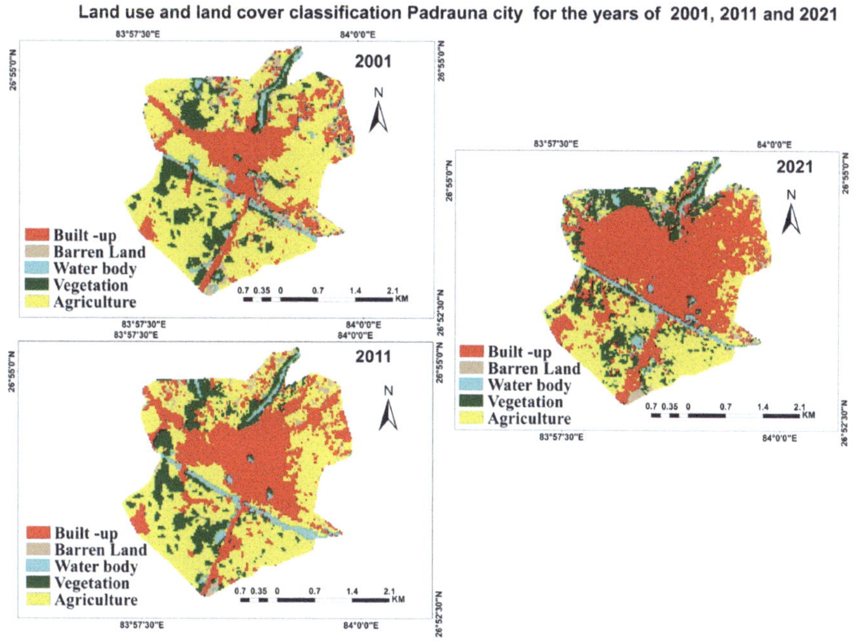

Fig. 4.5 Land use and land cover classification of Padrauna city for the years of 2001, 2011 and 2021. *Source* Based on remote sensing data (Landsat TM) for 2001–2021

Table 4.6 Land use/land cover change of Padrauna city, 2001–2021

Classes	Area in hectare			Area in percent			Percentage of relative change		
	2001	2011	2021	2001	2011	2021	2011–2001	2021–2011	2021–2001
Built- up	248.03	399.98	516.69	19.33	31.18	40.27	11.84	9.10	20.94
Barren land	48.85	46.06	35.84	3.81	3.59	2.79	– 0.22	– 0.80	– 1.01
Water body	160.67	151.67	142.38	12.52	11.82	11.10	– 0.70	– 0.72	– 1.43
Vegetation	329.34	286.58	225.38	25.67	22.34	17.57	– 3.33	– 4.77	– 8.10
Agriculture	496.11	398.71	362.71	38.67	31.08	28.27	– 7.59	– 2.81	– 10.40

Source Based on remote sensing data (Landsat TM) for 2001–2021

broader problems of urban growth, seeking to find balance between the need for development and the protection of natural resources and agricultural livelihoods.

Land use and land cover (LULC) changes in Tanda city from 2001 to 2021 portray major transformations, which comprise increasing the built up area at the expense of vegetation cover and agricultural land, as shown in Table 4.8 and Fig. 4.6. The urban growth took place and from 20.01% of built up area in 2001, the built up area expanded to 40.53% of built up area in 2021 (Table 4.8). On the contrary, vegetation cover experienced a decline from 33.29% in 2001 to 24.19% in 2021; agricultural

Table 4.7 LULC change in each class for Padrauna city, 2001–2021

Classes	2001–2011	2011–2021
	Area in hectare	Area in hectare
Agriculture to agriculture	361.71	295.13
Vegetation to agriculture	26.83	61.03
Water body to agriculture	10.17	6.55
Vegetation to vegetation	267.41	191.74
Agriculture to vegetation	19.17	33.64
Built up to built up	248.03	399.98
Agriculture to built up	106.4	55.5
Vegetation to built up	42.76	53.2
Barren land to built up	2.79	8.01
Agriculture to barren land	1.24	8.44
Barren land to barren land	34.65	25.11
Vegetation to barren land	10.17	2.29
Water body to water body	150.5	135.83
Vegetation to water body	1.17	6.55

Source Based on remote sensing data (Landsat TM) for 2001–2021

land decreased by 39.02–28.51% in the same period (Table 4.8). Such changes point to how urbanization is bearing down on the city's landscape.

Specific patterns of transformation of the LULC changes are found through detailed analysis as presented in Table 4.9. The main changes that occurred during the 2001–2011 period consisted of 89.81 ha of vegetation and 68.75 ha of agricultural land as converted to built up areas (Table 4.9). This trend continued in the next couple of decades when 96.95 ha of vegetation and 102.68 ha of agricultural land were turned into an urban use between 2011 and 2021. An additional important change was the transformation of agricultural land into barren land, at 7.6 ha in 2001–2011, and down to 2.7 ha in 2011–2021 (Table 4.9). This shift may be due tributes

Table 4.8 Land use/land cover change of Tanda city, 2001–2021

Classes	Area in hectare			Area in percent			Percentage of relative change		
	2001	2011	2021	2001	2011	2021	2001–2011	2011–2021	2001–2021
Built-up	368.82	533.74	747.23	20.01	28.95	40.53	8.95	11.58	20.53
Barren land	29.25	35.25	27.24	1.59	1.91	1.48	0.33	– 0.43	– 0.11
Water body	112.38	108.01	97.53	6.10	5.86	5.29	– 0.24	– 0.57	– 0.81
Vegetation	613.8	525.86	445.86	33.29	28.52	24.19	-4.77	-4.34	-9.11
Agriculture	719.28	640.67	525.67	39.02	34.75	28.51	-4.26	-6.24	-10.50

Source Based on remote sensing data (Landsat TM) for 2001–2021

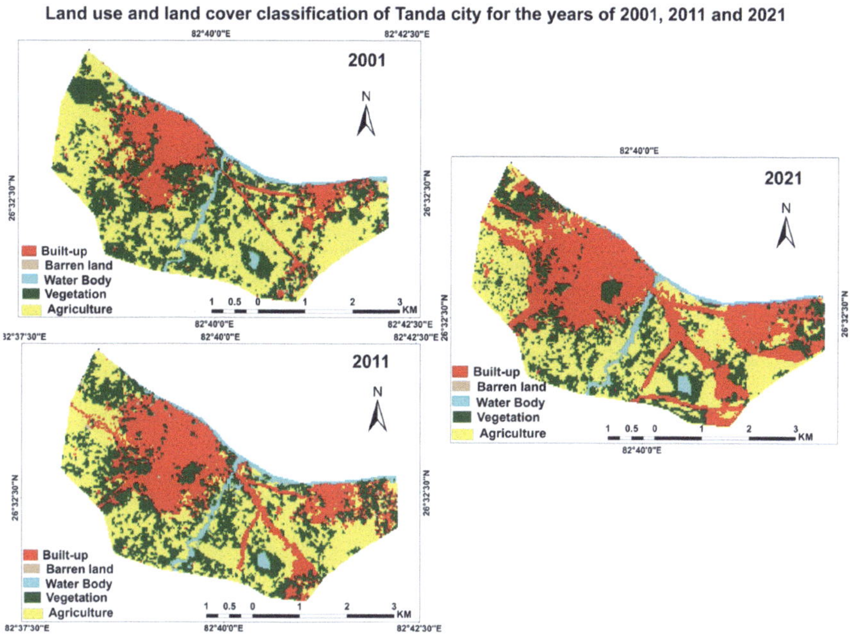

Fig. 4.6 Land use and land cover classification of Tanda city for the years of 2001, 2011 and 2021. *Source* Based on remote sensing data (Landsat TM) for 2001–2021

Land degradation or temporary land use transitions resulting from urban or industrial activity. In addition, a large amount of vegetation was converted to agricultural land and there was dynamic interaction between these two categories. The vegetation was transformed into agricultural land in 36.15 ha between 2001 and 2011 and it was 25.38 ha during 2011–2021 periods (Table 4.9). Such a pattern further implies that agricultural expansion in the city's peripheral areas is mainly due to population growth and a need for farming. Overall, over the 20 year period, we see that the changes to Tanda city's LULC are a rapid urbanization induced by expansion of built areas and the reduction in vegetation and agricultural land. These changes emphasize the requirement for balanced urban planning approach that includes environmental sustainability and preservation of green spaces and agrarian resources.

Temporal Patterns of Thermal Discomfort in Padrauna and Tanda City

The seasonal variations in Padrauna and Tanda city comfort and discomfort levels for the years 2001, 2011, and 2021 are a result of climatic factors, urbanization, and environmental changes and analyzed using the Discomfort Index (DI). In 2001

Table 4.9 LULC change in each class for Tanda City, 2001–2021

Classes	Area in hectare	
	2001–2011	2011–2021
Agriculture to agriculture	604.52	500.29
Vegetation to agriculture	36.15	25.38
Vegetation to vegetation	508.72	400.86
Agriculture to vegetation	17.08	45
Water body to water body	100.23	90.15
Vegetation to water body	6.74	6.03
Vegetation to barren land	4.76	3.15
Agriculture to barren land	7.6	2.7
Barren land to barren land	22.89	21.39
Built up to built up	368.82	533.74
Vegetation to built up	89.81	96.95
Agriculture to built up	68.75	102.68
Barren land to built up	6.36	13.86
Water body to agriculture	1.1	1.35

Source Based on remote sensing data (Landsat TM) for 2001–2021

and 2011, less than 50% of residents felt discomfort in the pre and post rainy sub season in Padrauna City (Table 4.10). However, during the rainy sub-season (May to September), there are more than 50% of the population that are uncomfortable to a certain degree. Between the period of 2018 and 2021, the discomfort became common as most residents experienced discomfort during this period. The reason for this is probably related to the rising temperatures, the urban heat island effect caused by more built up areas and higher relative humidity during the rainy season. However, the moderately warm January, February, November and December months experienced none of this, remaining very comfortable all three years with no record of discomfort from moderate temperatures.

In Tanda city, it was less than 50% of the population perceiving discomfort in 2001, 2011, and 2021 in the pre rainy season (Table 4.10). However, discomfort is more than 50% in all years during the rainy sub season (April to September) and was similar to Padrauna.

High humidity levels and urban densification have reduced vegetation and natural cooling to play major roles in generating this discomfort. Overall it was without pain as was Padrauna, however, Tanda had no discomfort toward the less cold months, since good weather prevailed. The discomfort experienced at certain time coincides with more than one cause. Compared to the dry and sandy sub season, high air humidity and temperatures, and especially rainy sub season, lead to dramatic amplification of discomfort, maintained under urban conditions of elevated densities. Urbanization and reduction in green cover has intensified the urban heat island effect, and now thermal discomfort is added to it at an even faster rate. Poor ventilation further

Table 4.10 Temporal pattern of thermal discomfort for the Padrauna and Tanda city for the year of 2001, 2011 and 2021

Month	Padrauna city			Tanda city		
	2001	2011	2021	2001	2011	2021
Jan	No discomfort	No discomfort	No discomfort	No discomfort	No discomfort	No discomfort
Feb	No discomfort	No discomfort	No discomfort	No discomfort	No discomfort	No discomfort
Mar	No discomfort	Less than 50% feels discomfort	Less than 50% feels discomfort	No discomfort	Less than 50% feels discomfort	Less than 50% feels discomfort
Apr	More than 50% feels discomfort	Less than 50% feels discomfort	Less than 50% feels discomfort	More than 50% feels discomfort	More than 50% feels discomfort	Less than 50% feels discomfort
May	More than 50% feels discomfort	More than 50% feels discomfort	More than 50% feels discomfort	More than 50% feels discomfort	More than 50% feels discomfort	More than 50% feels discomfort
Jun	More than 50% feels discomfort	More than 50% feels discomfort	Most of the population feels discomfort	More than 50% feels discomfort	More than 50% feels discomfort	More than 50% feels discomfort
Jul	More than 50% feels discomfort	More than 50% feels discomfort	Most of the population feels discomfort	More than 50% feels discomfort	More than 50% feels discomfort	More than 50% feels discomfort
Aug	More than 50% feels discomfort	More than 50% feels discomfort	Most of the population feels discomfort	More than 50% feels discomfort	More than 50% feels discomfort	More than 50% feels discomfort
Sep	More than 50% feels discomfort	More than 50% feels discomfort	Most of the population feels discomfort	More than 50% feels discomfort	More than 50% feels discomfort	More than 50% feels discomfort
Oct	Less than 50% feels discomfort	Less than 50% feels discomfort	More than 50% feels discomfort	Less than 50% feels discomfort	Less than 50% feels discomfort	Less than 50% feels discomfort
Nov	No discomfort	No discomfort	No discomfort	No discomfort	No discomfort	No discomfort
Dec	No discomfort	No discomfort	No discomfort	No discomfort	No discomfort	No discomfort
Annual	Less than 50% feels discomfort	Less than 50% feels discomfort	Less than 50% feels discomfort	Less than 50% feels discomfort	Less than 50% feels discomfort	Less than 50% feels discomfort

Source Calculated by authors

compounds the problem due to narrow streets and dense building. Moreover, climate insensitive infrastructure, including shaded area or heat reflective materials, aggravates discomfort when it is hot and humid. These seasonal discomfort trends would indicate that in both cities, less than 50% of the population had an annual discomfort in every year as of 2001, 2011, and 2021 (Table 4.10). Generally, comfortable conditions prevailed during the winter and post winter months, during which the temperatures were less, and heat stress was also low. Since these measures help to mitigate discomfort and increase green cover and focus on sustainable urban planning as well as the use of climate-resilient infrastructure are essential to keep up the urban livability. These could serve as incisive solutions to causes that are creating discomfort and improve the living quality of Padrauna and Tanda cities.

Discomfort zones (Pitman et al. 2015) are related closely to the quantity and quality of urban green spaces, which are key in reducing the impacts of these forces. Urban green areas such as parks, gardens and tree lined avenues help in lowering urban heat island effect, improving air quality, providing shade and are natural cooling agents. Dense urban areas (Gill et al. 2007) where the presence or degradation of urban green spaces is a great extent absent greatly increase discomfort. The decrease of vegetation cover between 2001 and 2021 is a clear tendency that can be observed in Padrauna and Tanda and has increased environmental stress and decreased the urban livability for residents. The implications for public health are enormous, and people who come in extended contact with unpleasant conditions run the risk of breathing problems, heat related illness as well as high stress levels to the mind (Gascon et al. 2015). Improving urban green spaces improves both the thermal comfort and encourages physical exercise, mental health and social cohesiveness (Elliott et al. 2020), and thus improving these spaces determines the creation of healthier, more resilient societies.

Environmental Concerns

The infrastructures between Tanda and Padrauna towns are so vast and it affects in terms of environmental degradation and sustainability. Based on pit systems, a total of 1590 units are in use and only twelve flush or pour flush toilets are available in Tanda, as shown in Table 4.10. Here, the reliance, however, is on pit systems so that if they are not maintained or sealed properly hold the potential for groundwater contamination. On the other hand, Padrauna has 1540 pit systems as opposed to (2803) flush or pour flush toilets (Table 4.10). Flush systems are more sanitary and do not require a well of water or adequate drainage to prevent some contamination of wastewater. The widespread usage of flush toilets in Padrauna, as well as its use of open drainage makes it prone to environmental difficulties resulting in the discharge of untreated waste into nature. Further, these two towns are differentiated by their drainage infrastructure. Not only is Padurana's open drainage system harmful to its environment and public health because it drowns, collects untreated wastewater, solid waste as well as runoff and subsequently attracts waterlogging, foul odour, and proliferation of diseases, but also at times it becomes a direct potential source for the outbreak of cases of cholera. But these pollutants can quite easily flow to local water bodies and cause damage to ecosystems. Tanda, with its mix of open and closed drainage systems, offers better containment and management of wastewater in certain areas.

Despite that, there still remains pollution. But these open drains which are in some parts of the town to say the least, unclean the land. A consequence of different levels of urbanization in the municipalities is the road infrastructure that reflects them. Padrauna is 49.61 km in total road length, whereas Tanda's road length is 71 km, as shown in Table 4.11. Improved road connectivity of Tanda encourages growth, but could lead to more automobile emissions, noise and heat island effects unless the development of the area is laced with a sustainable urban planning. Padrauna have 315 and Tanda 1032 road illumination spots (Table 4.11). Tanda's great lighting network might indicate new urban infrastructure, but it could also pretend the increased energy use which may or may not be accompanied by the reduction of pollution if renewable energy is not employed. Padrauna has fewer of illumination points, so if there is the dependency on inefficient or harmful temporary lighting sources, like diesel generators or kerosene lamp.

However, additional environmental issues are considered because of electricity usage patterns. Padrauna has 5242 home electrical and 108 industrial connections but in Tanda there is 7500 domestic and 143 industrial connections (Table 4.11). Tanda probably spends more energy, which may put a strain on resources and increase greenhouse gas emissions if the source of supply is not renewable. Energy demand in the business sector of Padrauna is evident from its higher number of commercial electricity connections (897) compared to Tanda (730), and will lead to carbon emissions if energy management is ineffective.

Both towns have environmental problems of similar if not more gravity. Padrauna's lack of closed drainage system, increased incidence of flush toilets and number of lit points are all causes for concern with regards to water pollution, waste management and energy efficiency. In particular, open drains are a scourge because untreated waste can pollute the environment through them. Groundwater pollution and air pollution risks are reflected in the extent use of pit systems and significant energy consumption caused by many electricity connections and road lighting points in Tanda. This calls for mitigating these issues which can be done through targeted interventions. For Tanda, better waste management systems and energy efficient infrastructure could be beneficial. Padrauna, on the other hand, may benefit from closed drainage system and good quality of wastewater treatment.

Discussion

Tanda alongside Padrauna witnessed major losses of urban green areas between 2001 and 2021 which exacerbated the environmental and health risks in rapidly developing parts of these areas. The growing population has depleted green spaces

Table 4.11 Status of sanitation, drainage, road and electrification in Tanda and Padrauna town in 2011

Towns	Sanitation			System of drainage	Road length (in km)	Electrification				
	Pit system	Flush/pour flush	Others			Domestic	Industrial	Commercial	Road lighting (points)	Others
Tanda	1590	12	1	Both Open and Closed	71	7500	143	730	1032	4
Padrauna	1540	2803	3	Open	49.61	5242	108	897	315	10

Source Census of India, 2011

which led to heat rise and magnified the urban heat island (UHI) effect alongside decreased urban life quality. Environment factors remain responsible for 24% of worldwide illness-related disease burden and 23% of mortality rates which affects heart disease alongside cancer and respiratory diseases and vector-borne illnesses (Guleria et al. 2019). Both cities have seen rising temperatures leading to heightened atmospheric development of dangerous chemicals which especially increases occurrences of ground-level ozone thus worsening respiration disorders including asthma and chronic bronchitis (Liu et al. 2023).

The combination of vegetation and water bodies in urban parks works as natural coolants to reduce LST by providing shade alongside plant water loss and evapotranspiration processes which both enhance sustainability and improve thermal comfort. Urban areas benefit from these green spaces because they both decrease high summer temperatures and defend against winter temperature loss thus serving as fundamental climate resilience elements for communities facing population growth while experiencing climate change impacts in Tanda and Padrauna. The presence of parks generates environmental benefits by trapping pollutants while increasing oxygen concentration which minimizes health risks stemming from heat-related disorders specifically affecting children and seniors.

Higher temperatures caused by the decline in Tanda and Padrauna's green spaces bring about UHI effects which impair heat regulation and lead to cardiovascular stress together with dehydration and respiratory disorders according to Singh et al. (2023). A lack of sufficient green spaces within urban areas prevents people from exercising while cutting off social opportunities and reducing mental health which leads to obesity and depression and anxiety conditions. This highlights the matter most pressing need of addressing health and environmental impacts of smaller green cover in urban areas because of growing demand for resources. To achieve good urban planning in Tanda and Padrauna, there must be a focus on building and preserving accessible, well designed parks that not only contribute to thermal comfort but to reduce health hazards and overall urban resilience. All of which are important for maximizing green space benefit such as HUI reduction and social well-being (Debbage and Shepherd 2015; Lopez 2020). However, as these cities expand over the coming decades, it is necessary that they plan urban green areas into such processes to ease this resource pressure and to keep their residents healthy and sustainable.

Conclusion

This is study of Tanda and Padrauna, two small and medium size centers in eastern Uttar Pradesh which are chosen to highlight the environmental and health problems which are characteristic of other such towns in Uttar Pradesh. The projected population trends indicate that these towns under study are consistency undergoing urban growth and urbanization which would result in increasing pressure on their basic infrastructure and resource base. The built up areas in the two towns have seen huge growth between 2001 and 2021, mostly at loss of agricultural land and vegetation cover. This urban green space transformation has resulted in the loss of considerable urban greens and consequently contributed to the increase in thermal discomforts and various health issues among residents. However, the UHI effect has been significantly amplified as a result of the loss of vegetation cover leading to higher land surface temperatures and comfortable living conditions. This Thermal stress has been related to the heat related illnesses such as heatstroke, dehydration and cardiovascular strain more especially of the vulnerable groups such as children and elderly. In addition, the green spaces have decayed, worsened air quality and led to respiratory diseases like asthma and chronic bronchitis. Many people don't have the opportunity for physical activity and relaxation of the mind which is reflected in the problems obesity, anxiety and depression. The environmental challenges that both towns face are linked to their inadequate infrastructure. Among our concerns in Padrauna are open drainage system, generic use of flush toilets and low road lighting. Especially open drains pose significant and bad impact over water quality and public health through untreated waste to the environment. However, a significant reliance on pit systems for sanitation and a greater number of electricity connection and road lighting points would be at odds with risks of water contamination and air pollution in Tanda. These challenges cannot be addressed unless there are targeted and strategic interventions. Improved waste management, integration of energy efficient infrastructure can be beneficial to Tanda, which would minimize the environmental impact. Environmental sustainability is very key for Padrauna to transition to a closed drainage system and to develop effective wastewater treatment practices. Both towns underscore the pressing need for urban planning that incorporates green infrastructure to enhance livability, mitigate the adverse effects of urbanization, and support sustainable development in eastern Uttar Pradesh.

References

Adger WN, Arnell NW, Tompkins EL (2005) Successful adaptation to climate change across scales. Glob Environ Chang 15(2):77–86

Ahmed M, Shuai C, Ahmed M (2022) Influencing factors of carbon emissions and their trends in China and India: a machine learning method. Environ Sci Pollut Res 29(32):48424–48437

Akram H, Hussain S, Mazumdar P, Chua KO, Butt TE, Harikrishna JA (2023) Mangrove health: a review of functions, threats, and challenges associated with mangrove management practices. Forests 14(9):1698

Amrit K, Pandey RP, Mishra SK (2021) Meteorological drought characteristics in eastern region of India. Hydrological extremes: river hydraulics and irrigation water management, pp 111–120

Ayers J, Huq S, Wright H, Faisal AM, Hussain ST (2017) Mainstreaming climate change adaptation into development in Bangladesh. In: Community-based adaptation. Routledge, pp 15–27

Battisti DS, Naylor RL (2009) Historical warnings of future food insecurity with unprecedented seasonal heat. Science 323(5911):240–244

Brundtland GH (1987) Our common future—call for action. Environ Conserv 14(4):291–294

Chakraborty K (2009) Vegetation change detection in Barak Basin. Current science, pp 1236–1242

Chaudhry D (2023) Climate change and health of the urban poor: the role of environmental justice. J Clim Change Health. https://doi.org/10.1016/j.joclim.2023.100277

Das S, Choudhury MR, Chatterjee B, Das P, Bagri S, Paul D et al (2024) Unraveling the urban climate crisis: exploring the nexus of urbanization, climate change, and their impacts on the environment and human well-being–A global perspective. AIMS Public Health 11(3):963

Debbage N, Shepherd JM (2015) The urban heat island effect and city contiguity. Comput Environ Urban Syst 54:181–194

Dewan AM, Kabir MH, Nahar K, Rahman MZ (2012) Urbanisation and environmental degradation in Dhaka Metropolitan Area of Bangladesh. Int J Environ Sustain Dev 11(2):118–147

Elliott H, Eon C, Breadsell JK (2020) Improving City vitality through urban heat reduction with green infrastructure and design solutions: a systematic literature review. Buildings 10(12):219

Feliciano D, Recha J, Ambaw G, MacSween K, Solomon D, Wollenberg E (2022) Assessment of agricultural emissions, climate change mitigation and adaptation practices in Ethiopia. Clim Policy 22(4):427–444

Gascon M, Triguero-Mas M, Martínez D, Dadvand P, Forns J, Plasència A, Nieuwenhuijsen MJ (2015) Mental health benefits of long-term exposure to residential green and blue spaces: a systematic review. Int J Environ Res Public Health 12(4):4354–4379

Gill SE, Handley JF, Ennos AR, Pauleit S (2007) Adapting cities for climate change: the role of the green infrastructure. Built Environ 33(1):115–133

Guleria R, Mathur V, Dhanuka A (2019) Health effects of changing environment. Natural resource management: ecological perspectives, 95–107

Gupta AK, Gavsker KK (2024) Regional urbanisation, rise and significance of census towns, and the challenges: a geographical study of Eastern Uttar Pradesh, India. Nagarok 56(2):135–157

IMD Annual Report (2021) India Meteorological Department, New Delhi. Accessed at https://metnet.imd.gov.in/docs/imdnews/ANNUAL_REPORT2021English.pdf

Izaguirre C, Losada IJ, Camus P, Vigh JL, Stenek V (2021) Climate change risk to global port operations. Nat Clim Chang 11(1):14–20

Jurgilevich A, Räsänen A, Groundstroem F, Juhola S (2017) A systematic review of dynamics in climate risk and vulnerability assessments. Environ Res Lett 12(1):013002

Leal Filho W, Azeiteiro UM, Balogun AL, Setti AFF, Mucova SA, Ayal D et al (2021) The influence of ecosystems services depletion to climate change adaptation efforts in Africa. Sci Total Environ 779:146414

Leppänen S, Saikkonen L, Ollikainen M (2014) Impact of climate change on cereal grain production in Russia: Mimeo. Agricultural goods and Bads: essays on agriculture and environmental externalities

Lipczynska-Kochany E (2018) Effect of climate change on humic substances and associated impacts on the quality of surface water and groundwater: a review. Sci Total Environ 640:1548–1565

Liu F, Chang-Richards A, Wang KIK, Dirks KN (2023) Effects of climate change on health and wellbeing: a systematic review. Sustain Dev 31(4):2067–2090

Lopez B (2020) Parks are critical urban infrastructure: perception and use of urban green spaces in NYC during COVID-19. Preprints. Accessed at https://www.preprints.org/manuscript/202008.0620/v2

Michel D, Eriksson M, Klimes M (2021) Climate change and (in) security in transboundary river basins. In: Handbook of security and the environment. Edward Elgar Publishing, pp 62–75

Mohan M, Gupta A, Bhati S (2013) A modified approach to analyze thermal comfort classification. Atmos Clim Sci 4(1):2014. Accessed at https://www.scirp.org/journal/paperinformation?paperid=40946

Mundhe NN, Jaybhaye RG (2014) Impact of urbanization on land use/land covers change using Geo-spatial techniques. Int J Geomatics Geosci 5(1):50–60

Ochoa JJ, Tan Y, Qian QK, Shen L, Moreno EL (2018) Learning from best practices in sustainable urbanization. Habitat Int 78:83–95

Panda A (2009) Assessing vulnerability to climate change in India. Econ Polit Wkly 44(16):105–107

Panda DK, AghaKouchak A, Ambast SK (2017) Increasing heat waves and warm spells in India, observed from a multiaspect framework. J Geophys Res Atmos 122(7):3837–3858

Pathak H (2023) Impact, adaptation, and mitigation of climate change in Indian agriculture. Environ Monit Assess 195(1):52

Pereira C, Flores-Colen I, Mendes MP (2024) Guidelines to reduce the effects of urban heat in a changing climate: green infrastructures and design measures. Sustain Dev 32(1):57–83

Pitman SD, Daniels CB, Ely ME (2015) Green infrastructure as life support: urban nature and climate change. Trans R Soc S Aust 139(1):97–112

Polydoros A, Cartalis C (2014) Assessing thermal risk in urban areas–an application for the urban agglomeration of Athens. Adv Build Energy Res 8(1):74–83

Roelofsen P (2016) A computer model for the assessment of employee performance loss as a function of thermal discomfort or degree of heat stress. Intell Build Int 8(4):195–214

Salunke AY (2021) Unravelling the genotoxic potential of agrochemicals on fish cell line, Doctoral dissertation, Maharaja Sayajirao University of Baroda (India). Accessed at http://okgw.hmlibrary.ac.in:8080/jspui/bitstream/123456789/1082/17/17.%20Synopsis.pdf

Schuurmans CJE (2021) The world heat budget: expected changes. In: Eisma D (ed) Climate change impact on coastal habitation. CRC Press, pp 1–15. Accessed at https://doi.org/10.1201/9781003069935

Seto KC, Dhakal S, Bigio A, Blanco H, Delgado GC, Dewar D, Huang L, Inaba A, Kansal A, Lwasa S, McMahon JE, Müller DB, Murakami J, Nagendra H, Ramaswami A (2014) Human settlements, infrastructure and spatial planning. In: Edenhofer O, Pichs-Madruga R, Sokona Y, Farahani E, Kadner S, Seyboth K, Adler A, Baum I, Brunner S, Eickemeier P, Kriemann B, Savolainen J, Schlömer S, von Stechow C, Zwickel T, Minx JC (eds) Climate change 2014: mitigation of climate change. Contribution of working group III to the fifth assessment report of the intergovernmental panel on climate change. Cambridge University Press, Cambridge, United Kingdom and New York, NY, USA

Shastry V, Mani M, Tenorio R (2016) Evaluating thermal comfort and building climatic response in warm-humid climates for vernacular dwellings in Suggenhalli (India). Archit Sci Rev 59(1):12–26

Singh S, Priyadarshni P, Pandey P (2023) Impact of urban heat Island: a local-level urban climate phenomenon on urban ecology and human health. In: Advanced remote sensing for urban and landscape ecology. Springer Nature Singapore, Singapore, pp 113–128

Sreenivasulu V, Bhaskar PU (2010) Change detection in land use and land cover using remote sensing and GIS techniques. Int J Eng Sci Technol 2(12):7758–7762

Turner BL, Moss RH, Skole DL (1993) Relating land use and global land cover change: a proposal for an IGBP-HDP core project. Stockholm: International Geosphere-Biosphere Programme: A Study of Global Change and the Human Dimensions of Global Environmental Change Programme

UN, (1992) Rio declaration on environment and development. United Nations, Rio de Janeiro

Weisheimer A, Palmer TN (2005) Changing frequency of occurrence of extreme seasonal temperatures under global warming. Geophys Res Lett 32(20)

Yadav MK, Singh RS, Singh KK, Mall RK, Patel CB, Yadav SK, Singh MK (2015) Assessment of climate change impact on productivity of different cereal crops in Varanasi, India. J Agrometeorol 17(2):179–184

Chapter 5
Urban Resilience in the Face of Climate Change: Lessons from Spain

Tarek Ben Hassen

Abstract With its Mediterranean climate, Spain is experiencing accelerated climate change impacts, including frequent heat waves, prolonged droughts, and extreme weather events that pose significant threats to urban infrastructure and public health. However, Spanish cities have had successful experiences with the implementation of resilience policies, particularly in adapting to climate change and addressing urban health risks. Accordingly, this chapter aims to explore the Spanish experience as a case study in integrating strategies for sustainable and resilient cities, focusing on how its urban planning, public health, and climate adaptation policies intersect. Based on policy analysis, case studies, and recent data, the chapter assesses Spain's response to these pressures, focusing on implementing climate-resilient infrastructure, promoting green spaces, and public health preparedness measures. A key focus of this case study is Spain's innovative approach to integrating health considerations into urban planning by promoting nature-based solutions and protecting vulnerable populations, such as the elderly, from heat-related illnesses and other climate-induced health risks. Particular attention is paid to successful initiatives such as Barcelona's "Superblocks" program and Madrid's urban forestry strategy. Finally, this chapter offers insights into Spain's successes, challenges, and lessons learned, providing valuable policy recommendations for other countries and regions with similar conditions, such as the MENA region, seeking to build resilient and sustainable cities.

Keywords Urban resilience · Climate change adaptation · Sustainable cities · Public health · Urbanization · Spain · Green infrastructure

---

T. Ben Hassen (✉)
Department of International Affairs, College of Arts and Sciences, Qatar University, Doha, Qatar
e-mail: thassen@qu.edu.qa

© The Author(s), under exclusive license to Springer Nature Switzerland AG 2025

A. I. Almulhim et al. (eds.), *Urbanization, Climate Change, and Health*, Climate Change Management, https://doi.org/10.1007/978-3-031-99132-5_5

Introduction

Climate change represents one of the most critical global challenges of the twenty-first century, exerting widespread effects on ecological systems, human societies, and economic frameworks, with certain regions and countries bearing a disproportionate burden (IPCC 2023). For instance, observation and model-based research have branded the Mediterranean area, including Spain, as one of the most notable climate change "hot spots." Indeed, since the 1980s, the Mediterranean area has seen faster climate warming than the global average (Christidis et al. 2023). If global greenhouse gas emissions are not significantly reduced, the Mediterranean area is projected to warm by 2 °C during the next 20 years. Under the worst-case emission scenario, warming might reach 4 °C by the end of the twenty-first century (Zittis et al. 2019). Likewise, elevated atmospheric evaporative demand, driven by rising temperatures, has intensified drought frequency, intensity, and duration, notably over the past five to six decades. This trend is projected to escalate in the future, exacerbating water scarcity in the Mediterranean region, as indicated by current climate projections (Vicente-Serrano et al. 2014). Extreme high temperatures, hot days, and nights are projected to rise significantly, with cooling needs in cities potentially increasing by 50–278% under a 2 °C global warming level (GWL) and 134–375% beyond 3 °C GWL. Urban heat island (UHI) effects are expected to amplify further these cooling demands (Coppola et al. 2021; Zinzi and Carnielo 2017).

Increasingly experiencing the effects of climate change, Spain is clearly on the frontlines of climate change in Europe. Increasing temperatures, extended heatwaves, less precipitation, and more frequent extreme weather phenomena, including severe storms and floods, present significant challenges for metropolitan regions (Lorenzo and Alvarez 2022). Spanish cities, where approximately 80% of the population resides, are particularly vulnerable to the escalating effects of global warming due to their diverse geographical and climatic conditions. These urban areas are increasingly experiencing prolonged droughts that strain water resources, intensifying heat waves that exacerbate the UHeffect, and acute water shortages that threaten both public health and economic stability. Consequently, combating climate change's impacts has become a top concern, with many communities implementing climate neutrality and adaptation strategies (Díaz-López et al. 2021). Madrid and Barcelona, two of Spain's most populous cities, are at the forefront of these challenges. Madrid is Spain's capital and biggest city. Madrid is a heavily populated urban region. As an inland city, it has an increasing UHI effect that aggravates rising summer temperatures, increasing its vulnerability (Suárez et al. 2024). On the other hand, Barcelona is dealing with both heat-related issues and water security issues, and its status as a coastline city is subject to increasing sea levels and floods (Chelleri 2018). These cities' dense populations, aged infrastructure, and need to balance urban development and climate adaption intensify their challenges. These environmental stresses highlight how urgently creative ideas catered to their particular setting are required. Accordingly, this chapter aims to explore the Spanish experience as a case study in integrating strategies for sustainable and resilient cities, focusing on how its urban planning, public health, and climate adaptation policies intersect.

Despite the growing body of research on climate change adaptation, a significant gap remains in examining how cities in Southern Europe combine urban resilience strategies with public health initiatives. The current body of literature predominantly emphasizes either the adaptation of technical infrastructure or broad climate policies, frequently neglecting the vital connection between climate risks—especially heatwaves—and health outcomes in densely populated urban settings. This is particularly significant in Mediterranean cities, which encounter severe climate challenges, including extreme heat, water scarcity, and air pollution. However, these cities also have sophisticated urban planning frameworks that could potentially serve as exemplary models for other regions.

This chapter's analytical focus on Madrid and Barcelona significantly fills a scholarly gap by offering a detailed examination of municipal-scale climate adaptation strategies explicitly designed to protect population health outcomes. It highlights innovative policies such as Barcelona's Superblocks and Climate Shelters, as well as Madrid's Urban Forest strategy, which are under-documented yet impactful examples of urban climate resilience with direct public health benefits. These initiatives demonstrate how integrating social equity, health protection, and environmental sustainability within urban planning frameworks can generate considerable health co-benefits. Additionally, the Spanish experience provides important empirical insights for countries in the Middle East and North Africa (MENA) region, where rapid urbanization, increased climate vulnerabilities, and weak health-adaptation integration present major governance challenges. The chapter analyzes these cases and fills a gap in international resilience and climate health studies. It offers transferable models that other cities can adapt to enhance climate resilience while safeguarding vulnerable populations.

This chapter is organized into four sections. The first outlines the methodological approach, followed by a section presenting the results under the theme "Climate Change, Urban Resilience, and Public Health in Spain: Lessons from Madrid and Barcelona." The third section provides a discussion of the key findings, while the final section offers the conclusion.

Methodology

This chapter applies a qualitative case study methodology to examine the urban resilience strategies of Spain in response to climate change, specifically highlighting the cases of Madrid and Barcelona. The choice of Spain is informed by its position as one of the most climate-vulnerable countries in Europe, providing significant insights into the interplay between urban planning, public health, and climate adaptation. The analysis is based on an extensive examination of secondary data, encompassing peer-reviewed journal articles, official reports, policy documents, and information from international organizations like the IPCC and the European Commission.

This includes Spanish municipal documentation and climate policy documents from Madrid and Barcelona, such as the "Climate Plan (2018–2030)" of the Ajuntament de Barcelona (regional government), and the "*Roadmap to Climate Neutrality by 2050" of* Madrid City Council, to gain a comprehensive insight into the initiatives and policies at the city level.

Moreover, the selection of Madrid and Barcelona as case studies is based on their substantial population, economic significance, and differing geographical settings, offering valuable insights into climate adaptation policies. Madrid, the capital of Spain, is situated at the heart of the country (40.42°N, 3.70°W). In 2022, it is home to over three million residents, 6.98 million when including the metropolitan area, making it the largest city in Spain, with a high population density of 5512 inhabitants per square kilometer (Eurostat 2022). Madrid has a continental Mediterranean climate, which makes it highly vulnerable to heatwaves and the urban heat island effect. Likewise, Barcelona, a major Mediterranean metropolis and capital of the Catalonia region, is characterized by its high population concentration, with more than 1.6 million residents occupying a relatively small urban area of just 100.4 km^2, with a population of 4.5 million in the metropolitan region. This results in a significant population density of approximately 15,873 inhabitants per square kilometer, making the city exceptionally compact and densely populated. As Spain's second-largest municipality and the sixth-most populous city in the European Union, Barcelona continues to grow, with its urban development extending beyond its official administrative borders (González et al. 2020). Barcelona faces the combined challenges of rising sea levels, air pollution, and heat.

Both cities have sophisticated urban climate resilience initiatives and house a sizable portion of Spain's population and economic activity. This chapter thoroughly examines innovative initiatives, including Barcelona's Superblocks and Climate Shelters and Madrid's Urban Forest strategy, evaluating their effectiveness, identifying challenges, and exploring their potential applicability to other regions.

Climate Change, Urban Resilience, and Public Health in Spain: Lessons from Madrid and Barcelona

The Climate-Health Nexus in Spain: A Comprehensive Overview

Spain is confronting the mounting consequences of climate change with increasing intensity. The country is grappling with rising temperatures, prolonged heatwaves, diminishing precipitation, and a growing frequency of extreme weather events, such as severe storms and devastating floods, which pose substantial challenges for its metropolitan areas (Fig. 5.1).

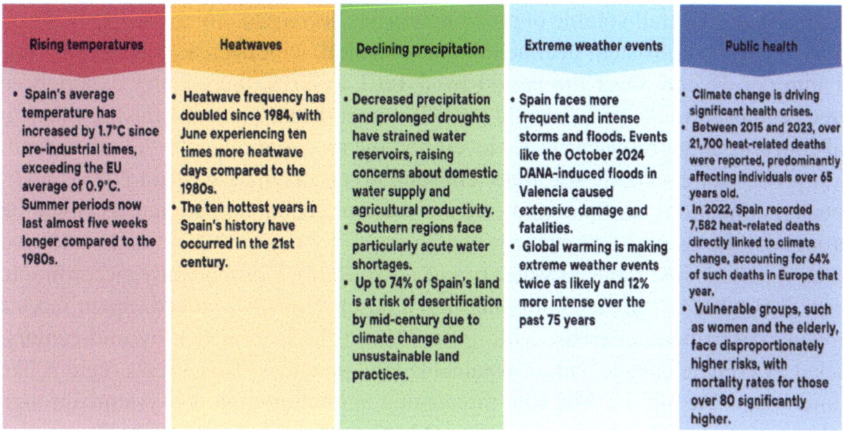

Fig. 5.1 Main climate change impacts on Spain

Firstly, Spain's average temperature has increased by 1.7 °C since pre-industrial times, compared to 0.9 °C for the rest of the EU. Between 1971 and 2022, temperatures in mainland Spain and the Balearic Islands rose by 3.27 °C. This increase significantly surpasses the global average of 1.19 °C and the Mediterranean regional average of 1.58 °C. The number of summer days—defined as days with maximum temperatures of 25 °C or above—increased significantly, rising from 82.4 in 1971 to 117.9 in 2022, marking a 43% increase. Over the past 50 years, summer has extended by an average of 36 days across the country. Simultaneously, the prevalence of tropical nights, where temperatures remain above 25 °C, surged dramatically, climbing from 1.73 in 1971 to 14.12 in 2022. Tropical nights are primarily concentrated in the Southern Plateau, the valleys of the Guadalquivir and Ebro rivers, and the Mediterranean coast, emphasizing the regional disparities in the warming trends' impact (IEA 2022; Roca et al. 2024). The temperature rise has been especially intense over the last decade. The ten hottest years in history in Spain have been experienced so far this century. For instance, in 2023, Spain experienced its hottest April day on record, hitting 38.8C. In some locations, records were beaten by a 5 °C margin (BBC 2023).

Secondly, the frequency of heatwaves in Spain has surged significantly, nearly doubling from less than one occurrence per year on average between 1971 and 1980 to almost two per year during the decade from 2013 to 2022. This sharp increase highlights the escalating impact of climate change on weather patterns. Moreover, the duration of these heatwaves has tripled, extending from an average of three days in the 1970s to a prolonged nine days in recent years. This intensification not only underscores the growing severity of extreme heat events but also exacerbates their adverse effects on public health and infrastructure. The prolonged nature of heatwaves increases the cumulative stress on urban centers, especially in densely populated regions like Madrid and Barcelona, where UHI effects further amplify these conditions (Lorenzo and Alvarez 2022; Roca et al. 2024).

Thirdly, the overall volume of precipitation has decreased. Since the 1950s, Spain has witnessed a consistent decline in annual rainfall, with decreases between 3 and 11 mm per decade. Over the period from 1961 to 2018, the country experienced three prolonged multi-year drought periods of significant intensity, raising substantial concerns about reservoir water levels and the potential impacts on both agricultural production and domestic water consumption (Greenpeace Spain 2023). Additionally, projections for Spain indicate a consistent decline in national precipitation throughout the twenty-first century, characterized by more frequent and prolonged drought periods and reduced water runoff (IEA 2022). Consequently, according to Roca et al. (2024), "Spain is slipping into a desert climate." Indeed, Spain faces a growing desertification crisis, with up to 74% of the land at risk by mid-century, fueled by climate change and unsustainable land practices. This affects regions like Almería and Castilla-La Mancha, threatening agriculture and ecosystems through soil degradation and reduced vegetation (Martínez-Valderrama et al. 2022).

Fourthly, the anticipated reduction in average annual precipitation does not preclude the possibility of extreme precipitation events. In fact, some regions of Spain may still experience increased torrential rainfall and flooding, highlighting the unpredictable and nuanced nature of climate change impacts. For instance, in October 2024, an isolated high-altitude depression (DANA) struck eastern Spain, causing torrential rainfall and devastating flash floods, severely impacting populated areas, human life, the environment, and infrastructure. Valencia, the third-largest city, was particularly affected, with at least 217 recorded fatalities. The floods are triggered by a deadly weather system in which cold and warm air mix and generate heavy rain clouds (European Commission 2024). However, a preliminary analysis by World Weather Attribution (WWA), a collection of worldwide scientists that examine the influence of global warming on severe weather, determined that over the past 75 years, extreme rainfall events during the September-December period in central and southeastern Spain have notably increased, with global warming causing these events to become roughly twice as likely and 12% more intense (World Weather Attribution 2024).

Fifthly, the impacts of climate change extend beyond environmental and economic dimensions, posing significant risks to public health (Iñiguez et al. 2021). Rising temperatures, prolonged heat waves, and deteriorating air quality exacerbate respiratory and cardiovascular conditions. At the same time, extreme weather events such as floods and storms heighten the risk of waterborne diseases and injuries. Indeed, the human cost of extreme temperatures in Spain has been substantial. According to Spain's daily mortality surveillance system (MoMo), the period from 2015 to 2023 saw more than 21,700 heat-related fatalities, with the majority of victims being individuals over the age of 65 (Euronews 2024). Climate change is, as well, driving significant public health crises in Spain, with extreme heat waves directly linked to thousands of deaths. Without rapid reductions in greenhouse gas emissions and comprehensive adaptation strategies, Spain will face even deadlier summers (Díaz et al. 2019).

For instance, the summer of 2022 is a stark example of escalating risks. Spain endured unprecedented high temperatures, with maximum anomalies of up to 8 °C above the seasonal norm in some regions. Cities like Madrid, Barcelona, and Valencia experienced daily temperatures exceeding Climate Shift Index levels of 5. This indicates that such extreme heat events would have been nearly impossible without the influence of human-induced climate change. Accordingly, research from the Barcelona Institute for Global Health (ISGlobal) highlights that in the summer of 2022, Spain recorded 7582 heat-related deaths attributable to climate change (Beck et al. 2024). This accounted for 64% of the 11,797 heat-related deaths during the period, with global warming doubling the likelihood of such fatalities compared to historical norms. Spain ranks as the second most affected European country after Italy in terms of heat-related mortality driven by climate change. The impact is particularly severe in Spanish southern regions, where global warming accounts for more fatalities than the European average. Vulnerable groups such as women and the elderly bear the brunt of these effects. Heat-related deaths among women attributable to climate change were 60% of the total, while those over 80 years old were disproportionately affected, with their mortality rate 13 percentage points higher than those under 64 (Beck et al. 2024).

In addition to these profound effects, climate change reshapes the urban landscape, presenting unique challenges for densely populated cities. To explore these localized impacts further, the following section examines the case of Madrid and Barcelona, offering insights into how these cities adapt to and mitigate climate change's effects.

The Climate Change Impact on Madrid and Barcelona

Spanish cities, home to 80% of the population, are especially exposed to the effects of global warming due to their geographical position, facing challenges such as droughts, heat waves, water shortages, and abrupt floods. Madrid and Barcelona, as important economic and demographic centers, confront a unique confluence of natural and human-caused climate threats. Urbanization, population density, tourism, and economic growth exacerbate these risks.

Based on the Köppen-Geiger classification, Madrid's climate is Mediterranean, bordering on semi-arid (Csa–BSk). Its elevation and inland location contribute to an intensified Mediterranean climate, marked by hot summers, cold winters, and precipitation concentrated mainly in spring and autumn (Núñez-Peiró et al. 2021; Rasilla et al. 2019). Summers in Madrid are notably hot, with an average maximum temperature of 32.1 °C in July, and extremely dry (Rasilla et al. 2019). The city features a diverse urban landscape with varied structural compositions, significant motorized transportation networks, and fragmented green spaces (Menéndez and Higueras García 2020). According to the Köppen-Geiger classification, Barcelona is classified under the "Csa" category, indicating a dry summer climate. Located near the coastline, Barcelona experiences a characteristic Mediterranean climate, featuring winters that are mild and moderately rainy and summers that are hot and relatively arid (Salvati et al. 2017).

Both Madrid and Barcelona face significant climate-related challenges exacerbated by their geographical locations and high levels of urbanization. These challenges include rising temperatures, frequent droughts, recurrent heatwaves, and increasing flood risks. In both cities, the UHI effect intensifies the already high summer temperatures, amplifying thermal stress for residents. Additionally, drought conditions strain water resources, jeopardizing the sustainability of urban water supplies and underscoring the urgent need for adaptive measures to enhance urban resilience (Table 5.1).

Firstly, in recent decades, both Madrid and Barcelona have experienced notable increases in average temperatures. Madrid's average annual temperature has risen by over 1.5 °C since the mid-twentieth century (Rasilla et al. 2019). Since 1780, Barcelona has experienced a gradual temperature increase of 0.08 °C per decade, translating to a cumulative rise of 1.8 °C. Climate projections indicate that if greenhouse gas emissions persist at their current trajectory, global average temperatures could increase by over 3° by 100 (Ajuntament de Barcelona 2024).

Secondly, this warming trend has led to more frequent and severe heatwaves, with summer temperatures often exceeding 35 °C and night-time temperatures remaining uncomfortably high. In Barcelona, heat waves have become increasingly frequent in recent years, with 2003 standing out as a particularly intense year, recording thirteen distinct heat wave events. The city has documented several heat waves since 2012. Future climate projections indicate that heat waves are expected to increase in frequency, severity, and duration. Night-time temperatures are anticipated to rise, and coastal neighborhoods will likely experience the most pronounced warming effects in 2100 (Ajuntament de Barcelona 2024). Likewise, Madrid has seen a rise in extreme hot days (EHD) frequency since the 1980s. Before 1980, such events were rare and typically lasted only one or two days, with some years recording no heatwaves. In contrast, post-1980 data shows a clear upward trend, with an average annual increase of 0.27 heatwave days, reflecting a stark shift in the city's summer climate patterns (Rasilla et al. 2019).

For instance, in 2024, both cities experienced prolonged periods of extreme heat, with 21 consecutive days under heat alerts (EL PAÍS 2024). The August 2024 heat wave was made at least five times more likely by human-induced climate change, according to the Climate Shift Index (CSI). During this period, Madrid recorded a maximum temperature of 36.3 °C with anomalies of 4.2 °C above average, while Barcelona experienced highs of 29.3 °C and anomalies of 2.4 °C. These anomalies exacerbate UHI effects in both cities, leading to extended discomfort and health risks (Climate Central 2024).

Thirdly, the UHI effect exacerbates high temperatures in both cities, leading to significant temperature differences between urban centers and surrounding rural areas. Previous research has shown that temperature differences in Madrid during the hottest summer periods can reach up to 8 °C, with variations depending on the city's urban structure (Sánchez-Guevara Sánchez et al. 2017). In both cities, the UHI effect intensifies already elevated summer temperatures, with recent years seeing recorded highs consistently above 40 °C (Sánchez-Guevara Sánchez et al. 2017). A comprehensive study on Barcelona's UHI effect, conducted by Martin-Vide

Table 5.1 Comparative impacts of climate change on Madrid and Barcelona

Climate impact	Madrid	Barcelona
Rising temperatures	– Average yearly temperatures have increased by over 1.5 °C since the mid-twentieth century, with more noticeable warming during summers – Heat anomalies during extreme events, such as the 2024 heatwave, exceeded 4.2 °C above average	– Cumulative temperature increase of 1.8 °C since 1780, with a gradual rise of 0.08 °C per decade – The 2024 heatwave saw anomalies of 2.4 °C above average
Increased heatwaves	– Extreme hot days (EHD) have increased significantly since the 1980s, with post-1980 data showing a sharp upward trend. Heatwaves now last longer, averaging up to 9 days	– Heatwaves are frequent and intense, with summer temperatures exceeding 35 °C and inadequate night-time cooling – Night-time temperatures are anticipated to rise, and coastal neighborhoods will likely experience the most pronounced warming effects in 2100
Urban heat island effect	– UHI intensifies heat effects, creating temperature differences of up to 8 °C between urban centers and rural surroundings	– UHI contributes to night-time temperatures averaging 2 °C higher in urban areas, especially compared to non-urban observatories like Barcelona Airport
Degraded air quality	– Heatwaves degrade air quality, with PM10 levels increasing by 60% in urban areas, often linked to Saharan dust and worsened by southerly winds	– Air quality remains a challenge, especially in high-traffic areas and during extreme heat, with persistent issues from vehicular emissions and reduced dispersion
Health impacts	– Heatwaves increase respiratory and cardiovascular disease hospitalizations, particularly among vulnerable populations – Approximately 4% of summer deaths in Madrid are attributed to high temperatures, with over 2000 deaths reported in the summer of 2024, primarily linked to the lack of climate shelters	– Vulnerable populations, including low-income households and the elderly, face significant health risks, with 24% of households experiencing energy poverty

Source Compiled by the authors based on reviewed literature

and Moreno-Garcia (2020), revealed significant temperature variations between the city and surrounding areas. By analyzing data from 2004 to 2013, they discovered that Barcelona's night-time air temperatures were, on average, nearly 2 °C higher than those recorded at the Barcelona airport, which (serves as an ideal non-urban observatory due to its location and elevation).

Fourthly, heatwaves in Madrid significantly degrade air quality, with notable increases in pollutant concentrations, particularly PM10 levels, which rise by over 60% in urban areas compared to less than 25% in suburban regions. PM10 exceedances during heatwaves are up to 12 times more likely than on typical summer days, especially in the city center, driven by southerly winds displacing urban plumes and contributions from Saharan dust outbreaks. Air quality indices shift to worse categories during heatwaves, particularly at urban traffic stations. Saharan dust is identified as a major contributor to elevated PM10 levels, exacerbating the air quality challenges associated with heat waves (Rasilla et al. 2019).

Fifthly, the health impacts of heatwaves are a growing concern in Madrid and Barcelona, particularly for vulnerable groups like the elderly and those with underlying medical disorders. In Madrid, long-term exposure to heat has been linked in studies to higher hospitalizations for respiratory and cardiovascular conditions (Sánchez-Guevara Sánchez et al. 2017). For instance, approximately 4% of summer deaths in Madrid are attributed to high temperatures, estimated to result from the UHI effect (Gallo et al. 2024). During the summer of 2024, more than 2000 deaths were attributed to high temperatures, the majority occurring in the province of Madrid, which still lacks a network of climate shelters.

Moreover, climate change disproportionately affects vulnerable populations, exacerbating existing social and economic inequalities. Factors such as physical health, age, gender, socioeconomic status, and living and working conditions influence the extent to which individuals are affected. Consequently, marginalized groups, such as low-income families, the elderly, and women, are often the most impacted due to limited access to resources, inadequate housing, and reduced capacity to adapt to extreme weather events like heat waves and floods. Additionally, the severity of climate change impacts varies across city districts, influenced by building quality, vegetation cover, impervious surface percentages, urban density, heat source distribution, and local socioeconomic conditions.

For instance, heatwaves are increasingly affecting Madrid and Barcelona, with their impact varying significantly across different districts due to a combination of environmental and sociodemographic factors. This variation highlights the critical role of income levels, access to cooling systems, age distribution, and urban green spaces in shaping vulnerability to extreme heat. In Barcelona, an estimated 24% of households face energy poverty, with women and individuals born in lower-middle-income countries disproportionately affected, according to the Public Health Agency of Barcelona. This issue is most prevalent in the districts of Ciutat Vella, Sants-Montjuïc, and Nou Barris, highlighting the unequal territorial distribution of energy poverty across the city (Ajuntament de Barcelona 2024). The following box summarizes key findings from a study analyzing the interplay of these factors in determining heatwave-related mortality in Madrid (Box 1).

Box 1: Heatwave Impacts and the Role of Sociodemographic Factors in Madrid

- Vulnerable populations, including low-income groups and those without adequate cooling systems, bear a disproportionate burden of heatwave-related mortality. Socioeconomic inequalities exacerbate these vulnerabilities.
- Disparities in Heatwave Impact: Income level was the most significant determinant, with lower-income districts facing higher mortality risks. Wealthier districts showed a reduced impact, likely due to better access to cooling systems and healthcare resources.
- The presence of air conditioning units significantly mitigated the impact of heatwaves. However, even in households with air conditioning, usage was limited in lower-income groups due to energy costs, reflecting the influence of energy poverty.
- Increased green zones were associated with reduced heatwave impacts in univariate analyses. However, their significance diminished in multivariate models, indicating that urban green spaces alone may not suffice to offset risks in vulnerable districts.
- Surprisingly, districts with higher percentages of elderly residents did not exhibit a statistically significant heatwave impact. This was attributed to the higher income levels in these districts, which provided resources for adaptive measures.

Source: López-Bueno et al. (2020).

Urban Resilience Framework in Madrid and Barcelona

Madrid and Barcelona have implemented innovative frameworks to strengthen urban resilience and adapt to the growing challenges of climate change. These cities have embraced strategies such as enhancing public health systems, expanding green infrastructure, and promoting community-driven initiatives. Their approaches offer valuable insights into creating sustainable and climate-resilient urban environments.

Madrid has established itself as a pioneering city in urban innovation and structural transformation, consistently developing progressive public policies that align with European economic and environmental commitments. The city has particularly distinguished itself by advancing towards a low-carbon, circular, and ecosystem-friendly economy, with a notable emphasis on Nature-Based Solutions (NBS) as a strategic approach to enhancing urban resilience in the context of climate change (Oquendo Di Cosola et al. 2020). In March 2021, Madrid's city council released the Roadmap to Climate Neutrality by 2050, an ambitious strategic document that outlines a comprehensive approach to reducing greenhouse gas emissions (Madrid City Council 2021). The roadmap aims to achieve a 65% reduction in emissions within the city's boundaries by 2030, with the ultimate goal of reaching climate neutrality by 2050. This policy framework builds upon Madrid's previous decarbonization efforts over the past 15 years and explicitly aligns with international climate initiatives, specifically the Paris Climate Agreement and the European Green Deal. The roadmap represents the city's strategic response to climate change and sustainability challenges, proposing a diverse set of policy initiatives and potential scenarios to guide Madrid's environmental transformation (Ciambra et al. 2023).

Further, Madrid's urban environmental strategy is embodied in two complementary initiatives: the Madrid + Natural Plan and Plan A: Air Quality and Climate Change Plan. The Madrid + Natural Plan represents a comprehensive approach to developing green urban infrastructure strategically organized across three scales—building, neighborhood, and city-wide. Its multi-faceted objectives include improving building energy performance, regulating rainwater flow, creating green spaces, mitigating the UHI effect, enhancing air quality, and implementing systematic monitoring to assess the cost–benefit ratio of these environmental solutions (Ciambra et al. 2023). Plan A builds upon this foundation, serving as a dynamic tool to reduce air pollution, prevent climate change, and define adaptive strategies. Structured around four key areas—sustainable mobility, urban regeneration, climate change adaptation, and public awareness—the plan adopts a holistic ecological vision. This approach encompasses diverse urban domains, including mobility, public awareness, building rehabilitation, local energy production, water management, and urban renaturation. The plans reflect Madrid's sophisticated and integrated approach to urban environmental resilience and sustainability (Oquendo Di Cosola et al. 2020).

For instance, in 2019, Madrid adopted the Urban Forestry Strategy. A cornerstone of this strategy is the Madrid Green Belt ("Bosque Metropolitano"). This 75-km forest belt encompasses a total projected area of 32,035 ha, with 81% of existing natural spaces and the remainder designated for planting on wasteland and reclaimed landfills. The plan includes planting two million trees over the next decade on 2300 ha of peripheral land, home to approximately 1.2 million residents, half privately owned. Initially envisioned as a fully public urban infrastructure with an estimated cost of €3.6 million, the project has garnered significant private sector support, with businesses sponsoring planting activities through the "Foro de Empresas por Madrid" initiative. These partnerships underscore the collaborative nature of the project, combining public and private efforts to enhance urban resilience, mitigate climate impacts, and provide vital green spaces for the city's inhabitants (Regalado Zamora et al. 2021; Romero-Muñoz et al. 2023).

Another example is the "Refúgiate en la Cultura," an initiative by Madrid's City Council to offer residents and visitors a respite from high summer temperatures by engaging in cultural activities. During July and August 2024, the program featured 46 brief flamenco performances in major museums such as the Prado, Reina Sofía, Thyssen-Bornemisza, and the Royal Collections Gallery. These 15-min shows took place between 3:00 PM and 5:00 PM, the hottest part of the day, providing an opportunity to enjoy art in cool environments (Veranos de la Villa 2024). Additionally, municipal museums and libraries hosted monologues by comedian Sergio Pazos, further enriching the cultural offerings. The program also collaborated with cinemas across the city to offer discounted tickets for afternoon screenings before 5:00 PM, encouraging people to seek refuge from the heat in air-conditioned spaces while enjoying cultural content (Gacetín Madrid 2024).

On the other hand, Barcelona has consistently demonstrated leadership in addressing resilience and climate change, implementing forward-thinking initiatives such as the Citizen Commitment for a More Sustainable Barcelona, formally declaring a Climate Emergency, and aligning its strategies with the 2030 Agenda. In response to a series of critical events between 2006 and 2008—including extreme drought, railway service disruptions, and electrical blackouts—Barcelona initiated urban resilience projects focused on infrastructure and service improvements. These efforts were designed to systematically address and mitigate the vulnerabilities exposed by these challenging episodes (Urban Resilience Hub 2022). Accordingly, Barcelona established a governance-centered framework integrating public–private partnerships, technical expertise, and international collaboration. Central to this model is the creation, in 2009, of the Taula de Infraestructuras I Serveis Urbans (TISU), which coordinated over 40 resilience projects and engaged diverse stakeholders. The city successfully leverages infrastructure improvements, global networks (e.g., UN-Habitat and Rockefeller 100 Resilient Cities), and strategic urban planning initiatives to address risks like flooding, drought, and social inequities (Chelleri 2018).

So, as part of the policies for changing the model to make the city a greener, fairer, more efficient, and healthier place, various government measures and strategic plans have been drawn up over the last few years to achieve the climate goals. In 2018, the regional government adopted the Climate Plan (2018–2030), a rigorous, well-grounded roadmap that aims to mitigate and adapt to the effects of climate change in Barcelona. It has a budget of over €1.8 billion to accelerate the climate transition and six programs with specific actions (Table 5.2).

For instance, the Barcelona City Council has created a network of more than 350 climate shelters where citizens can shelter from the heat during the summer and the cold during the winter. Most of them are open all year round (Box 2).

Box 2: Characteristics of the Climate Shelters in Barcelona

- A network of 350 venues
- These areas serve a dual purpose, providing refuge from extreme temperatures while preserving their original functional characteristics.
- These climate-resilient spaces encompass both indoor and outdoor environments. Indoor locations such as libraries and civic centers offer shelter, while outdoor spaces like parks and block interiors also provide relief from temperature extremes.
- These climate refuge spaces are universally accessible at no cost, except municipal swimming pools, which offer entry at regulated public-sector rates and include subsidized pricing options.

Table 5.2 The main elements of the climate plan (2018–2030) of Barcelona

Mobility plan	**Energy transition plan**
• Improving public transport (L8, L9, Rodalies trains, Tram) • Electrifying the fleets and spreading the Endolla Barcelona charger network • Transforming public space to foster sustainable mobility on foot, by bicycle or on public transport	• Multiplying the amount of renewable and local energy generated by five • Expanding the heating and cooling networks in 22@ and Zona Franca and opening a new area in La Sagrera • Stepping up the energy renovation of buildings
Heat plan	**Neighbourhood climate plan**
• Expanding the climate shelter network and adding water play areas in public spaces • School climate plan and residencies climate plan to work towards equipping these spaces with air conditioning • Shade plan • Improving the response to heat emergencies	• Including the climate perspective as a working area in the neighbourhood plan • Defining the health and climate change strategy • Implementing the local and interior spaces programme (PEPI)
Resilience plan	**Let's change for the climate plan**
• Adapting to constant drought and rolling out an alternative water resources plan • Preparing the coastline • Lowering the risk of floods and forest fires • Enlarging and improving green areas	• Setting up a climate advisory council • Strengthening the more sustainable Barcelona Network and fostering a results-orientated approach to the city's challenges • Announcing more calls for climate subsidies • Holding the 2nd citizen climate assembly • Developing the carbon neutral city council programme • See the complete government measure for the climate plan here

Source Ajuntament de Barcelona (2018)

- These spaces prioritize accessibility and comfort, ideally featuring seating areas like chairs or benches and providing complimentary drinking water to enhance visitor comfort and usability.
- For indoor spaces, it is advised to set air conditioning systems to 26 °C during summer and 21 °C during winter.
- Some of these spaces offer activities or reading materials to make the visit more enjoyable.
- By 2030, the city aims to ensure that every resident in Barcelona has access to a climate shelter within a 5-min walk, with identified and mapped optimal routes.

Source: Ajuntament de Barcelona (2024), Pede (2024)

Another example is the "Superblocks" initiative, which represents a transformative urban planning strategy to reduce traffic, improve air quality, and enhance public health. It was adopted in 2016, initially in the Poblenou neighborhood. These car-free zones group streets into pedestrian-friendly areas, reallocating space once dominated by vehicles to green spaces, play areas, and walking or cycling paths (Zografos et al. 2020). A study by the Barcelona Institute for Global Health estimates that full implementation of all 503 planned Superblocks could save 667 lives annually by reducing NO2 levels by 24%, mitigating traffic noise and heat island effects, and encouraging physical activity. Initial resistance has given way to widespread support as communities enjoy improved quality of life, increased local business activity, and enhanced public spaces, positioning the model as a pioneering example of sustainable urban development (Palència et al. 2020). Also, as part of the Superblock model, the Barcelona City Council is rolling out the "Eixos Verds" Plan to extensively green urban streets, with many potential mental health benefits (Vidal Yañez et al. 2023).

Discussion

This chapter explores the diverse approaches Madrid and Barcelona took to address the growing challenges of climate change and enhance urban resilience. The journey toward urban resilience in Madrid and Barcelona exemplifies the complexities of addressing climate change in densely populated, economically vibrant cities. However, the path is fraught with challenges that require immediate attention to ensure sustainable progress.

Firstly, one of the most significant challenges is the stark socioeconomic disparities that influence the ability of residents to adapt to climate risks. Vulnerable populations, including low-income families, the elderly, and those in marginalized districts, are disproportionately affected by heatwaves, air pollution, and inadequate infrastructure (López-Bueno et al. 2020).

Secondly, Madrid and Barcelona encounter significant challenges in harmonizing efforts across various departments and stakeholders. The fragmentation of responsibilities often delays project execution, misaligned priorities, and inefficient resource utilization. For instance, the Poblenou superblock project in Barcelona faced significant opposition driven by everyday political struggles over municipal authority and competing visions for the city's future. The project revealed that urban transformation often hinges not just on environmental or quality-of-life benefits but also on short-term political gains and competitive urbanism. This case underscores that resistance to transformational adaptation can stem not only from concerns about material changes but also from debates over who holds the authority to define and act on "the common good" (Zografos et al. 2020).

Thirdly, the aging infrastructure in both cities poses a considerable challenge. Upgrading buildings for improved energy efficiency, enhancing stormwater management systems to cope with extreme weather, and providing equitable access to green spaces demand significant financial and logistical resources. The urban heat island effect alsontensifies the need for innovative infrastructure strategies to combat rising temperatures (Suárez et al. 2024). Furthermore, Barcelona and Madrid have adopted different yet complementary strategies to address the challenges posed by climate change. Barcelona's Climate Shelters program effectively responds to extreme heat, providing secure environments for at-risk groups during heatwaves and fostering community engagement and inclusivity. In contrast, Madrid has prioritized its urban forestry initiatives to enhance green infrastructure to mitigate the UHI effect and improve air quality.

The case studies expose complex implementation dynamics spanning institutional, social, and political aspects that extend beyond technical solutions. For instance, Madrid's Urban Forest initiative offers a potentially effective strategy for tackling the urgent need for comprehensive nature-based interventions at the city's edges. However, its success depends on navigating conflicting land-use priorities, addressing administrative jurisdictional challenges, and guaranteeing fair access for residents in socioeconomically disadvantaged areas. The way green infrastructure investments are spread out often reflects current socio-spatial inequalities, which could reinforce environmental privilege unless intentionally designed with equity as a core principle. Likewise, Barcelona's Climate Shelter network focuses on demographically vulnerable populations but faces considerable operational challenges. These include the methodological complexities of accurately identifying and mapping at-risk demographic segments, setting appropriate vulnerability thresholds that balance inclusivity and resource limitations, and ensuring ongoing financial support for long-term maintenance and community engagement efforts. Establishing these shelters uncovers conflicts between centralized planning methods and the unique requirements of neighborhoods, emphasizing the significance of participatory governance strategies that integrate local insights and preferences.

These case studies indicate that effective urban resilience strategies must fundamentally intertwine with social equity considerations during their conceptualization, implementation, and evaluation phases. Climate adaptation interventions that do not explicitly address equity dimensions may worsen existing socioeconomic disparities, potentially leading to a situation where the "resilience dividends" are primarily beneficial to privileged populations. At the same time, vulnerable communities remain inadequately protected against escalating climate hazards. Indeed, this possibility of exacerbated inequities is extensively documented in the literature on urban climate adaptation, distinguishing between several types of adaptation injustice (Shi et al. 2016). As Anguelovski et al. (2016) highlighted, urban adaptation injustices can be categorized into two kinds: acts of commission, when interventions adversely impact or displace disadvantaged populations, and acts of omission, wherein they safeguard and favor elite groups to the detriment of the urban poor. The adaptation strategies should also explore the beneficial effects for socioeconomically vulnerable populations, such as the reduction of urban heat island effects through increased tree canopy coverage and high-albedo surfaces, as well as the possible negative socioeconomic consequences linked to these interventions. This includes the observed relationship between greater vegetation density and rising property values, which may lead to housing affordability issues and the displacement of economically marginalized communities due to gentrification.

Finally, the chapter has some limitations. Firstly, the analysis focuses solely on Madrid and Barcelona as case studies, which may limit the applicability of the findings to other urban areas in Spain or the wider Mediterranean region that could have different socio-political, economic, and environmental traits. Secondly, the methodological approach primarily depends on secondary data sources, including municipal policy frameworks, institutional reports, and peer-reviewed literature, while lacking the incorporation of primary empirical data collection methods, such as qualitative interviews with key stakeholders or direct observational studies. These limitations indicate the necessity for more thorough and diverse research methodologies in upcoming studies focused on urban climate resilience within Mediterranean settings.

Conclusion

This chapter has explored the climate adaptation strategies applied by Madrid and Barcelona, focusing specifically on how they incorporate public health considerations and the potential for these strategies to be applied in other contexts within the Mediterranean and MENA regions. The analysis uncovers important insights that can enhance urban climate resilience in similar climatic and urban environments. The experiences of Barcelona and Madrid demonstrate that addressing climate change requires a multi-faceted approach that combines innovation, inclusivity, and adaptability. These experiences serve as a valuable reference for other European cities and provide important lessons for regions like the MENA, which encounter similar climatic challenges yet frequently lack effective adaptation strategies (Ben Hassen and Hageer 2024; Eltinay and Egbu 2024). For cities worldwide, particularly those in the MENA region, Spain's examples of Climate Shelters and urban forestry offer practical and scalable solutions. However, these must be adapted to local contexts, focusing on equitable implementation to ensure no community is left behind in the fight against climate change. Additionally, these experiences highlight how carefully context-specific factors—such as urban morphological density, socioeconomic stratification, and governance fragmentation—must be systematically addressed within comprehensive climate adaptation plans.

However, further research and analysis are needed to evaluate the long-term effectiveness of both cities' approaches and identify best practices for enhancing urban resilience in climate change. This research should focus on assessing the social, economic, and environmental impacts of different strategies and exploring innovative solutions to address the complex challenges of climate change. Sharing knowledge and collaborating across cities and regions can accelerate the transition towards more resilient and sustainable urban environments.

Acknowledgements In preparing this chapter, Grammarly was used solely for grammar refinement, minor rephrasing to enhance clarity, and correcting English errors. This tool was employed to improve the readability and technical accuracy of the text while ensuring linguistic precision. Its use was limited to language enhancement and did not influence the substantive content, analysis, or arguments presented. The intellectual contributions, critical analysis, and overall structure of the chapter remain entirely the author's, developed through rigorous research and scholarly engagement with the subject matter.

References

Ajuntament de Barcelona (2018) Climate plan (2018–2030). https://www.barcelona.cat/barcelona-pel-clima/sites/default/files/climate_plan_maig.pdf

Ajuntament de Barcelona (2024) Barcelona for climate. https://www.barcelona.cat/barcelona-pel-clima/en/specific-actions/heat-waves?utm_source=chatgpt.com

Anguelovski I, Shi L, Chu E, Gallagher D, Goh K, Lamb Z, Reeve K, Teicher H (2016) Equity impacts of urban land use planning for climate adaptation. J Plan Educ Res 36(3):333–348. https://doi.org/10.1177/0739456X16645166

BBC (2023) Climate change: Spain breaks record temperature for April. https://www.bbc.com/news/science-environment-65403381

Beck TM, Schumacher DL, Achebak H, Vicedo-Cabrera AM, Seneviratne SI, Ballester J (2024) Mortality burden attributed to anthropogenic warming during Europe's 2022 record-breaking summer. Npj Clim Atmos Sci 7(1):245. https://doi.org/10.1038/s41612-024-00783-2

Ben Hassen T, Hageer Y (2024) Urban climate resilience in MENA region: opportunities and challenges of nature-based solutions. In: Leal Filho W, Nagy GJ, Ayal DY (eds) Handbook of nature-based solutions to mitigation and adaptation to climate change. Springer International Publishing, pp 1–23. https://doi.org/10.1007/978-3-030-98067-2_161-1

Chelleri L (2018) Barcelona experience in resilience: an integrated governance model for operationalizing urban resilience, pp 111–127. https://doi.org/10.1007/978-3-319-75798-8_6

Christidis N, Mitchell D, Stott PA (2023) Rapidly increasing likelihood of exceeding 50 °C in parts of the Mediterranean and the Middle East due to human influence. Npj Clim Atmos Sci 6(1):45. https://doi.org/10.1038/s41612-023-00377-4

Ciambra A, Stamos I, Siragusa A (2023) Localizing and monitoring climate neutrality through the sustainable development goals (SDGs) framework: the case of Madrid. Sustainability 15(6):4819. https://doi.org/10.3390/su15064819

Climate Central (2024) Climate change-fueled heat impacts Spain amid ongoing wildfire threat in the Mediterranean. https://www.climatecentral.org/climate-shift-index-alert/spain-and-mediterranean-august-21-2024

Coppola E, Nogherotto R, Ciarlo' JM, Giorgi F, van Meijgaard E, Kadygrov N, Iles C, Corre L, Sandstad M, Somot S, Nabat P, Vautard R, Levavasseur G, Schwingshackl C, Sillmann J, Kjellström E, Nikulin G, Aalbers E, Lenderink G et al (2021) Assessment of the European climate projections as simulated by the large EURO-CORDEX regional and global climate model ensemble. J Geophys Res Atmos 126(4). https://doi.org/10.1029/2019JD032356

Díaz J, Sáez M, Carmona R, Mirón IJ, Barceló MA, Luna MY, Linares C (2019) Mortality attributable to high temperatures over the 2021–2050 and 2051–2100 time horizons in Spain: adaptation and economic estimate. Environ Res 172:475–485. https://doi.org/10.1016/j.envres.2019.02.041

Díaz-López C, Verichev K, Holgado-Terriza JA, Zamorano M (2021) Evolution of climate zones for building in Spain in the face of climate change. Sustain Cities Soc 74:103223. https://doi.org/10.1016/j.scs.2021.103223

EL PAÍS (2024) Las ciudades no pueden ser hornos invivibles. https://elpais.com/opinion/2024-08-16/las-ciudades-no-pueden-ser-hornos-invivibles.html?utm_source=chatgpt.com

Eltinay N, Egbu C (2024) Urban resilience and climate change in the MENA region. Routledge. https://doi.org/10.4324/9781003363224

Euronews (2024) Spain has suffered 22,000 heat-related deaths in the last 8 years. How will a new map help? https://www.euronews.com/green/2024/05/17/spain-has-suffered-22000-heat-related-deaths-in-the-last-8-years-how-will-a-new-map-help

European Commission (2024) Flash floods in Spain: joining forces for rapid recovery. https://civil-protection-humanitarian-aid.ec.europa.eu/news-stories/stories/flash-floods-spain-joining-forces-rapid-recovery_en

Eurostat (2022) Annual—population on the 1st of January, total—Madrid. https://db.nomics.world/Eurostat/urb_lpop1?dimensions=%7B%22cities%22%3A%5B%22ES001F%22%5D%7D&tab=list

Gacetín Madrid (2024) Madrid lanza "Refúgiate en la Cultura": flamenco y monólogos en museos durante julio y agosto, además de descuentos en cine. https://gacetinmadrid.com/2024/06/26/madrid-refugiate-cultura-flamenco-monologos-museos-julio-agosto/?utm_source=chatgpt.com

Gallo E, Quijal-Zamorano M, Méndez Turrubiates RF, Tonne C, Basagaña X, Achebak H, Ballester J (2024) Heat-related mortality in Europe during 2023 and the role of adaptation in protecting health. Nat Med 30(11):3101–3105. https://doi.org/10.1038/s41591-024-03186-1

González A, Gabàs A, Cardoso MA, Brito RS, Pereira C, Russo B, Martínez M, Velasco M, Domínguez JL, Sánchez-Muñoz D, Pardo M, Monjo R, Martinez E, Guerrero M, Forero E, Pagani G, Fourniere H, Locatelli L (2020) Barcelona Resilience Action Plan. . . In Resilience Action Plans of the RESCCUE Cities. D6.2 RESCCUE Project. https://toolkit.resccue.eu/wp-content/uploads/2020/11/Barcelona-Resilience-Action-Plan_Toolkit.pdf

Greenpeace Spain (2023) Race against the climate clock: climate change and extreme weather events in Spain. https://es.greenpeace.org/es/sala-de-prensa/informes/informe-race-against-the-climate-clock-climate-change-and-extreme-weather-events-in-spain/

IEA (2022) Climate resilience policy indicator. https://www.iea.org/reports/climate-resilience-policy-indicator

Iñiguez C, Royé D, Tobías A (2021) Contrasting patterns of temperature related mortality and hospitalization by cardiovascular and respiratory diseases in 52 Spanish cities. Environ Res 192:110191. https://doi.org/10.1016/j.envres.2020.110191

IPCC (2023) Climate change 2022—impacts. Adapt Vulnerability. https://doi.org/10.1017/9781009325844

López-Bueno JA, Díaz J, Sánchez-Guevara C, Sánchez-Martínez G, Franco M, Gullón P, Núñez Peiró M, Valero I, Linares C (2020) The impact of heat waves on daily mortality in districts in Madrid: the effect of sociodemographic factors. Environ Res 190:109993. https://doi.org/10.1016/j.envres.2020.109993

Lorenzo MN, Alvarez I (2022) Future changes of hot extremes in Spain: towards warmer conditions. Nat Hazards 113(1):383–402. https://doi.org/10.1007/s11069-022-05306-x

Madrid City Council (2021) Roadmap to climate neutrality by 2050. https://www.madrid.es/UnidadesDescentralizadas/Sostenibilidad/EspeInf/EnergiayCC/06Divulgaci%C3%B3n/6cDocumentacion/6cNHRNeutral/Ficheros/RoadmapENG2022.pdf

Martínez-Valderrama J, del Barrio G, Sanjuán ME, Guirado E, Maestre FT (2022) Desertification in Spain: a sound diagnosis without solutions and new scenarios. Land 11(2):272. https://doi.org/10.3390/land11020272

Martin-Vide J, Moreno-Garcia MC (2020) Probability values for the intensity of Barcelona's urban heat island (Spain). Atmos Res 240:104877. https://doi.org/10.1016/j.atmosres.2020.104877

Menéndez EP, Higueras García E (2020) Urban sustainability versus the impact of covid-19: a Madrid case study. DisP Plann Rev 56(4):64–81. https://doi.org/10.1080/02513625.2020.1906059

Núñez-Peiró M, Sánchez-Guevara Sánchez C, Neila González FJ (2021) Hourly evolution of intra-urban temperature variability across the local climate zones. The case of Madrid. Urban Clim 39:100921. https://doi.org/10.1016/j.uclim.2021.100921

Oquendo Di Cosola V, Sánchez-Reséndiz JA, Olivieri L, Olivieri F (2020) Actions for adaptation and mitigation to climate change: Madrid case study. Revista Facultad de Ingeniería Universidad de Antioquia. https://doi.org/10.17533/udea.redin.20200795

Palència L, León-Gómez BB, Bartoll X, Carrere J, Díez E, Font-Ribera L, Gómez A, López MJ, Marí-Dell'Olmo M, Mehdipanah R, Olabarría M, Pérez G, Puig-Ribera A, Rico M, Rojas-Rueda D, Vásquez-Vera H, Pérez K (2020) Study protocol for the evaluation of the health effects of superblocks in Barcelona: The "Salut Als Carrers" (health in the streets) project. Int J Environ Res Public Health 17(8):2956. https://doi.org/10.3390/ijerph17082956

Pede EC (2024) Heat waves and urban vulnerability: climate shelters, public services and innovative solutions. Lessons from Barcelona. Urban Res Pract 17(3):465–471. https://doi.org/10.1080/17535069.2024.2329058

Rasilla D, Allende F, Martilli A, Fernández F (2019) Heat waves and human well-being in Madrid (Spain). Atmosphere 10(5):288. https://doi.org/10.3390/atmos10050288

Regalado Zamora G, Martínez Díaz E, Sánchez Guardia MR (2021) Espacios de oportunidad y política municipal madrileña de parques y espacios abiertos públicos. X Congreso Internacional de Ordenación Del Territorio: Recuperación, Transformación y Resiliencia. El Papel Del Territorio, 2021, ISBN 978-84-9133-414-9, pp 853–860. https://dialnet.unirioja.es/servlet/articulo?codigo=8629331&info=resumen&idioma=SPA

Roca J, Arellano B, Zheng Q (2024) Spain: towards a drier and warmer climate? https://doi.org/10.5194/ems2024-932

Romero-Muñoz S, Alméstar M, Sánchez-Chaparro T, Muñoz Sanz V (2023) The impact of institutional innovation on a public tender: the case of Madrid Metropolitan Forest. Land 12(6):1179. https://doi.org/10.3390/land12061179

Salvati A, Coch Roura H, Cecere C (2017) Assessing the urban heat island and its energy impact on residential buildings in Mediterranean climate: Barcelona case study. Energy Build 146:38–54. https://doi.org/10.1016/j.enbuild.2017.04.025

Sánchez-Guevara Sánchez C, Núñez Peiró M, Neila González FJ (2017) Urban heat island and vulnerable population. The case of Madrid. In: Sustainable development and renovation in architecture, urbanism and engineering. Springer International Publishing, pp 3–13. https://doi.org/10.1007/978-3-319-51442-0_1

Shi L, Chu E, Anguelovski I, Aylett A, Debats J, Goh K, Schenk T, Seto KC, Dodman D, Roberts D, Roberts JT, VanDeveer SD (2016) Roadmap towards justice in urban climate adaptation research. Nat Clim Chang 6(2):131–137. https://doi.org/10.1038/nclimate2841

Suárez M, Rieiro-Díaz AM, Alba D, Langemeyer J, Gómez-Baggethun E, Ametzaga-Arregi I (2024) Urban resilience through green infrastructure: a framework for policy analysis applied to Madrid, Spain. Landscape Urban Plann 241:104923. https://doi.org/10.1016/J.LANDURBPLAN.2023.104923

Urban Resilience Hub (2022) Resilience profile Barcelona. https://urbanresiliencehub.org/publication/profilebarcelona/

Veranos de la Villa (2024) Refúgiate en la Cultura. https://www.veranosdelavilla.com/es/evento/refugiate-en-la-cultura?utm_source=chatgpt.com

Vicente-Serrano SM, Lopez-Moreno J-I, Beguería S, Lorenzo-Lacruz J, Sanchez-Lorenzo A, García-Ruiz JM, Azorin-Molina C, Morán-Tejeda E, Revuelto J, Trigo R, Coelho F, Espejo F (2014) Evidence of increasing drought severity caused by temperature rise in southern Europe. Environ Res Lett 9(4):044001. https://doi.org/10.1088/1748-9326/9/4/044001

Vidal Yañez D, Pereira Barboza E, Cirach M, Daher C, Nieuwenhuijsen M, Mueller N (2023) An urban green space intervention with benefits for mental health: a health impact assessment of the Barcelona "Eixos Verds" Plan. Environ Int 174:107880. https://doi.org/10.1016/j.envint.2023.107880

World Weather Attribution (2024) Extreme downpours increasing in southeastern Spain as fossil fuel emissions heat the climate. https://www.worldweatherattribution.org/extreme-downpours-increasing-in-southern-spain-as-fossil-fuel-emissions-heat-the-climate/

Zinzi M, Carnielo E (2017) Impact of urban temperatures on energy performance and thermal comfort in residential buildings. The case of Rome, Italy. Energy Build 157:20–29. https://doi.org/10.1016/j.enbuild.2017.05.021

Zittis G, Hadjinicolaou P, Klangidou M, Proestos Y, Lelieveld J (2019) A multi-model, multi-scenario, and multi-domain analysis of regional climate projections for the Mediterranean. Reg Environ Change 19(8):2621–2635. https://doi.org/10.1007/s10113-019-01565-w

Zografos C, Klause KA, Connolly JJT, Anguelovski I (2020) The everyday politics of urban transformational adaptation: struggles for authority and the Barcelona superblock project. Cities 99:102613. https://doi.org/10.1016/J.CITIES.2020.102613

Chapter 6
Climate Change and Public Health in Senegal: An In-Depth Analysis of Impacts and Sustainable Urban Solutions

Cheikh Cisse

Abstract This article studies the complex and growing impacts of climate change on public health in Senegal, focusing on direct and indirect effects such as the increase in respiratory, waterborne and nutritional diseases. Based on a multidisciplinary methodology combining a systematic literature review and qualitative interviews with health professionals, the study demonstrates the aggravation of health vulnerabilities in urban and peri-urban areas—particularly in Dakar—due to extreme climatic events and poor urban planning. It also identifies promising adaptive responses such as urban greening, sustainable mobility initiatives and improved waste management systems. However, institutional, financial and socio-political constraints limit the scalability and sustainability of these solutions. The article calls for integrated urban planning and multi-sectoral governance, supported by community participation and innovative financing mechanisms, to build resilient and health-inclusive cities in Senegal.

Keywords Climate change · Public health · Urban resilience · Senegal · Sustainable planning

Introduction

Climate change is currently one of the most pressing threats to public health and sustainable development at the global level. This phenomenon, which manifests itself, among other things, through rising average temperatures, intensification of extreme weather events, and degradation of ecosystems, exacerbates pre-existing vulnerabilities, particularly in underdeveloped countries.

C. Cisse (✉)
Department of Urbanism, Lecturer at Ecole Supérieure des Professions Immobilières (ESPI), Paris, France
e-mail: c.cisse@groupe-espi.fr

© The Author(s), under exclusive license to Springer Nature Switzerland AG 2025

A. I. Almulhim et al. (eds.), *Urbanization, Climate Change, and Health*, Climate Change Management, https://doi.org/10.1007/978-3-031-99132-5_6

In Africa, and more specifically in Senegal, these impacts are amplified by structural and institutional challenges. Phenomena such as recurrent flooding in Dakar, desertification in rural northern regions like Fouta (Matam), and coastal erosion in coastal cities along the western, southwestern, and northwestern coasts (Bargny-Dakar, Saint-Louis, and the Petite Côte—Thiès, Fatick) compromise food security, degrade health infrastructure, and worsen socio-economic disparities. These realities make it essential to adopt systemic solutions, such as sustainable urban planning policies that integrate health and climate dimensions.

Moreover, rapid and poorly managed urbanization—with urban populations exceeding 50% in major regions—aggravates climate and health vulnerabilities. Highly populated areas such as Pikine and Guédiawaye, in the suburbs of Dakar, are particularly exposed to the effects of flooding and increasing air pollution, which raises the prevalence of respiratory and waterborne diseases. These challenges underscore the need to explore the potential of sustainable urban development models to strengthen the resilience of populations in the face of climate and health crises. The following map (see Fig. 6.1) shows the location of the different regions of Senegal mentioned in the text.

In light of these various issues, a central question arises: how does climate change impact public health in Senegalese cities, and what is the potential of sustainable urban planning to mitigate these effects?

To support this line of inquiry, the article is structured into several complementary sections. First, a literature review presents existing work on the links between climate change, health challenges, and urbanism, while identifying gaps in the literature. Then, the research methodology combines a primarily qualitative and occasionally

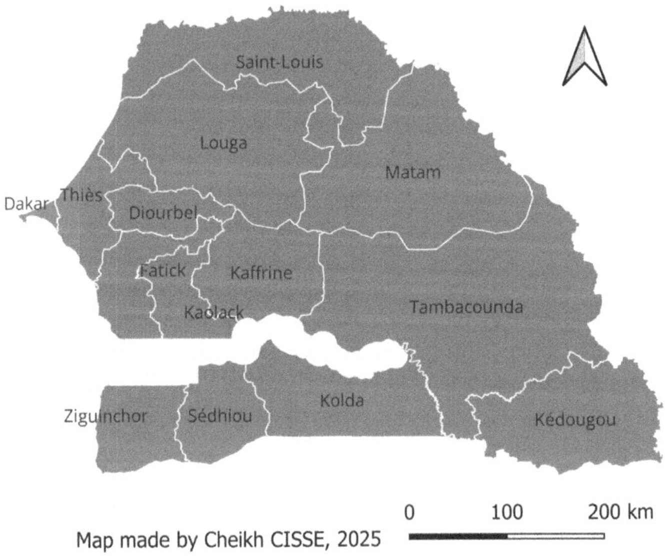

Fig. 6.1 Situation of the regions of Senegal

quantitative analysis based on secondary data and surveys with several health professionals, in order to capture the complexity of the issues studied. The results offer an analysis of the impacts of climate change on public health in the Senegalese context, particularly regarding respiratory, waterborne, and nutritional diseases, as well as emerging adaptive responses.

In the discussion section, these results are critically analyzed to assess the effectiveness of current strategies, identify institutional and financial challenges, and propose avenues for improvement. Finally, recommendations are provided to strengthen urban and health resilience, while opening perspectives for a sustainable and inclusive future in Senegal.

Literature Review

This part presents an analytical and critical review of the main scientific and institutional publications on the links between climate change, public health and urban trends, particularly in the Senegalese and African context. It aims to identify theoretical and empirical advances on the subject, while exposing existing gaps in the literature. The aim is to situate the article in the scientific field, by emphasizing its contribution to a better understanding of the intersecting impacts of climate and urbanization on health in Senegal.

State of the Art

The impacts of climate change on public health are numerous and complex, affecting both direct and indirect dimensions. McMichael et al. (2003) showed that climate change influences public health by disrupting the economy, degrading the environment, and causing interruptions in development following extreme weather events. These impacts include the worsening of respiratory, cardiovascular, infectious, and waterborne diseases, particularly in vulnerable urban and rural contexts. Watts et al. (2015) also demonstrated that heatwaves significantly increase hospitalizations related to heat stress and cardiovascular illnesses, while floods intensify the spread of waterborne diseases such as cholera. In Senegal, Fall and Dramani (2023) emphasized that recurrent flooding in urban areas such as Dakar exacerbates cases of diarrhea and malaria, particularly in suburbs like Pikine and Guédiawaye. Descroix and Lambert (2018), at the Sahel level, showed that coastal erosion affects the livelihoods of communities living near the coasts, increasing their socio-economic vulnerability.

The health status of populations is also linked to the effectiveness of climate policies. Haines and al. (2006) argued that current measures to reduce climate risks are insufficient, making the fight against climate-sensitive diseases more difficult. Fayissa and Gutema (2005) state that health in Sub-Saharan Africa is heavily influenced by factors such as GDP per capita, literacy rate, and food availability, while

Nocera and Zweifel (1998) emphasize the role of healthcare costs, education, and wealth in the demand for health services.

Specific observations in Senegal reveal that climate variability enhances health inequalities. Mbaye (2015) showed that floods during the rainy season and rising temperatures contribute to an increase in acute respiratory infections, identified by Cissé et al. (2019) in Dakar suburbs such as Pikine, along with diarrheal diseases and hypertension. Furthermore, Thiam et al. (2017) found that high temperatures (above 36 °C) and cumulative rainfall beyond 57 mm increase cases of diarrhea, with a more pronounced impact in rural than urban areas.

Vector-borne diseases such as malaria also show a strong correlation with climate change. Patz et al. (2005) demonstrated that climatic variations influence the geographic distribution of vector-borne diseases, as confirmed by Sy and al. (2016), who noted a significant increase in malaria incidence in Senegalese urban areas during the rainy season. Sy et al. (2022) also established a link between heatwaves and mortality in Dakar, particularly affecting the most vulnerable groups, such as the elderly and children.

In terms of urban planning, the work of Newman and Jennings (2008) underlines the importance of green infrastructure and sustainable mobility in mitigating climate impacts while improving urban health. Niang et al. (2010) showed that sustainable urban drainage systems play a key role in reducing flood risks, offering an example of effective local solutions. Frumkin et al. (2008) stressed the importance of urban planning in promoting public health in the face of climate change.

Despite these initiatives, institutional and financial obstacles persist. Thiam et al. (2017) identify a lack of intersectoral coordination and resources as major challenges to implementing effective climate policies. Bowen and al. (2014) emphasize the need to integrate health and climate policies to improve urban resilience. Sy et al. (2022) highlight the importance of multisectoral and participatory approaches to build sustainable and resilient cities in the face of climate crises. Finally, Watts et al. (2021) underscore the urgency of coordinated global action to mitigate the health impacts of climate change.

In summary, research findings converge to highlight the urgency of integrating climate and health considerations into urban planning in Senegal. This integration holds significant potential that deserves to be explored in this article.

Literature Gap and Scientific Contribution

Research on the impacts of climate change and its implications for public health has significantly evolved in recent years. However, the majority of studies focus on global or regional contexts, without accounting for local specificities. For instance, the work of Watts et al. (2015) explores the links between heatwaves and cardiovascular diseases, while that of Descroix and Lambert (2018) focuses on the impact of coastal erosion on the livelihoods of rural populations in the Sahel region.

In Senegal, although some studies like that of Fall and Dramani (2024) expose the health consequences of floods, a strictly local and integrative analysis linking climate, public health, and urban planning remains rare. Rapid urbanization and specific challenges in Senegal, such as ineffective drainage systems and increasing air pollution, call for a contextual approach that has not yet been fully explored in the literature.

This article aims to partially fill this gap by adopting a three-dimensional analysis that combines climate, health, and urban data. It adds value by integrating the voices of local actors, particularly health professionals, in the analysis of vulnerabilities and potential solutions. This approach relies on a cross-analysis of qualitative and quantitative data to ensure a scientific interpretation.

Methodology

The article is based on an integrative methodology that combines documentary analysis and qualitative data to analyze the impacts of climate change on public health in Senegal and adaptive responses in terms of sustainable urban planning. Particular attention was paid to methodological rigor to ensure scientific validity and the reproducibility of the analysis.

The approach combines a systematic literature review, based on PRISMA standards, and an exploratory qualitative survey through semi-structured interviews with health professionals. This dual approach allows the study of complex interactions between climate change, public health, and urban dynamics in the Senegalese context.

The documentary research strategy consisted of querying the *Cairn, ScienceDirect,* and *Google Scholar* databases between the end of 2024 and early 2025. Keywords used, in French and English, included "climate change," "public health," "Senegalese cities," "air pollution," "floods," "malaria," "urban resilience," and "heatwaves." The search initially identified 48 publications. The inclusion criteria were: (1) published between 2005 and 2024, (2) providing specific data on climate impacts in Senegal or in similar urban contexts, and (3) being peer-reviewed documents or from recognized organizations (WHO, UNICEF, IPCC, World Bank). After reviewing titles, abstracts, and applying scientific quality criteria (level of evidence, contextual relevance, methodological rigor), a total of 28 references were selected for analysis.

In parallel, a qualitative survey was conducted in January 2025 with 7 Senegalese health professionals (doctors, epidemiologists, public health officers, local health officials), recruited through purposive sampling. The semi-structured interviews, averaging 30 min, were conducted remotely via *WhatsApp.* The interview guide focused on five areas: direct health impacts, vulnerable populations, institutional challenges, infrastructure needs, and envisioned solutions. The interviews were recorded with the participants' informed consent, then fully transcribed.

Data analysis mainly relied on qualitative methods. For the documentary data, a thematic content analysis was carried out manually, following the steps of open, axial, and selective coding, based on Bardin's (2013) principles of content analysis.

This approach organizes meaning units extracted from the documents into successive thematic categories to identify the underlying logics of the studied phenomena. The identified themes are: air pollution, waterborne diseases, institutional vulnerability, sustainable mobility, and climate governance.

The interview data were processed using inductive coding, with the support of *NVivo 12 software*. An Excel table was also used to cross-reference responses by thematic axis and respondent profile. This approach allowed the emergence of empirical categories, such as institutional barriers to climate action or perceptions of the link between health inequalities and urbanization.

Data triangulation was performed using a methodological approach of source convergence. The results from the interviews were compared and validated by data from official reports (World Health Organization—WHO, Senegalese Ministry of Health, Senegalese National Agency for Statistics and Demography–ANDS) and scientific publications. This triangulation ensured the coherence of the results and identified discrepancies between institutional discourse and field realities.

The analytical framework of the study is structured around three interdependent dimensions: (1) climate data (heatwaves, pollution, floods), (2) health data (respiratory, waterborne, and nutritional morbidity), (3) institutional data (adaptive capacities, public policies, urban governance). This framework enables a systemic and contextualized reading of the impacts of climate change on public health in Senegal.

Finally, the reliability of the conclusions was ensured through cross-validation of all sources and careful attention to the internal consistency of the results. Convergent results were highlighted, while divergences were analyzed to identify critical reflections and formulate contextualized recommendations.

Results

Climate change in Senegal has varied impacts on public health, encompassing both direct and indirect dimensions. Extreme climatic conditions exacerbate pre-existing health vulnerabilities, disproportionately affecting the most fragile urban and rural populations. This section outlines the main diseases associated with climate change while presenting the challenges faced by Senegal's health systems and the necessary adaptive responses.

Impacts of Climate Change on Health

The health effects of climate change are manifested by an increase in respiratory, waterborne, and nutritional diseases. These already prevalent pathologies are enhances by environmental factors such as air pollution, heatwaves, floods, and declining agricultural yields.

The Surge in Respiratory and Cardiovascular Diseases

In Dakar, concentrations of suspended particles (PM10 and PM2.5) largely exceed the annual thresholds set by the WHO, reaching critical levels, with exceedances of 6–9 times the limit for PM10 and 2–5 times for PM2.5. Consequently, the Air Quality Index (AQI) ranges from "poor" ($100 \leq AQI < 200$) to "very poor" ($AQI \geq 200$) during the dry season, which extends from November to May, worsening respiratory diseases such as asthma and chronic bronchitis. These pollutants mainly stem from biomass combustion and the open burning of solid waste, intensifying asthma attacks and cardiac failures, especially during heatwaves.

Biomass combustion, used by more than 75% of households in Sub-Saharan Africa, dominates the residential sector in Senegal, accounting for 83% of consumed energy and nearly 48% of national energy consumption (IEA 2019; Tchanche 2018). Burned in open fires or rudimentary stoves, these biofuels emit significant fine particles. Similarly, the informal burning of solid waste—including plastics, tires, and glass—also contributes to 43.25% of soot carbon emissions and 12.9% of NO_x, worsened by a limited waste collection rate of 35%. The Mbeubeuss landfill, in eastern Dakar, receives 2000 tons of unsorted waste per day, increasing pollutant dispersion. These practices led to a 25% increase in hospitalizations for respiratory failure and a 30% rise in asthma attacks between 2015 and 2020, especially affecting children and the elderly (Seck 2021; Toure-Badiane 2019).

Moreover, recurrent floods caused by heavy rainfall increase the contamination of drinking water systems and the spread of pathogens such as Salmonella and Vibrio cholerae, responsible for waterborne diseases that cause about 842,000 deaths annually, including 361,000 among children under 5 (WHO 2015). These events also raise the risks of foodborne illnesses, worsened by the use of contaminated water in food processing, contributing to a global burden of 33 million Disability-Adjusted Life Years (DALYs) in 2010, the highest in Africa (Cissé 2019). In the urban and peri-urban areas of Dakar, where interviewed health professionals report inadequate health infrastructure, these challenges are intensified. Initiatives such as urban micro-gardening aim to stabilize soils, improve water resource management, and reduce climatic and health vulnerabilities, but their implementation remains limited (French Development Agency 2017; IPCC 2018).

The Proliferation of Waterborne Diseases

Recurrent floods and insufficient wastewater management in vulnerable areas such as Pikine and Guédiawaye intensify the spread of waterborne and vector-borne diseases, notably cholera, diarrhea, and malaria. During the 2005 cholera outbreak, Dakar recorded 5474 cases, representing 34.6% of the national total, largely due to intense rainfall reaching 277 mm in one week, which favored stagnant water and contamination by *Vibrio cholerae* (Constantin de Magny et al. 2012). These epidemics were accelerated by the use of unprotected water stored in barrels due to periodic water shortages, especially in densely populated neighborhoods. At the same time, the failure of health infrastructure aggravates the burden on health facilities, particularly during climatic crises. Children and pregnant women, already among the most vulnerable populations, are especially affected by the rise in infant mortality and complications linked to these diseases.

Malaria remains a major public health concern in Senegal, with its impacts intensified in the flooded urban areas of Dakar and its suburbs. Since 2005, recurrent floods have created favorable larval habitats for *Anopheles arabiensis*, the primary malaria vector in the region. Between 2012 and 2014, increased resistance of *Vibrio cholerae* strains to several insecticides, notably pyrethroids and DDT, was identified, attributed to specific genetic mutations (L1014F and L1014S) and metabolic mechanisms involving *GST and CYP450* enzymes. These genetic adaptations, combined with climatic factors such as heavy precipitation and high temperatures, favored the resurgence and persistence of cholera epidemics in certain regions, as observed in Senegal in 2005 (Moore et al. 2005). Only organophosphate molecules remain lethal. This presents a growing challenge in a context of increasingly intense and prolonged seasonal rainfall due to climate change, which lengthens larval habitat duration and urban transmission.

Despite progress through periodic distribution campaigns of insecticide-treated nets and indoor spraying by the Senegalese Ministry of Health with support from the World Health Organization, resistance management remains a priority to mitigate health impacts in these densely populated areas.

Food Insecurity: A Real Threat

Recurrent droughts and climate fluctuations in the arid and semi-arid regions of northern and eastern Senegal have significantly reduced agricultural yields. Rainfall has decreased by 30% in recent decades, while temperatures have increased by 0.9 °C, leading to increasing aridification, particularly in the North and Center of the country (UNEP 2010). These conditions have led to a significant drop in agricultural yields, notably for millet and peanuts, which fell by 20% between 2015 and 2020 (UNEP 2010). This decline in agricultural production and reliance on expensive food imports exacerbate malnutrition in certain departments in the Fouta region (Podor, Kanel, Ranérou Matam), where food insecurity affected 245,000 people in 2018 (Action Against Hunger 2018).

These challenges are compounded by the destruction of agricultural and coastal erosion control ecosystems stabilized by past initiatives, such as the reforestation of more than 400 ha in the 1950s in coastal areas of Dakar (Malika, Retba, and Cambérène). Over 2000 ha of these areas have been urbanized, undermining their role in food security and increasing dependency on costly food imports, thereby worsening socio-economic inequalities (Rouchiche 2001). These factors underscore the need for integrated policies to strengthen the resilience of agricultural systems to climate change.

These intertwined health and environmental dynamics, visually summarized in Fig. 6.2, which highlights the main links between climate risks and public health

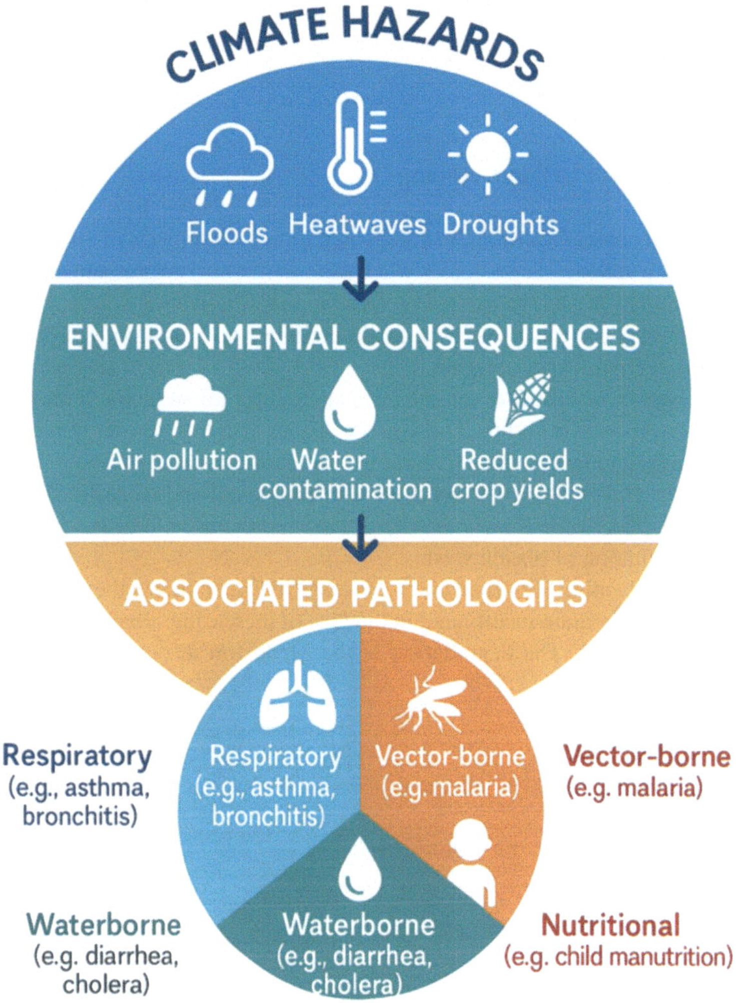

Fig. 6.2 The impact of climate risks on public health. Realized by Cheikh CISSE (2025)

outcomes in Senegal, underscore the urgent need to develop robust and integrated policies that combine health, social, and environmental responses to effectively mitigate the growing impacts of climate change on public health.

Adaptive Solutions

To mitigate the impacts of climate change on public health in Senegal, several adaptive initiatives have been implemented or are under development. This section analyzes the main responses in the fields of urban greening, sustainable mobility, and waste management, while highlighting their potential for improvement and expansion.

The Externalities of Urban Greening

Urban greening constitutes a major strategy to mitigate the impacts of climate change on public health in Senegal, particularly in Dakar, where heat islands and air pollution affect the population. As shown in Fig. 6.3, the main remaining forest areas in the Dakar region play a critical role in regulating climate and promoting urban health. Spaces such as Hann Park, covering 60 ha, absorb up to 15 tons of fine particles per hectare per year (Constructive Voices 2023) and reduce urban temperatures through shading and evapotranspiration. These green spaces also improve air quality, create cooling islands, and help reduce respiratory and cardiovascular diseases. Likewise, the cornices of Dakar, along the coastlineoffer ecological and recreational benefits, although their effectiveness is limited by lack of maintenance, illegal constructions, and the accumulation of plastic waste.

However, with only 15% urban green coverage in Dakar, below the recommended 30%, these efforts remain insufficient in the face of increasing urbanization pressure and lack of resources. The degradation of the Mbao forest, once 770 ha and now

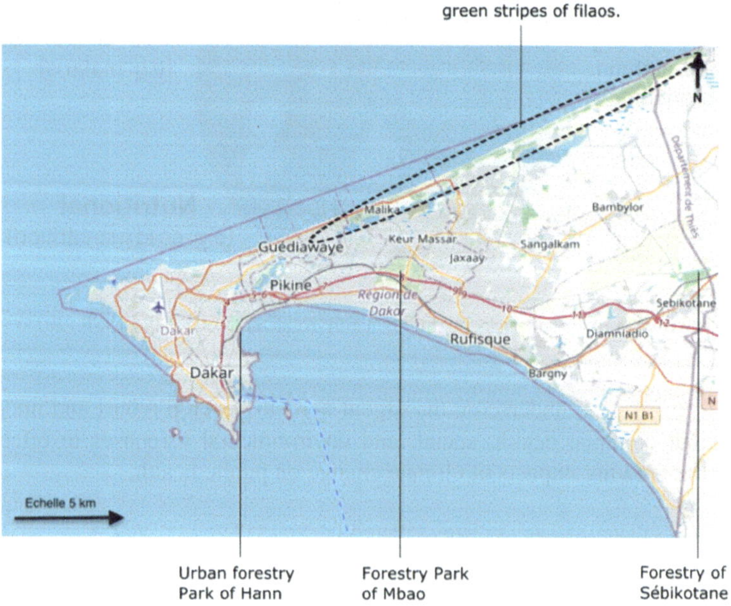

Fig. 6.3 Main forest areas in the Dakar region. *Source* Open Street Map, Cheikh CISSE rehabilitation, 2024

reduced to 500 ha, and the urbanization of more than 2000 ha of reforested coastal areas illustrate these challenges. Nevertheless, small-scale reforestation and urban greening initiatives continue to show measurable impacts on urban heat.

Promoting Sustainable Mobility

The Bus Rapid Transit (BRT) project, launched in 2023 in Dakar, represents a major step toward more sustainable urban transport. This system aims to reduce road congestion and CO_2 emissions while improving access to transport for low-income populations. With 18.5 km of dedicated lanes and a capacity of 300,000 passengers per day, it is expected to reduce 2 million tons of CO_2, valued at 60 million euros (World Bank Group 2020). According to estimates by the World Bank Group (2020), the BRT could reduce urban transport-related greenhouse gas emissions by 25% by 2030 and transport more than 300,000 passengers daily.

Despite these advantages, the project faces significant challenges. Some densely populated peripheral areas of Dakar remain underserved, limiting the inclusivity and social impact of the BRT (Pikine, Keur Massar, etc.). Thus, although the BRT improves urban resilience by allocating 15% of its budget to improving drainage systems to reduce flood risks, assessing its long-term impacts on air quality and reducing mobility inequalities in Dakar's peripheral areas remains a challenge.

The Impacts of Waste Management

Waste management in Senegal presents critical challenges for public health and the environment, particularly in densely populated urban areas such as Dakar and Thiès. With an annual production of 250,000 tons of plastic waste, only 12% is collected for recycling, with the majority accumulating in uncontrolled dumps or being burned in the open. These practices release toxic substances such as dioxins and furans, worsening respiratory and waterborne diseases. The Mbeubeuss landfill, in the Dakar region, receives 2000 tons of unsorted waste daily on its 75 ha, illustrating the scale of the problem, with pollutant dispersion affecting soils, groundwater, and surrounding populations.

Community and national initiatives such as the *sétal sunu reew* day, launched by President Bassirou Diomaye Faye since taking office in April 2024, on the first Saturday of each month, although limited, offer promising prospects in terms of sanitation and public health, while generating economic opportunities, particularly in the waste recovery sector. However, waste management in Senegal remains marked by regional disparities and uneven implementation of existing strategies, reinforcing socio-economic and health inequalities between urban and rural areas. Figure 6.4 also visually synthesizes the adaptive solutions explored in this study, particularly in terms of sustainable planning, public health, and climate resilience.

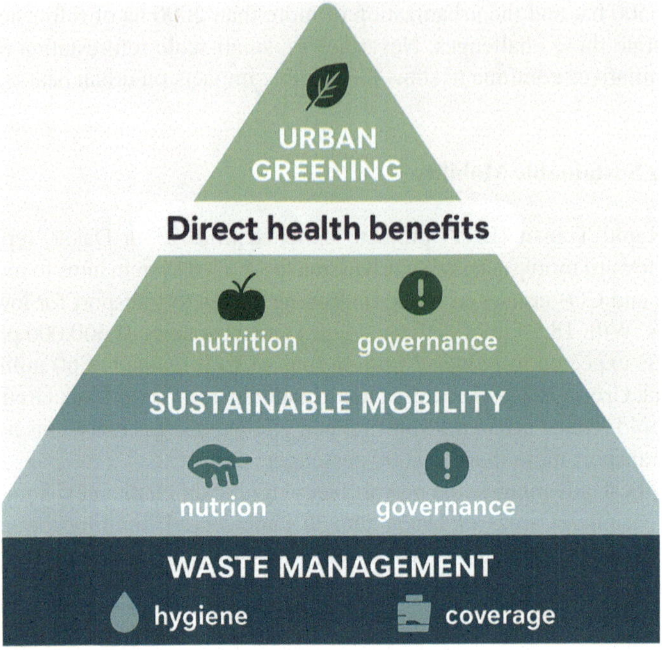

Fig. 6.4 Adaptable solutions. Realized by Cheikh CISSE (2025)

Discussion: Toward a Coherent and Contextualized Approach to Climate and Health Resilience in Senegal

The impacts of climate change on public health and sustainable urban planning solutions in Senegal raise several discussion points.

A Critical Analysis of the Resilience Strategies Implemented

The results confirm that certain initiatives, such as urban greening and the implementation of the BRT in Dakar, have measurable effects on public health and the environment. Greening, for example, helps mitigate the effects of heatwaves and reduce urban heat islands. However, these strategies remain limited in terms of territorial coverage and accessibility for the most vulnerable groups. Implementation is fragmented, hindered by a lack of intersectoral coordination between the concerned ministries.

Despite their potential, the analyzed initiatives suffer from a lack of scalability. The BRT, for example, although effective in certain areas, still excludes the most disadvantaged peripheries of Dakar (Pikine, Keur Massar), limiting its impact on mobility inequalities and overall emissions. Likewise, waste management projects face chronic dependence on external funding, which undermines their sustainability.

Institutional, Financial, and Socio-political Challenges

Institutional and financial constraints are major obstacles. The inefficiency of regulatory frameworks, the lack of human and technical resources at the local level, and the low capacity to capture available climate funding (Green Climate Fund, carbon markets) limit the scaling-up of projects. Moreover, adaptation policies are often designed without significant consultation with local communities, which affects their social acceptability.

Triangulation of sources has shown a lack of synergy between public, private, and community actors. Local institutions lack the levers to operationalize national adaptation strategies, which are often too generic. Consequently, multi-level governance remains ineffective in addressing such complex challenges.

Toward Contextualized and Sustainable Anchoring of Solutions

To strengthen the feasibility and scalability of the proposed solutions, it seems relevant to design governance frameworks adapted to the Senegalese context, based on intersectoral and inter-municipal coordination platforms. These platforms would align priorities between actors, pool resources, and standardize actions in health, urban planning, and climate.

On the financial front, it is advisable to develop hybrid financing mechanisms (public–private, participatory, or resilience fund-based). The strategic mobilization of other innovative financing instruments, adapted to local realities, can ensure the sustainability of initiatives.

Finally, the success of projects depends on strong community ownership. The qualitative data collected confirm that awareness programs, neighborhood committees, and partnerships with local NGOs are powerful levers for transformation. By involving citizens in the design and evaluation of projects, their territorial anchoring and social relevance are strengthened.

The results of this study thus highlight the need for a paradigm shift in urban planning in Senegal, moving from a sectoral and top-down approach to integrated governance that is sensitive to local specificities.

Conclusion and Recommendations

Climate change exacerbates pre-existing vulnerabilities in public health in Senegal. Extreme weather events such as heatwaves, floods, and droughts have direct and indirect effects on the prevalence of respiratory, waterborne, vector-borne, and nutritional diseases. These impacts, worsened by rapid, poorly planned, and often unequal urbanization, reveal the urgency of a systemic and coordinated response.

Initiatives such as urban greening, the development of sustainable transport, or the improvement of waste management show positive effects but remain limited by institutional, financial, and social obstacles. Thus, to strengthen their effectiveness and ensure their sustainability, it is essential to mobilize a coherent set of action levers.

First, strengthening institutional frameworks is a priority. The establishment of intersectoral coordination mechanisms involving the ministries of health, environment, urban planning, and finance is essential. An interministerial committee dedicated to urban resilience, endowed with resources and steering capacity, could play a structuring role.

Next, mobilizing financial resources requires innovative solutions. Public–private partnerships (PPPs) should be promoted through clear legal frameworks, tax incentives, and equitable risk-sharing mechanisms. The creation of a fund dedicated to urban adaptation, co-financed by the State, local authorities, and private investors, would accelerate the realization of sustainable projects.

Finally, the involvement of local communities remains a determining factor. Environmental awareness, citizen participation in project design, and the creation of neighborhood-based initiatives (community gardens, recycling, educational programs) are essential to foster social ownership of proposed solutions.

Beyond technical approaches, this study also emphasizes the need to strengthen the production, dissemination, and use of reliable climate and health data. These data are indispensable for informing public decisions, anticipating risks, and evaluating implemented policies.

References

Journal Articles

Bowen S et al (2014) Relative efficacy of mindfulness-based relapse prevention, standard relapse prevention, and treatment as usual for substance use disorders: a randomized clinical trial. JAMA Psychiatry 71(5):547–556.https://doi.org/10.1001/jamapsychiatry.2013.4546

Cissé G (2019) Food-borne and water-borne diseases under climate change in low- and middle-income countries: further efforts needed for reducing environmental health exposure risks. Acta Trop 194:181–188. https://doi.org/10.1016/j.actatropica.2019.03.012

Descroix L, Lambert L (2018) Changements climatiques et essor djihadiste au Sahel: une approche critique pour des solutions adaptées. Regards géopolitiques

Fall ND, Dramani L (2023) « Impact du changement climatique sur la santé de la population au Sénégal ». Rev Fr d'Economie Gestion 4(7):446–471. https://zenodo.org/records/8188140

Fayissa B, Gutema P (2005) Estimating a health production function for Sub-Saharan Africa (SSA). Appl Econ 37(2):155–164. https://doi.org/10.1080/0003684042000290095

Frumkin H et al (2008) Climate change: the public health response. Am J Public Health 98(3):435–445 https://doi.org/10.2105/AJPH.2007.119362

Haines A et al (2006) Climate change and human health: impacts, vulnerability, and mitigation. Lancet 367(9528) :2101–2109.https://doi.org/10.1016/S0140-6736(06)68933-2

Mbaye I (2015) Perception des impacts du changement climatique... VertigO 15(1). https://doi.org/10.4000/vertigo.16557

McMichael AJ et al (2003) Global climate change and health. Environ Health Perspect.https://doi.org/10.1289/ehp.4302

Moore S et al (2005) Cholera outbreak in Senegal in 2005: was climate a factor? Am J Trop Med Hyg 73(5):996–1002.https://doi.org/10.4269/ajtmh.2005.73.996

Patz JA et al (2005) Impact of regional climate change on human health. Nature. https://doi.org/10.1038/nature04188

Sy I et al (2016) Climate variability and malaria transmission in Senegal: a review. IJERPH 13(11):1234. https://doi.org/10.3390/ijerph13111234

Tchanche B (2018) Analyse du système énergétique du Sénégal. Rev Energies Renouvelables 21(1):73–88. https://www.cder.dz/download/Art21-1_9.pdf

Thiam S et al (2017) Association between childhood diarrhoeal incidence... Mbour, Senegal. https://doi.org/10.1371/journal.pntd.0005614

Watts N et al (2015) Health and climate change: policy responses to protect public health. Lancet. https://doi.org/10.1016/S0140-6736(15)60854-6

Watts N et al (2021) The 2020 report of The Lancet countdown. Lancet 397(10269):129–170. https://doi.org/10.1016/S0140-6736(20)32290-X

Books and Book Excerpts

Bardin L (2013) L'analyse de contenu. P.U.F., Collection, Paris

Newman P, Jennings I (2008) Cities as sustainable ecosystems: principles and practices. Island Press, Washington, DC. https://parfikh.wordpress.com/wp-content/uploads/2012/01/cities-as-sustainable-ecosystem.pdf

Nocera S, Zweifel P (1998) The demand for health: an empirical test of the grossman model using panel data. In: Zweifel P (eds) Health, the medical profession, and regulation. Developments in health economics and public policy, vol 6. Springer, Boston, MA. https://doi.org/10.1007/978-1-4615-5681-7_2

Sy I et al (2022) Heatwaves and Health Risks in The Northern Part of Senegal : analysing the distribution of temperature-related diseases and associated risk factors. Environ Sci Pollut Res Int (55):83365–83377. https://doi.org/10.1007/s11356-022-21205-x

Reports, Official Documents and Institutional Sources

Action Contre la Faim (2018) Près d'un enfant sur cinq atteint de malnutrition aiguë. https://www.actioncontrelafaim.org/a-la-une/pres-dun-enfant-sur-cinq-atteint-de-malnutrition-aigue/

Agence Française de Développement (2017) Food for cities. https://www.afd.fr/en/ressources/food-cities-what-roles-local-governments-global-south

Constructive Voices (2023) Comment mettre en œuvre des solutions urbaines fondées sur la nature. https://constructive-voices.com/fr/comment-mettre-en-oeuvre-des-solutions-urbaines-fondees-sur-la-nature/

IEA (2019) World energy outlook 2019. https://www.iea.org/reports/world-energy-outlook-2019

IPCC (2018) Global warming of 1.5°C. https://www.ipcc.ch/sr15/

Rouchiche S (2001) La foresterie urbaine. FAO. https://www.fao.org/4/x3994f/X3994F01.htm

Seck M (2021) Analyse des effets liés à la pollution atmosphérique au Sénégal. Mémoire de maitrise en environnement de l'Université de Sherbrooke. https://savoirs.usherbrooke.ca/bitstream/handle/11143/18798/seck_moussa_MEnv_2021.pdf

Toure-Badiane NO (2019) Les vagues de chaleur et les maladies respiratoires et cardiovasculaire. Atelier de Formation sur les Vagues de Chaleur et leurs Impacts sur la Santé Dakar, Sénégal. https://www.cpc.ncep.noaa.gov/products/international/climatehealth/training/2019/day1/vagues_de_chaleurs_et_maladies_respiratoires_et_cardiovasculaires.pdf

Dr. Cheikh Cisse is an urban planner and geographer specializing in sustainable urban development and climate challenges in African cities. He holds a PhD in spatial planning and works as a lecturer in planning and real estate development at the Ecole Supérieure des Professions Immobilières (ESPI) Paris and as an affiliated researcher at the Leiden Centre for African Studies.

Part II
Integrating Health, Climate, and Social Systems in Urban Policy and Design

Chapter 7
Salutogenic Urbanism: Bridging Climate Challenges to Healthy Cities

Borhan Sepehri, Alan Dilani, Mohammad Anvar Adibhesami, Alex Opoku, Marjan Mehrmanesh, and Rosalina Gadyuchkova

Abstract With increasing global warming recently and, parallel to it, health-threatening challenges such as pandemics, cities as population centers have become hot spots to deal with these two challenges. There is a clear need to address climate change and health challenges simultaneously, through consistent policies and approaches. In this context, the Salutogenic approach emphasizes prioritizing human health needs while promoting sustainable environmental design. Salutogenic design focuses on creating environments that promote health and well-being rather than merely preventing illness. Antonovsky's salutogenesis theory, which emphasizes the factors that support human health and well-being, serves as the foundation for this concept. However, except for a limited number of articles, the development of the Salutogenic approach has always been on an architectural scale, and there is a need to develop it further on an urban scale. Therefore, the current study uses a systematic review of the literature on the Salutogenic design to critically examine the relationship between this approach and urbanism. Additionally, it seeks to analyze the gaps and potential of this approach for urbanism. Key findings reveal a notable

B. Sepehri (✉)
Department of Urban Planning and Design, Faculty of Arts and Architecture, Tarbiat Modares University, Tehran, Iran
e-mail: borhanspehri@gmail.com

A. Dilani (✉)
International Academy for Design and Health, Stockholm, Sweden
e-mail: alan@dilani.org

M. A. Adibhesami
School of Architecture and Environmental Design, Iran University of Science and Technology, Narmak, Tehran, Iran

A. Opoku
Department of Architectural Engineering, College of Engineering, University of Sharjah, Sharjah, UAE

M. Mehrmanesh
Department of Urban Planning, Kermanshah Branch, Islamic Azad University, Kermanshah, Iran

R. Gadyuchkova
Bartlett School of Planning, University College London, London, UK

© The Author(s), under exclusive license to Springer Nature Switzerland AG 2025
A. I. Almulhim et al. (eds.), *Urbanization, Climate Change, and Health*, Climate Change Management, https://doi.org/10.1007/978-3-031-99132-5_7

increase in publications on Salutogenic urbanism from 2019–2020 to 2023, most likely due to the COVID-19 pandemic. The study points out significant trends: mental health as a primary concern, a change in research emphasis from rural to urban, and omissions in addressing physical health and environmental change. A co-occurrence analysis clearly highlights connections between urban planning, public health, and well-being. The study proposes a framework for Salutogenic urbanism based on three components: comprehensible (e.g., green spaces, walkability), manageable (e.g., noise reduction, air quality), and meaningful (e.g., social cohesion, biodiversity). Results clarify that using solutions including waste management systems, energy-efficient designs, active transportation networks, and smart city technologies enables Salutogenic urbanism to solve health and environmental issues as well. These results provide direction in creating and developing sustainable healthy cities consistent with the Sustainable Development Goals.

Keywords Salutogenic · Salutogenic urbanism · Climate change · Healthy cities · Sustainable cities

Introduction

In the modern world, the sudden increase in the population of cities has brought about the need for rapid and cheap transportation and a large number of private cars (Bozhkov et al. 2021; Zipperer and Pickett 2012). As a result, cities have expanded their non-standard development without considering environmental and human indicators such as land and altitude (Sepehri et al. 2025). According to UN projections, more than 55% of the world's population resided in urban areas by 2020, and by 2025, this figure is predicted to rise to 68% (United Nations 2018). The two main problems resulting from these developments are climate change and the concomitant endangerment of citizens' health (Baer 2016; Giles-Corti et al. 2016; Hou et al. 2022).

Climate change has resulted in elevated sea levels, flooding, heightened natural disasters, and global warming (Kumar et al. 2021). Citizens' health is jeopardized by air and water pollution, inadequate waste management, unsafe housing conditions, traffic congestion, insufficient green areas, and detrimental urban lifestyles. The World Health Organization estimates that environmental factors related to urbanization account for approximately 23% of global deaths (Noncommunicable Diseases, Rehabilitation and Disability 2025). This could indicate an urgent need for sustainable urbanization that simultaneously prioritizes increasing urban climate resilience and improving the health and well-being of citizens.

Furthermore, the COVID-19 pandemic presents a pivotal opportunity to reassess the design and planning of cities (Afrin et al. 2021). Subsequently, some cities adopted policies and approaches that have achieved positive results in terms of both climate and urban health goals. For example, cities such as Bologna, which promote energy efficiency and use renewable energy sources, not only reduce climate change but also improve indoor air quality, which is beneficial to public health (Nuvolari-Duodo et al. 2024). They also contribute to healthier living environments and reduce energy consumption by encouraging green building standards and retrofitting existing structures (Nuvolari-Duodo et al. 2024). In the area of transport, cities have helped promote green transport and improved the design of urban spaces in favor of active mobility (Becvarik et al. 2024; Khreis et al. 2019). This has led to a decrease in the use of private cars and, consequently, a reduction in air pollution and CO_2 emissions (Becvarik et al. 2024; Khreis et al. 2019). Furthermore, policies to expand parks and, in general, green infrastructure in cities have had positive effects on reducing heat islands and improving citizens' mental and physical health of citizens (Mullenbach and Wilhelm Stanis 2024; Nuvolari-Duodo et al. 2024). A comprehensive look at these policies and approaches shows that cities around the world need to place humans and nature at the center of their design, planning, and management policies in an integrated manner.

In the last few decades, designers and researchers have begun using the Salutogenic design approach (SDA), which has led to positive results in architecture based on real-world examples (Yeang and Dilani 2022). The SDA emphasizes the development of surroundings that promote health and well-being, rather than concentrating exclusively on illness avoidance. This design philosophy, which has its roots in the concept of salutogenesis, developed by Aaron Antonovsky, emphasizes the importance of understanding and enhancing the factors that support human health (Yeang and Dilani 2022).

Salutogenic Urbanism (SU) can be defined as an urban planning and design approach that prioritizes the creation of environments promoting physical, mental, and social well-being while addressing challenges such as climate change and public health that reduce community stress as much as possible. It integrates principles of sustainability, equity, and resilience to foster healthy, livable cities that enhance citizens' quality of life through comprehensible, manageable, and meaningful design elements.

Despite the potential mentioned in some research and published documents for SDA as an urbanism approach, the design frameworks for this approach at the urban scale are vague and do not have a specific place in the existing literature. Therefore, there is a need to conduct a general review of the SDA literature. This study addressed this issue. In this study, we aim to review the literature on SDA and understand the extent of their development at the urban planning scale. Our findings and analyses answer the following questions.

1. What are the most popular topics and gaps in the literature?
2. What are the most important factors of the Salutogenic Urbanism?
3. What suggestions and possibilities can be considered for developing the Salutogenic urbanism approach, especially emphasizing climate and health challenges?

The results of this study not only provide a better understanding of how the SDA has changed over time in the research field, but we also think that they can help the growth of Salutogenic urbanism (SU), which means that cities can be built that can handle climate and health problems.

Background

Antonovsky's Salutogenic theory serves as the foundation for SDA. This theory was introduced as a philosophy that health should be achieved by promoting healthy factors rather than by treating and neutralizing pathogenesis. Dilani first applied this approach to the realm of design and architecture (Dilani 2005, 2008). According to Cecon et al. (2021) and Eriksson et al. (2024), Dilani's work is particularly concerned with Dilani is interested in understanding the factors that influence affect health and well-being, especially in the area of "sense of coherence" (SOC) (Cecon et al. 2021). SOC refers to an individual is a person's ability to perceive life as comprehensible, manageable, and meaningful. The main goal of Salutogenic design is to create a built environment that embraces these factors. In this regard, the SDA has become a valuable framework for various scales and multiple purposes.

For instance, healthcare design has always incorporated the SDA as a fundamental characteristic. Studies have shown that implementing SDA in healthcare design is beneficial by addressing both the physical and psychological needs of patients. In this regard, Golembiewski also adds that designing healthcare spaces based on SDA can reduce stress and promote the well-being of patients and healthcare providers alike (Golembiewski 2017). Especially in the design of hospitals, SAD helped the patient's recovery and satisfaction (Ghaffari et al. 2023). Researchers have found that adding things such as natural light, green spaces, and well-thought-out space arrangements can make healthcare facilities healthier, which ultimately leads to better health outcomes (Essa and Jabbari 2020; Mazuch 2017).

In addition, especially in recent years, SAD has undergone initial development at the urban scale. The application of SDA in urbanism has been to enhance the health and improve the quality of life of citizens. In fact, most research and practice of Salutogenic urbanism have been around creating a policy of designing healthier cities, especially for vulnerable people in society, to improve mental and physical health (Brunelli et al. 2022; Capolongo et al. 2018). One of the main characteristics of SDA (both in architecture and urban planning) is the use of biophilic design principles (Stoltz and Schaffer 2018; Yeang and Dilani 2022). This refers to the close interaction between urban structure and nature, as well as the creation of urban gardens and parks, which can not only improve health, but also address climate and ecological challenges (Lee and Kim 2021).

This study examined what has already been written about SDA in cities and considered how this approach might help solve health and climate problems in cities.

Methodology

This study evaluated the most important literature on SDA and urbanism. The first step was to create a comprehensive search string appropriate to the objectives of the present study and Salutogenic concept. As shown in Fig. 7.1, we selected keywords and concepts from four main areas to include as many articles as possible in our analysis.

In this search string, the first part aimed to identify all the articles that were directly related to Salutogenic. Therefore, we used three keywords: Salutogenic, Salutogenesis, and Sense of Coherence. Next, a long sentence was created, which, as shown in Fig. 7.1, was intended to identify all Salutogenic articles on an urban scale that were also related to urban health or climate:

(("Salutogenic" OR "Salutogenesis" OR "Sense of Coherence") AND ("urbanism" OR "urban design" OR "urban planning" OR "town design" OR "town planning" OR "city design" OR "city planning" OR "Spatial planning" OR " spatial design" OR "urban development" OR "city development" OR "town development" OR "spatial development" OR "biophilic urbanism" OR "biophilic urban design" Or "biophilic urban planning" OR "biophilic urban development" OR "biophilic town design" OR "biophilic town planning" Or "biophilic city design" OR "biophilic city planning" OR "biophilic spatial design" OR "biophilic spatial planning" OR "nature inspired urban design" OR "nature inspired urban planning" OR "nature inspired town design" OR "nature inspired town planning" OR "nature inspired spatial design" OR "nature inspired urbanism" OR "ecological urbanism" OR "ecological urban design" OR "ecological urban planning" OR "ecological town design" OR "ecological town planning" OR "ecological spatial design" OR "ecological spatial planning" OR "green urbanism" OR "green urban design" OR "green urban planning" OR "green town design" OR "green town planning" OR "green spatial design" OR "green spatial planning" OR "sustainable urbanism" OR "sustainable urban design" OR "sustainable urban planning" OR "sustainable town design" OR "sustainable town planning" OR "sustainable spatial design" OR "sustainable spatial planning" OR "environmental urbanism" OR "environmental urban design" OR "environmental urban planning" OR "environmental town design" OR "environmental town planning" OR "environmental spatial design" OR "environmental spatial planning" OR "health* city" OR "health* cities" OR "health* urbanism" OR "health* urban design" OR "health* urban planning" OR "health* town design" OR "health* town planning" OR "health* spatial design" OR "health* spatial planning" OR "urban* health*" OR "urban* mental health*" OR "urban* well-being" OR "eco-friendly urbanism" OR "eco-friendly urban design" OR "eco-friendly urban planning" OR "eco-friendly town design" OR "eco-friendly town planning" OR "eco-friendly spatial design" OR "eco-friendly spatial planning" OR "climate-responsive urbanism" OR "climate-responsive urban design" OR "climate-responsive urban planning" OR "climate-responsive town design" OR "climate-responsive town planning" OR "climate-responsive spatial design" OR "climate-responsive

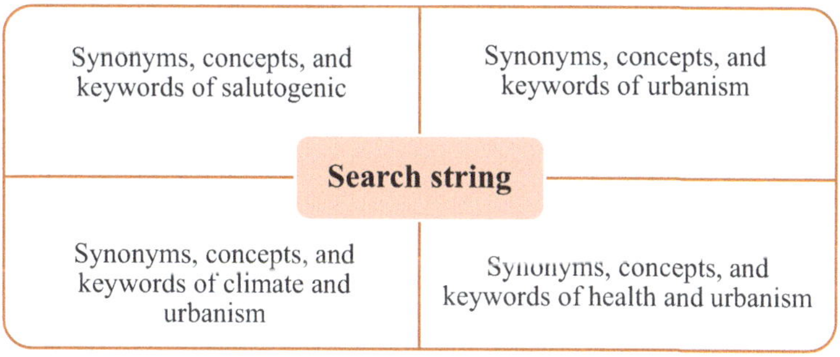

Fig. 7.1 Synonyms, concepts, and keywords for creating the search strings

spatial planning" OR "low-carbon urbanism" OR "low-carbon urban design" OR "low-carbon urban planning" OR "low-carbon town design" OR "low-carbon town planning" OR "low-carbon spatial planning" OR "low-carbon spatial design" OR "evidence-based urbanism" OR "evidence-based urban design" OR "evidence-based urban planning" OR "evidence-based town planning" OR "evidence-based town design" OR "evidence-based spatial design" OR "evidence-based spatial planning"))

After finalizing the search string, we selected the Web of Science and Scopus databases to identify the articles. These two databases are very important for literature reviews because they contain a lot of scientific content, can track citations well, offer advanced search options, and are trusted sources of academic research. We used the PRISMA checklist to create a unified, comprehensive, and understandable search.

On January 17, 2025, the search string was applied to the Title, Abstract, and Keywords sections of Scopus and to the Topic section of Web of Science to identify only English articles. Figure 7.2 illustrates the identification of 64 articles in Scopus and 59 articles in the Web of Science. We downloaded 123 article records from Scopus and the Web of Science after verifying their relativity. During the screening stage, 44 articles were identified as duplicates and irrelevant, leading to their removal. Ultimately, 69 articles were selected as the final database for this study. We used Bibliomtrix and VOSviewer software to analyze these articles. We also performed in-depth and critical analyses to group the topics into groups and identify factors that are beneficial for urbanism.

Results and Discussion

Scientific Production

The Salutogenic approach has likely advanced on an urban scale in recent years, with the hypothesis that an increasing number of publications will explore it owing to the emergence of COVID-19 and the escalating issues of urban health. Figure 7.3 illustrates the publication patterns of articles from 1988 to 2024. Primarily between 2014 and 2016, SU drew interest. Several elements could explain this. The 10th and 11th World Congresses on Design and Health took place in Toronto and Hong Kong, respectively, in 2014 and 2015. Consistently supporting the Salutogenic design philosophy, these conferences have greatly helped to increase attention on Salutogenic urbanization. The adoption of global frameworks such as the United Nations Sustainable Development Goals (SDGs) in 2015 brought into focus the importance of sustainable cities (Goal 11) (Reich 2022). Stimulated research on health-promoting urban policies using health-enhancing ideas corresponding to SDG 11.7, which emphasizes inclusive and accessible green areas. Figure 7.3 also shows a significant increase in publications related to Salutogenic urbanism from 2019–2020 to 2023. This increase could be attributed to the onset of the COVID-19 pandemic, further highlighting the need for healthy cities.

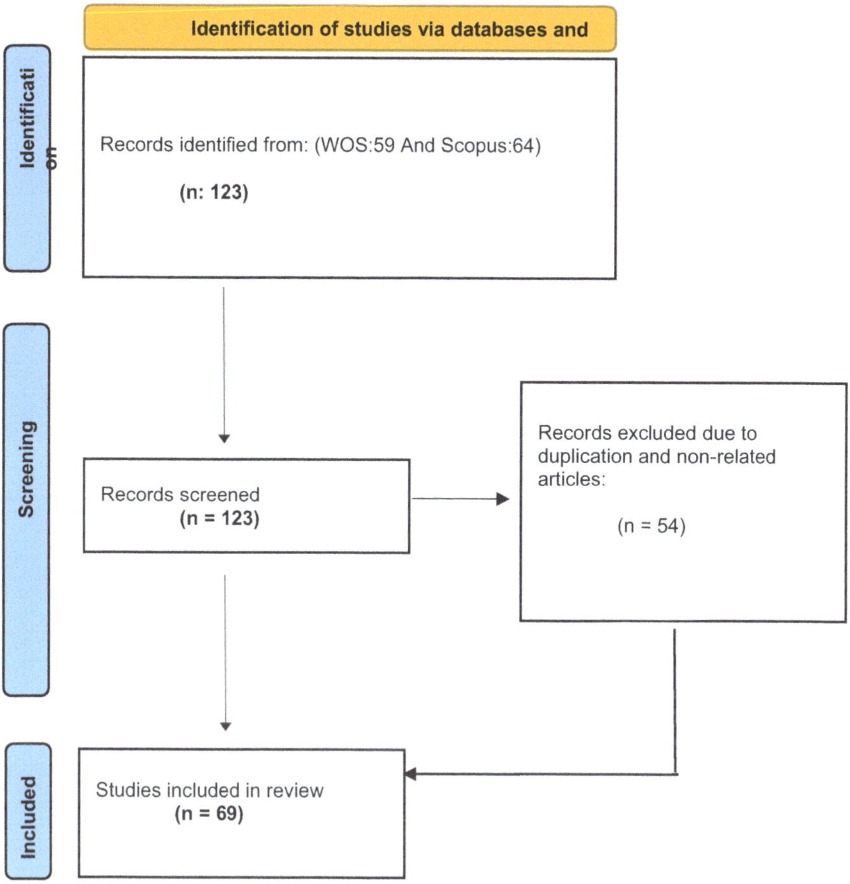

Fig. 7.2 Procedures for bibliometric review based on PRISMA checklist (Page et al. 2021)

Keywords and Topics

Figure 7.4 shows the most important keywords for the identified articles. Public health was the most frequent keyword. This aligns with the Salutogenic concept's primary objective of prioritizing public health. The importance of human keywords also emphasizes the main concept of salutogenesis, which is that humans should be at the center of urban design and planning.

Figure 7.4 shows several important keywords. The keyword Europe indicates the geographic focus of the SU. European countries and researchers have played important roles in the development of this approach. Decision making is a crucial keyword. This suggests that urban decision making plays a crucial role in promoting health in cities, as viewed through the lens of the Salutogenic approach. Figure 7.4, on the other hand, shows only a few keywords that are related to environmental

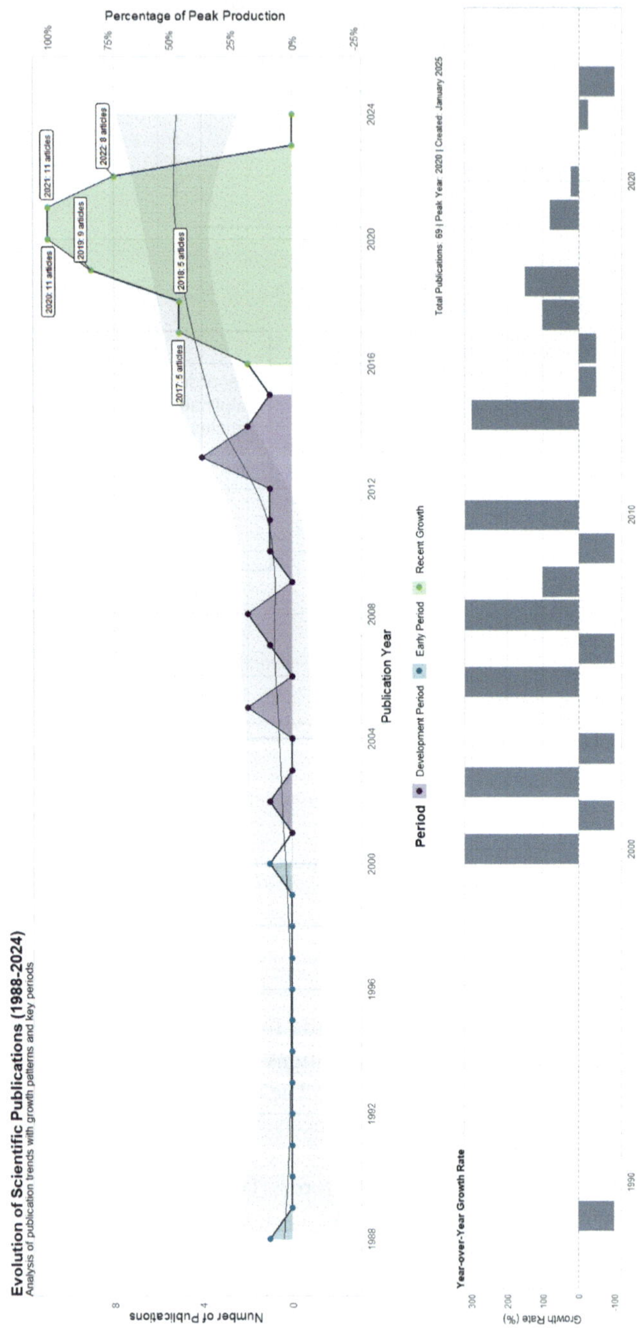

Fig. 7.3 Scientific production

Fig. 7.4 Word cloud of most relevant words

problems such as climate change and air quality. This means that these keywords are new or have not been studied as extensively in studies on Salutogenic urbanism.

We analyzed trending topics in Fig. 7.5 using the Bibliometrix software, highlighting significant points. Mental health has been a trending topic for several years. This shows that the focus of research has been more on mental health because the nature of the Salutogenic approach and sense of coherence has been developed in line with mental health and stress reduction. Figure 7.5 does not include the topic of physical health, which suggests a potential thematic gap. A healthy city is designed and planned in line with the physical health of citizens in addition to promoting public health. In addition, the two topics of cities and rural areas are worth considering because they trended in two different periods. Figure 7.5 shows that the rural topic showed a trend from 2004 to 2012, while the topic showed a trend from 2020 to 2024. This shows that the focus of the research has changed from rural to urban areas. This change can be attributed to the fact that there have been more health challenges in cities in recent years, and as a result, the need to implement the Salutogenic approach has also increased.

Important References

The review revealed the six publications most commonly cited among the sample of SU studies. The table below lists the six most frequently cited works relating to Salutogenics.

Kaplan's book (1989) is highly influential to Salutogenic research. Throughout the literature on Salutogenics, authors often turn to Kaplan's (1989) book, Addressing

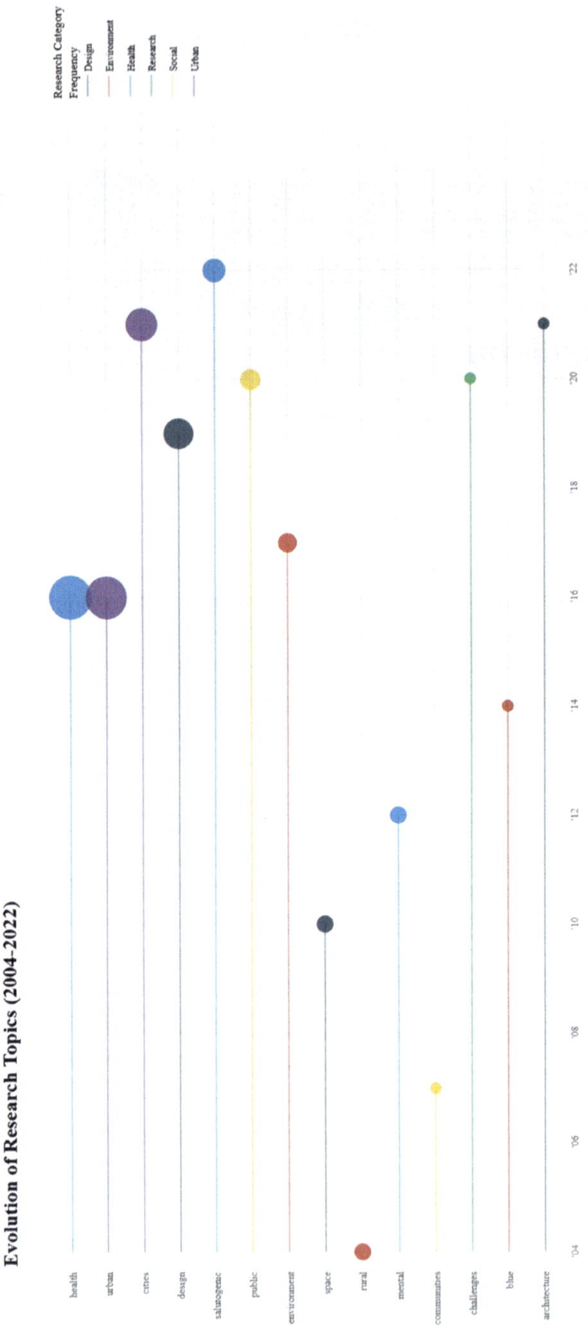

Fig. 7.5 Trend topics

the Mindful Self, to understand the mechanisms underlying the beneficial effects of interaction with nature. His Attention Recovery Theory (ART) shows how being in nature contributes to stress relief and the renewal of concentration and works with Salutogenics' emphasis on what maintains or improves health. Kaplan's book (1989) is useful for analyzing how interconnections contribute to favorable impacts on people's mental well-being (Kaplan 1989).

Antonovsky's (1979) book is likely to be considered a significant reference. Antonovsky's (1979) book Health, Stress, and Coping is critical to Salutogenic research since it introduced the concept of Salutogenesis and studied factors that promote well-being instead of just preventing illness (Antonovsky 1979). Antonovsky developed the concept of a "sense of coherence," which forms the basis for understanding how people cope with stress, build resilience, and thrive. This text has become widely recognized as the foundation for research exploring the environmental factors that support positive health. This book provides the foundation for investigations into methods that help people stay well and adapt to challenges. The work lines up with research into healthy and resilient settings, stress management approaches, and mental health strategies (Antonovsky 1979).

Dilani points out how integral the built environment is to improving the well-being of communities in her edited volume from 2004, "Design & Health III." The presentation of initiatives like vegetated neighborhoods, pedestrian-friendly areas, convenient services, and close proximity to civic life emphasizes the importance of environmental design (Dilani 2004). The method promotes creating built environments that reflect local needs to minimize common sources of stress—pollution and noise—and support increased physical activity, improved mental health, and closer community ties. Central to its approach is making urban design an indispensable tool used for fostering better health by uniting architecture, public health, and cities. Dilani also highlights how environmental design plays an important role in promoting public health by encouraging such urban policies as green spaces, walkable neighborhoods, easily accessible services, and opportunities for people to interact socially. Salutogenic design suggests using architectural design to lessen environmental stress factors such as air pollution or street noise and at the same time promoting physical activity, mental health, and community bonding (Dilani 2004). Connecting urbanism, architecture, and public health makes it possible to understand the major role that the built environment plays for individual and community health. The publication serves as an important and innovative guide to the field of SU.

More important references are shown in Table 7.1.

Co-occurrence Analysis

A co-occurrence analysis was conducted using VOSviewer software, as illustrated in Fig. 7.6. Co-occurrence analysis is valuable because it identifies patterns and relationships among words, concepts, or variables within a dataset, illustrating their connections and temporal variations (Leydesdorff et al. 2013; Van Eck and Waltman

Table 7.1 Most important references

Reference	Topic	Relation to Salutogenic urbanism
Antonovsky (1979)	Introduction of the Salutogenic model focusing on health rather than illness	This foundational work introduces the concept of Salutogenesis, emphasizing the importance of understanding health development, which is crucial for urban design that promotes well-being through supportive environments
Antonovsky (1987)	Stress management and health maintenance strategies	The insights into how individuals cope with stress contribute to designing urban spaces that enhance resilience and promote mental well-being, aligning with Salutogenic principles in urban planning
Kaplan (1989)	Psychological benefits of nature exposure	This work supports SU by highlighting the mental health benefits derived from natural environments, advocating for their integration into urban designs to foster well-being and reduce stress in city settings
Dilani (2004)	Salutogenic design theory Integrating Antonovsky's Salutogenesis into architecture For the first time at the Karolinska Institute, Medical University, Stockholm, this innovative approach has been developed. applied primarily in healthcare design	Recently, it has evolved into a systematic tool—Matrix Design and Health—enabling architects and planners to implement the principles of Salutogenesis across various sectors of architecture and urban design
Warburton et al. (2006)	Evidence-based benefits of physical activity for health	The findings underscore the necessity of creating urban environments that encourage physical activity through accessible parks and recreational areas, which are essential components of Salutogenic urban design aimed at enhancing public health outcomes
Hale et al. (2011)	Relationship between food environments and health through aesthetics and community gardening	This research illustrates how community gardens can enhance social interaction and access to healthy food within urban settings, promoting a Salutogenic approach by fostering community resilience and well-being through aesthetic and functional design in public spaces
Conradson (2017)	Exploration of places designed for relaxation and retreat in urban contexts	The concept of creating spaces for stillness and retreat aligns with Salutogenic principles by emphasizing the need for restorative environments in urban planning that contribute to mental well-being and overall health in densely populated areas

2014). It is extensively utilized in bibliometrics, text mining, and network analysis to discern research patterns, delineate knowledge systems, and facilitate decision making (Leydesdorff et al. 2013; Van Eck and Waltman 2014).

Figure 7.6 shows that the red cluster had the largest co-occurrence of words. In this cluster, the words "urban planning," "public health," and "well-being" had the highest co-occurrence. This could refer to research addressing the role of urban planning in the public health and well-being of citizens (Capolongo et al. 2018; Reich 2022). In addition, in the red cluster, the words urban area, urbanization, and environmental health have a high co-occurrence, which could indicate the importance of environmental health in urban areas (EUPHA-URB 2023; Souter-Brown 2023). Because of the increase in population and urbanization, environmental challenges that endanger health have also increased every year. Creating an integrated framework, such as Salutogenic urban development, can reduce the environmental health challenges in cities in line with their needs. In addition, in the red cluster, the word climate change is located on the edge of the cluster and does not have high co-occurrence. This could indicate that the relationship between Salutogenic urban development and climate issues are still in its early stages; previous research has not addressed this issue. However, as Table 7.2 shows, SU has a strong connection with green development and privileges urban design and planning based on nature-based approaches. This could create the strongest connection between SU and climate challenges.

In the green cluster, the terms "human", "health promotion," "sense of coherence," and "urban design" had the highest co-occurrence. The centrality of "human" in this

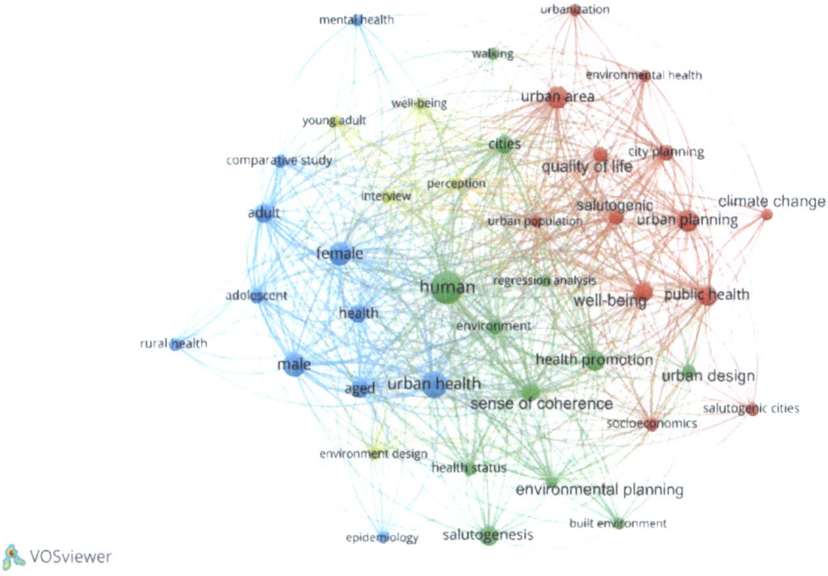

Fig. 7.6 Co-occurrences analysis

Table 7.2 SU factors based on literature review

SOC elements	SU factors	Definition/Description	Documents
Comprehensible	Green spaces	Access to parks and natural environments that promote mental and physical well-being	Capolongo et al. (2018), D'Alessandro et al. (2023), Sarkar et al. (2015)
	Walkability	Design of pedestrian-friendly pathways to encourage physical activity	Appolloni et al. (2020), Paglione et al. (2024), Sarkar et al. (2015)
	Urban comfort	Design strategies to enhance thermal comfort and reduce urban heat islands	Tamaskani Esfehankalateh et al. (2021), Ulpiani (2021)
	Accessibility to services	Proximity to healthcare, education, and recreational facilities	Buffoli and Rebecchi (2022), Capolongo et al. (2020)
	Safety and security	Urban design that ensures safety and reduces crime	Alderton et al. (2020), Quesada-García et al. (2023)
Manageable	Noise reduction	Strategies to minimize urban noise pollution for better mental health	Quesada-García et al. (2023), Ulpiani (2021)
	Air quality	Measures to improve air quality, such as reducing traffic emissions	Tamaskani Esfehankalateh et al. (2021), Ulpiani (2021)
	Resilience to climate change	Urban planning that mitigates climate risks and enhances adaptability	D'Alessandro et al. (2023), Ulpiani (2021)
	Health-promoting infrastructure	Design of buildings and public spaces that support physical and mental health	Quesada-García et al. (2023), Souter-Brown (2023)
Meaningful	Social cohesion	Urban design that fosters community interaction and social support	Alderton et al. (2020), Franceschini et al. (2018)
	Biodiversity	Integration of diverse flora and fauna to enhance ecosystem services	Chen et al. (2019), Mhuireach et al. (2021)
	Public participation	Involving citizens in urban planning to ensure inclusivity and equity	Buffoli and Rebecchi (2022), Maass et al. (2022)
	Therapeutic landscapes	Urban spaces designed to provide restorative and healing experiences	Souter-Brown (2023), Voelker and Kistemann (2013)

(continued)

Table 7.2 (continued)

SOC elements	SU factors	Definition/Description	Documents
	Equitable access to nature	Ensuring all socio-economic groups have access to green and blue spaces	Buffoli and Rebecchi (2022), Chen et al. (2019)
	Cultural ecosystem services	Spaces that provide cultural, spiritual, and recreational benefits	Chen et al. (2019), Souter-Brown (2023)

cluster reminds us that the concept of Salutogenics places humans at the center of attention, and that promoting human health is important regardless of pathogenic factors. The fact that the words "urban design" and "health promotion" are used together also shows the importance of urban design to the health and well-being of people in SU. Brunelli et al. emphasize this fact in their study (Brunelli et al. 2022). This study is useful because it connects the theory of Salutogenics with actual urban designs. It provides useful information on how to create places that promote health and well-being, especially for older people and other vulnerable groups (Brunelli et al. 2022). In addition, the co-occurrence of the terms "sense of coherence" and "environmental planning" in the green cluster is significant. The idea of salutogenesis places a high value on fostering a sense of coherence so that people feel less stressed. Planning the environment can help achieve this (Yeang and Dilani 2022). An (urban) environment that is Salutogenic will have the least health challenges, and as a result, the stress level of society and citizens will be at its lowest (Yeang and Dilani 2022). In addition, the co-occurrence of "walking" is small, but important in this cluster. This could refer to the promotion of walking in cities, the effects of which have been proven on mental and physical health, and is in line with recent concepts, such as walkable cities and X-minute cities (Sepehri and Sharifi 2025; Zhang et al. 2023). Although walking has rarely been addressed in the field of SU, Fig. 7.6 shows that this discussion is expanding (Mazuch 2017; Paglione et al. 2024; Sarkar et al. 2015).

Given the significant co-occurrence of the terms "male," "female," and "aged," aged, we can link the blue cluster to the demographic topic. In particular, the term "aged" shows that research has paid more attention to the elderly segment of society, who are also exposed to more health and climate challenges (Brunswick and Merzel 1988; Yeang and Dilani 2022). However, the blue cluster marginalizes these two prominent terms. The first is "rural health," which has been marginalized despite having the links above. This is in line with the findings of Fig. 7.5, which showed that rural topics have not been a trend in recent years (Forbes and Janzen 2004). The second is "mental health". Since the idea of salutogenesis was first developed with mental health in mind, the fact that this word was not included in the co-occurrence analysis came as a bit of surprise. Nevertheless, research in SU and mental health has continued to attract the attention of researchers (Altwaijri et al. 2023; Langeland et al. 2007; Tsuno and Yamazaki 2012). Perhaps one of the terms that was expected to be present in the analysis of Fig. 7.6 was "physical health," which was not found in the analysis, as in the analysis of Fig. 7.5. Although physical health can be included within the scope of the terms health and public health, this is not an area that should be further investigated, as other studies have shown that healthy cities have a particularly positive impact on the physical health of citizens (Sallis et al. 2016).

Salutogenic Urbanism

A thorough review of the literature identified important parts of SU, which are shown in Table 7.2, according to the three parts of Sense of Coherence (SOC): Comprehensible, Manageable, and Meaningful. This extensive framework illustrates how urban design and planning components may facilitate health promotion and well-being.

Current research on SU lacks a unified framework, similar to the works of Yeang and Dilani regarding Salutogenic architecture (2022). An integrated strategy is crucial for effectively addressing the health and climate challenges in urban environments (see Fig. 7.7).

Future Directions for Studies in SU

Research on SU must be expanded to address the issues of emerging cities and new aspects of healthy, ecologically viable urban environments. Key future research areas for SU consist of waste management in urban environments, energy efficiency, a cycling-friendly city environment, urban public transportation in a sustainable way, flexible public spaces in an urban setting, green roofs and garden roofs, and intelligent urban technologies incorporated into smart cities.

Research should be conducted to develop a waste management system that causes minimal harm to the environment and improves public health. Integrating waste-to-energy technology with a circular economy will reduce landfill burden while minimizing greenhouse gas emissions. Similarly, energy-efficient urban designs, including passive building techniques and the use of renewable energy, can reduce the effects of climate change, improve air quality, and decrease urban heat islands. Researchers should study ways to increase the availability and fairness of sustainable practices in all communities. For example, active transportation networks, including bike-friendly infrastructure and environmentally sustainable public transit systems, are crucial to reduce emissions and increase physical activity.

Future research needs to be directed at building accessible systems of transport usable for all sections, including the elderly and disabled. It can now examine the integration of intelligent technologies, such as real-time transit monitoring and bike-sharing applications, that could enhance user experience and encourage widespread adoption. Likewise, a safe, connected, and visually appealing urban design in prioritizing cycling and pedestrian routes is paramount. The design of urban public spaces needs to be flexible, incorporating pop-up parks, modular seating, and multivariate plaza uses to allow for community interaction and resilience. Green and garden roofs

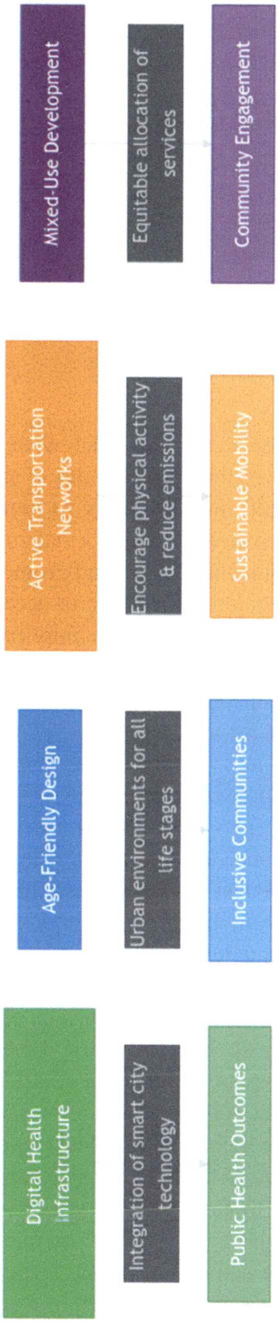

Fig. 7.7 Urban health and sustainability framework

enhance biodiversity in cities, reduce the heat island effect, and provide psychological benefits.

Further research needs to be directed at how green infrastructure can be scaled up and integrated into mainstream urban planning. In addition, we want to explore smart city technologies, such as IoT sensors and AI-powered urban management systems, to enhance resource distribution for better health surveillance and to increase the quality of general urban living. The concentration of future research may be able to find a linkage between SU and the SDGs while creating cities that are good for all health, environmental, and belonging perspectives. We will create urban environments with thorough buildings to improve the well-being of the present and future generations.

Recommendations and Implementations

Urban planners, policymakers, and designers that wish to include Salutogenic design into their work and advance the public health of individuals with the help of constructed environments would find this study to be rather enlightening. Our review of the literature has revealed eight techniques that guide unified urban design, simultaneously addressing environmental issues and promoting public health.

Green spaces and walkability require investing in accessible parks, communal gardens, and well-designed pedestrian networks to amplify a sense of coherence as well as increase activity levels within the city. Such strategies help alleviate heat in cities and increase fresh air by incorporating trees and green corridors. Planning policies must enhance green space proximity for every community and design walkable streets with integrated shade and rainwater management relying on ecological design. Furthermore, efficient responses to environmental health challenges involve using different nature-based strategies to lessen the effects of urban air pollution and loud noise. All tall buildings must have vertical gardens and green roofs by law, along with a network of IoT-enabled air quality sensors. The future of urban mobility depends on creating electric micro-mobility zones and incentives for emissions-free travel that align with WHO's air quality guidelines, which will help reduce carbon emissions in cities.

Engaging the public in collaborative design processes becomes essential to those aspects of SOC that relate to purpose and satisfaction. Communities should participate in the creation of therapeutic public spaces and multicultural hubs using inclusive workshop sessions. These places are designed to be welcoming and accessible to all citizens. Combining community urban farms and preservation areas effectively enhances community camaraderie, as these zones also support valuable ecosystem functions such as habitat connections and carbon capture.

Prioritizing people over cars through dedicated bike lanes and pedestrian-only spaces is the best way to achieve active transportation. X-minute cities should be reinforced by transit-oriented development enhanced by mobile apps that link convenience to multiple travel modes. These systems yield dual benefits: health hazards in respiratory disease incidence are mitigated by lowered concentrations of PM2.5

while reducing transportation-related emissions by 22–35%. Improving the number of green roofs and using surfaces that allow water to soak through is now crucial for making cities better prepared for climate challenges in their planning rules. Data on land height and local climate forecasts can help find the best spots for buildings that can resist flooding and guide efforts to restore wetlands, which can protect against storms and create enjoyable outdoor spaces.

Incorporating AI-driven energy and waste management techniques can help cities use their resources more efficiently. Testing in certain areas that combine sensor-based water monitoring and blockchain-powered carbon tracking will allow these solutions to grow through data sharing between government and private companies. To ensure these technologies are used fairly, strong policies must be created to prevent excluding marginalized groups, as the benefits of these innovations are fully realized. Also, equitable distribution of natural spaces requires analyzing disparities in access to green areas across communities. The results of poverty mapping for parks should guide the allocation of green spaces to regions with the highest need, along with the implementation of mobile green infrastructure in densely populated neighborhoods. Urban policymakers should recognize access to urban nature as a fundamental human right, adopting inclusive zoning policies that hold developers accountable for maintaining nearby parks.

Ultimately, Adaptive public space design is essential due to it allows spaces to adapt with outstanding flexibility via portable community seating arrangements and small, movable plazas for changing functions. These areas should be designed to capture renewable energy and stormwater on-site so that they can become living laboratories for circular economy initiatives.

Conclusion

SU adopts approaches focused on developing urban settings that promote and improve the well-being of its residents. The study emphasizes the value of techniques that tackle the difficulty of integrating health promotion and climate change adaptation. An integrated urban planning strategy includes making use of resources such as green infrastructure, advanced technology, and nature-based options, all of which contribute to improving a city's flexibility and ability to overcome various difficulties.

The results of the research indicated that there were few examples of having applied Salutogenic concepts to urbanism. Although Salutogenic approaches have consistently yielded positive outcomes in healthcare and architecture, their application in the field of urbanism remains incomplete. Constructing and evaluating suitable strategies would enable cities to enhance their opportunities in SU and thereby improve both their people's health and their environment. Nonetheless, there is a need for further investigation into the outcomes of SU practices applied in various scenarios across many locations. There is currently not enough research on how to implement SU principles effectively in public policies, hindering progress in this field. We must create flexible assessment tools and utilize intelligent technologies to evaluate how SU strategies are being implemented in various cities and urban environments.

According to the most recent studies, making essential improvements to energy efficiency, implementing strategies that support cycling, and enhancing public transportation are critical steps cities should take to attain greater environmental sustainability and the promotion of positive health in their residents. City planners can achieve a more livable and significant city by incorporating a "sense of coherence" into their urban designs. Implementing these processes when shaping cities has been shown to encourage residents to form stronger bonds and enjoy higher levels of health.

Studying how breakthrough technologies can strengthen the ability for cities to both advance sustainability efforts and reach the UN Sustainable Development Goals is a crucial step for the development of urban design in the coming years. Implementing eco-friendly forms of waste management and circular economy initiatives in urban planning may reduce emissions and limit the amount of rubbish dumped in landfills. Implementing these solutions will benefit the environment and improve the living conditions in cities.

Integrating Salutogenic principles into urbanism is an important and ambitious action that cities should take to address the challenges of climate change and improve the well-being of their people. Implementing these environmental planning guidelines will have an important effect on developing cities that can provide equal access to resources for their residents. To provide healthier environments and confront the impact of climate change, cities must adapt and change their construction methods.

AI Use Acknowledgment

No potential conflict of interest was reported by the author(s). During the preparation of this work, the author(s) used Grammarly to enhance grammar, punctuation, and clarity. Following the use of this tool, the author(s) thoroughly reviewed, edited, and approved the final content, taking full responsibility for the publication's material.

References

Afrin S, Chowdhury FJ, Rahman MM (2021) COVID-19 pandemic: rethinking strategies for resilient urban design, perceptions, and planning. Front Sustain Cities 3:668263. https://doi.org/10.3389/frsc.2021.668263

Alderton A, Henry N, Foster S, Badland H (2020) Examining the relationship between urban liveability and gender-based violence: a systematic review. Health Place 64:102365. https://doi.org/10.1016/j.healthplace.2020.102365

Altwaijri YA, Al-Subaie AS, Al-Habeeb A, Galea S, Akkad M, Naseem MT, Bilal L (2023) Urbanization and mental health: a perspective from Riyadh City, Kingdom of Saudi Arabia. Int J Soc Psychiatry 69(5):1121–1133. https://doi.org/10.1177/00207640231152202

Antonovsky A (1979) Health, stress, and coping. In: New perspectives on mental and physical well-being. pp 12–37. Available on: https://cir.nii.ac.jp/crid/1573105974919935616. Accessed on 15 Jan

Antonovsky A (1987) How people manage stress and stay well. In: Unraveling the mystery of health. pp 127–139

Appolloni L, Giretti A, Corazza MV, D'Alessandro D (2020) Walkable urban environments: an ergonomic approach of evaluation. Sustainability 12(20):8347. https://doi.org/10.3390/su12208347

Baer HA (2016) Private cars as environmental health hazards: the critical need for public transit in the era of climate change. In: A companion to the anthropology of environmental health. pp 458–482.https://doi.org/10.1002/9781118786949.ch22

Becvarik ZA, White LV, Lal A (2024) The health and wellbeing co-benefits of policies and programs to address climate change in urban areas: a scoping review. Environ Res Lett. https://doi.org/10.1088/1748-9326/ad78ef

Bozhkov S, Tonkov G, Boinovski V, Chernev P (2021) Experimental results of the hybrid electric vehicle energy efficiency in urban transportation. Trans Motauto World 6(2):43–46. Available on: https://stumejournals.com/journals/tm/2021/2/43.full.pdf. Accessed on 15 Jan 2025

Brunelli L, Smith H, Woolrych R (2022) A Salutogenic urban design framework: the case of UK local high streets and older people. Health Promot Int 37(5). https://doi.org/10.1093/heapro/daac102

Brunswick AF, Merzel CR (1988) Health through three life stages: a longitudinal study of urban black adolescents. Soc Sci Med 27(11):1207–1213. https://doi.org/10.1016/0277-9536(88)90350-4

Buffoli M, Rebecchi A (2022) The proximity of urban green spaces as urban health strategy to promote active, inclusive and Salutogenic cities. In: International conference on technological imagination in the green and digital transition. pp 1017–1027.https://doi.org/10.1007/978-3-031-29515-7_90

Capolongo S, Rebecchi A, Dettori M, Appolloni L, Azara A, Buffoli M, Capasso L, Casuccio A, Oliveri Conti G, D'Amico A (2018) Healthy design and urban planning strategies, actions, and policy to achieve Salutogenic cities. Int J Environ Res Public Health 15(12):2698. https://doi.org/10.3390/ijerph15122698

Capolongo S, Rebecchi A, Buffoli M, Appolloni L, Signorelli C, Fara GM, D'Alessandro D (2020) COVID-19 and cities: from urban health strategies to the pandemic challenge. A decalogue of public health opportunities. Acta Bio Med Atenei Parmensis 91(2):13. https://doi.org/10.23750/abm.v91i2.9515

Cecon N, Pfaff H, Lee S, Dresen A (2021) A Salutogenic model predicting the need for psycho-oncological care and its utilisation—the role of generalised resistance resources and sense of coherence. Eur J Cancer Care 30(1):e13335. https://doi.org/10.1111/ecc.13335

Chen X, de Vries S, Assmuth T, Dick J, Hermans T, Hertel O, Jensen A, Jones L, Kabisch S, Lanki T (2019) Research challenges for cultural ecosystem services and public health in (peri-) urban environments. Sci Total Environ 651:2118–2129. https://doi.org/10.1016/j.scitotenv.2018.09.030

Conradson D (2017) The experiential economy of stillness: places of retreat in contemporary Britain. In: Therapeutic landscapes. Routledge, pp 33–48. https://doi.org/10.4324/9781315551166

D'Alessandro D, Rebecchi A, Appolloni L, Brambilla A, Brusaferro S, Buffoli M, Carta M, Casuccio A, Coppola L, Corazza MV (2023) Re-thinking the environment, cities, and living spaces for public health purposes, according with the COVID-19 lesson: the LVII Erice charter. Land 12(10):1863. https://doi.org/10.3390/land12101863

Dilani A (ed) (2004) Design and health III: health promotion through environmental design. Int Acad Des Health (IADH). Available on: https://dilani.org/Books/Design%20and%20Health%20IV%20Book.pdf. Accessed on 15 Jan 2025

Dilani A (2005) Psychosocially supportive design. In: Dilani A (ed) Design and health IV: future trends in healthcare design. pp 13–22. Available on: https://designhealth.academy/onewebmedia/Design%20and%20Health%20IV%20Book.pdf#page=14. Accessed on 15 Jan 2025

Dilani A (2008) Psychosocially supportive design: a Salutogenic approach to the design of the physical environment. Des Health Sci Rev 1(2):47–55. Available on: https://www.researchgate.net/profile/Alan-Dilani/publication/265349464_Psychosocially_Supportive_Design_A_Salutogenic_Approach_to_the_Design_of_the_Physical_Environment/links/54b7be500cf2e68eb28047d0/Psychosocially-Supportive-Design-A-Salutogenic-Approach-to-the-Design-of-the-Physical-Environment.pdf. Accessed on 15 Jan 2025

Essa SS, Jabbari AN (2020) The impact of Salutogenic factors on the process of patient's recovery Case study; Erbil city hospitals. Anbar J Eng Sci 11(2):137–153. https://doi.org/10.37649/aengs.2023.176831

Forbes DA, Janzen BL (2004) Comparison of rural and urban users and non-users of home care in Canada. Can J Rural Med 9(4):227. Available on: https://www.proquest.com/openview/60a566f151e87a98d65457d2ef23e6c0/1?cbl=45825&pq-origsite=gscholar. Accessed on 15 Jan

Franceschini MC, de Andrade EA, Cimmino K (2018) Healthy cities, healthy settings and education: how do they work together to promote sustainable development? In: Lifelong learning and education in healthy and sustainable cities. pp 207–218.https://doi.org/10.1007/978-3-319-69474-0_12

Ghaffari F, Shabak M, Norouzi N, Nayyeri Fallah S (2023) Hospital Salutogenic public spaces: a conceptual framework of effective perceptional environment quality components on patients' satisfaction. Int J Build Pathol Adapt 41(5):965–987. https://doi.org/10.1108/IJBPA-05-2021-0071

Giles-Corti B, Vernez-Moudon A, Reis R, Turrell G, Dannenberg AL, Badland H, Foster S, Lowe M, Sallis JF, Stevenson M (2016) City planning and population health: a global challenge. Lancet 388(10062):2912–2924. Available on: https://viewpointvancouver.ca/wp-content/uploads/2016/12/lancet-city-planning.pdf. Accessed on 15 Jan

Golembiewski JA (2017) Salutogenic architecture in healthcare settings. In: The handbook of Salutogenesis. pp 267–276.https://doi.org/10.1007/978-3-319-04600-6

Hale J, Knapp C, Bardwell L, Buchenau M, Marshall J, Sancar F, Litt JS (2011) Connecting food environments and health through the relational nature of aesthetics: gaining insight through the community gardening experience. Soc Sci Med 72(11):1853–1863. https://doi.org/10.1016/j.socscimed.2011.03.044

Hou L, Wang Y, Zheng Y, Zhang A (2022) The impact of vehicle ownership on carbon emissions in the transportation sector. Sustainability 14(19):12657. https://doi.org/10.3390/su141912657

Kaplan R (1989) The experience of nature: a psychological perspective. Cambridge University Press. Available on: http://psichenatura.it/fileadmin/img/R._Kaplan__S._Kaplan_The_Experience_of_Nature__Introduction_.pdf. Accessed on 15 Jan

Khreis H, Sudmant A, Gouldson A, Nieuwenhuijsen M (2019) Transport policy measures for climate change as drivers for health in cities. In: Integrating human health into urban and transport planning: a framework, pp 583–608.https://doi.org/10.1007/978-3-319-74983-9_28

Kumar A, Nagar S, Anand S (2021) Climate change and existential threats. In: Global climate change. Elsevier, pp 1–31. https://doi.org/10.1016/B978-0-12-822928-6.00005-8

Langeland E, Wahl AK, Kristoffersen K, Nortvedt MW, Hanestad BR (2007) Sense of coherence predicts change in life satisfaction among home-living residents in the community with mental health problems: a 1-year follow-up study. Qual Life Res 16:939–946. https://doi.org/10.1007/s11136-007-9199-z

Lee S, Kim Y (2021) A framework of biophilic urbanism for improving climate change adaptability in urban environments. Urban Forest Urban Greening 61:127104. https://doi.org/10.1016/j.ufug.2021.127104

Leydesdorff L, Rafols I, Chen C (2013) Interactive overlays of journals and the measurement of interdisciplinarity on the basis of aggregated journal–journal citations. J Am Soc Inform Sci Technol 64(12):2573–2586. https://doi.org/10.1002/asi.22946

Maass R, Lillefjell M, Espnes GA (2022) Applying Salutogenesis in towns and cities. In: The handbook of Salutogenesis. pp 361–370. Available on: https://library.oapen.org/bitstream/handle/20.500.12657/52407/978-3-030-79515-3.pdf?sequence=1#page=359. Accessed on 15 Jan

Mazuch R (2017) Salutogenic and biophilic design as therapeutic approaches to sustainable architecture. Archit des 87(2):42–47. https://doi.org/10.1002/ad.2151

Mhuireach GA, Wilson H, Johnson BR (2021) Urban aerobiomes are influenced by season, vegetation, and individual site characteristics. EcoHealth 18:331–344. https://doi.org/10.1007/s10393-020-01493-w

Mullenbach LE, Wilhelm Stanis SA (2024) Climate change adaptation plans: inclusion of health, equity, and green space. J Urban Aff 46(4):701–716. https://doi.org/10.1080/07352166.2022.2091449

Noncommunicable Diseases, Rehabilitation and Disability (2025). Retrieved 10 Jan 2025, from https://www.who.int/teams/noncommunicable-diseases/integrated-support/environmental-risk-factors-and-ncds

Nuvolari-Duodo I, Dolcini M, Buffoli M, Rebecchi A, Giuliano D, Monticelli C, Vertua C, Brambilla A, Capolongo S (2024) Integrated climate change mitigation and public health protection strategies: the case of the city of Bologna, Italy. Int J Environ Res Public Health 21(11):1457. https://doi.org/10.3390/ijerph21111457

Stefano Capolongo (EUPHA-URB), J.L. (EUPHA-P) (2023). 5.E. workshop: Salutogenic cities: European experiences in Urban Health. Eur J Public Health 33(Supplement_2):160–288. https://doi.org/10.1093/eurpub/ckad160.288

Page MJ, McKenzie JE, Bossuyt PM, Boutron I, Hoffmann TC, Mulrow CD, Shamseer L, Tetzlaff JM, Akl EA, Brennan SE (2021) The PRISMA 2020 statement: an updated guideline for reporting systematic reviews. Bmj 372.https://doi.org/10.1136/bmj.n71

Paglione L, Donato MA, Cofone L, Sabato M, Appolloni L, D'Alessandro D (2024) The healthy city reimagined: walkability, active mobility, and the challenges of measurement and evaluation. Urban Sci 8(4):157. https://doi.org/10.3390/urbansci8040157

Quesada-García S, Valero-Flores P, Lozano-Gómez M (2023) Towards a healthy architecture: a new paradigm in the design and construction of buildings. Buildings 13(8):2001. https://doi.org/10.3390/buildings13082001

Reich SLJ (2022) Assessing the health potential of urban green space in an urban planning and design context: a comparative case study in the city of Stockholm on differences in accessibility, quality, and inclusivity. Available on: https://www.diva-portal.org/smash/get/diva2:1664131/FULLTEXT01.pdf. Accessed on 15 Jan

Sallis JF, Cerin E, Conway TL, Adams MA, Frank LD, Pratt M, Salvo D, Schipperijn J, Smith G, Cain KL (2016) Physical activity in relation to urban environments in 14 cities worldwide: a cross-sectional study. Lancet 387(10034):2207–2217. https://doi.org/10.1016/S0140-6736(15)01284-2

Sarkar C, Webster C, Pryor M, Tang D, Melbourne S, Zhang X, Jianzheng L (2015) Exploring associations between urban green, street design and walking: results from the Greater London boroughs. Landsc Urban Plan 143:112–125. https://doi.org/10.1016/j.landurbplan.2015.06.013

Sepehri B, Sharifi A (2025) X-minute cities as a growing notion of sustainable urbanism: a literature review. Cities 161:105902. https://doi.org/10.1016/j.cities.2025.105902

Sepehri B, Adibhesami MA, Parvanehdehkordi A, Karimi H (2025) Unlocking the power of snailstep application: a smartphone-based way to improve walking and urban health. J Transp Health 41:102001. https://doi.org/10.1016/j.jth.2025.102001

Souter-Brown G (2023) Urban health: applying therapeutic landscape design. Methods, design strategies and new scientific approaches. In: Therapeutic landscape design: methods, design strategies and new scientific approaches. pp 1–12. https://doi.org/10.1007/978-3-031-09439-2_1

Stoltz J, Schaffer C (2018) Salutogenic affordances and sustainability: multiple benefits with edible forest gardens in urban green spaces. Front Psychol 9:2344. https://doi.org/10.3389/fpsyg.2018.02344

Tamaskani Esfehankalateh A, Ngarambe J, Yun GY (2021) Influence of tree canopy coverage and leaf area density on urban heat island mitigation. Sustainability 13(13):7496. https://doi.org/10.3390/su13137496

Tsuno YS, Yamazaki Y (2012) Relationships among sense of coherence, resources, and mental health in urban and rural residents in Japan. BMC Public Health 12:1–9. https://doi.org/10.1186/1471-2458-12-1107

Ulpiani G (2021) On the linkage between urban heat island and urban pollution island: three-decade literature review towards a conceptual framework. Sci Total Environ 751:141727. https://doi.org/10.1016/j.scitotenv.2020.141727

United Nations (2018) 68% of the world population projected to live in urban areas by 2050, says UN. Retrieved 10 Jan 2025, from https://www.un.org/uk/desa/68-world-population-projected-live-urban-areas-2050-says-un. Accessed on 15 Jan 2025

Van Eck NJ, Waltman L (2014) Visualizing bibliometric networks. In: Measuring scholarly impact: methods and practice. Springer, pp 285–320. https://doi.org/10.1007/978-3-319-10377-8_13

Voelker S, Kistemann T (2013) Reprint of: "I'm always entirely happy when I'm here!" Urban blue enhancing human health and well-being in Cologne and Düsseldorf, Germany. Soc Sci Med 91:141–152. https://doi.org/10.1016/j.socscimed.2013.04.016

Warburton DER, Nicol CW, Bredin SSD (2006) Health benefits of physical activity: the evidence. CMAJ 174(6):801–809. https://doi.org/10.1503/cmaj.051351

Yeang K, Dilani A (2022) Ecological and Salutogenic design for a sustainable healthy global society. Cambridge Scholars Publishing. Available on: https://designhealth.academy/onewebmedia/ecologicalSalutogenicfinal.pdf. Accessed on 15 Jan

Zhang S, Wu W, Xiao Z, Wu S, Zhao Q, Ding D, Wang L (2023) Creating livable cities for healthy ageing: cognitive health in older adults and their 15-minute walkable neighbourhoods. Cities 137:104312. https://doi.org/10.1016/j.cities.2023.104312

Zipperer WC, Pickett STA (2012) Urban ecology: patterns of population growth and ecological effects. ELS. https://doi.org/10.1002/9780470015902.a0003246.pub2

Chapter 8
Conceptualizing the Integration of Sexual and Reproductive Health and Rights and Climate Change in Early Warning Systems

Brenda Nyeverwai Rumutsa and Boemo Mmandu Sekgoma

Abstract This paper presents a conceptual Early Warning System that integrates climate change and sexual and reproductive health and rights (SRHR) endorsing multi-sectoral coordination. A complete and effective integrated system requires the inclusion of all essential components, working together to form a protective safety net. It aims to shield at-risk populations by eliminating gaps and reducing uncertainty, ensuring comprehensive risk preparedness. A qualitative methodology guided by Grounded Theory, with purposive sampling to select relevant literature was adopted in this study. Articles were sourced from databases such as Scopus, Web of Science, and Google Scholar. Open coding allowed key themes to emerge and guide the refinement of search terms. This study recommends the development of an integrated early warning system that functions like an orb-web carefully woven with multiple interconnected strands. Just as an orb-web connects each thread to strengthen the whole structure, the early warning system should link all relevant components and actors in a coordinated way to ensure timely protection from vulnerabilities. This system must be built upon established concepts, models, and frameworks. Importantly, it should be a safety–critical tool designed to detect and respond to climate change and sexual and reproductive health and rights vulnerabilities.

Keywords Hazard · Disaster · Early warning system · Hazard · Risk reduction · Sexual and reproductive health and rights

Background

In recent times, the international community has witnessed a surge in initiatives aimed at bolstering Early Warning Systems (EWS), catalysing a profound shift towards the cultivation of a culture centred on proactive risk prevention (Tan 2024). This

B. N. Rumutsa (✉) · B. M. Sekgoma
SADC Parliamentary Forum—SRHR Sweden Funded Project, Windhoek, Namibia
e-mail: brendarumucha@yahoo.com

© The Author(s), under exclusive license to Springer Nature Switzerland AG 2025

A. I. Almulhim et al. (eds.), *Urbanization, Climate Change, and Health*, Climate Change Management, https://doi.org/10.1007/978-3-031-99132-5_8

strategic shift is in alignment with a resolution passed by the United Nations General Assembly, which underscored the urgent need for a collaborative global endeavour to harness existing scientific, technical, and indigenous knowledge within each nation. In the wake of tragic events in 2005, the absence of integrated early warning system was glaringly evident in the aftermath. United Nations Secretary-General Mr. Kofi Annan emphasized the importance of comprehensively assessing worldwide capacities and gaps in Early Warning Systems (EWS) (UNISDR 2006). A visionary goal fuelled this initiative to establish universally accessible EWS for all natural hazards, including those induced by climate change.

Moreover, the emergence of the COVID-19 pandemic (biological Hazard) served as a stark awakening call to communities worldwide, highlighting the pressing necessity for robust early warning systems. The pandemic's rapid spread and devastating impact underscored the crucial role of proactive measures in mitigating the effects of disasters, including those as a result of climate change as well as emerging sexual and reproductive health and rights (SRHR) issues. Segal and Giudice (2022) reveal that sexual and reproductive health and rights (SRHR) issues are also being exacerbated by climate change. Communities worldwide have been prompted to recognize the importance of crafting comprehensive integrated Early Warning Systems (EWS) that can effectively detect and respond to not only natural disasters but also emerging challenges, as buttressed by Joakim et al. (2021). The global crisis brought up by climate change has further reinforced the urgency of bolstering integrated disaster risk reduction (DRR) strategies that encompass SRHR issues.

The intersection of climate change and sexual reproductive health and rights (SRHR) remains largely unexplored in academic frameworks though the link has been established. Most of the existing early warning system frameworks that have been developed concentrate on climate change induced hazards such as cyclones, floods and droughts while neglecting human vulnerabilities relating to SRHR. The oversight is critical given that SRHR challenges often escalate during crises and it should be well-known that nature does not pause for a crisis, women and girls will continue to menstruate, and people will have sex even in most difficult conditions. This converges with Bhadra (2022) who affirms that sexuality is an important part of a person.

This study aims to conceptualise an integrated early warning system (IEWS) that links Climate change as well as sexual and reproductive health and rights (SRHR) interconnected crises. Most studies have developed fragmented and hazard-specific EWS that often prioritise technical forecasting while undermining social, reproductive and gender concatenated vulnerabilities. Most of the early warning systems prove to be typically mono-sectoral lacking a multi-sectoral inclusive, and systems thinking approach. A system should consider multi-sectorial coordination as a central mechanism for management. The authors sought to advance existing knowledge by developing a detailed early warning system framework for biological hazards that integrates climate change and SRHR. This framework offers a new perspective on designing safety–critical systems aimed at preventing hazards from escalating into disasters, using the orb web as a conceptual illustration.

Methods

This study used a qualitative methodology, complementing Purposive Sampling and Grounded Theory. The main methodological framework for the literature selection and analysis was grounded theory. Using broad keywords derived from the study's title and focus area, articles were first found through methodical searches on scholarly databases like Scopus, Web of Science, and Google Scholar. As enunciated by the tenets of grounded theory abstracts and article content were initially open-coded, enabling the inductive emergence of important themes and categories (Charmaz 2017; Oktay 2012). Search terms were improved as these categories evolved, and theoretical sampling, a fundamental part of the Grounded Theory methodology, was used to find more literature. After defining the thematic categories, the study used Purposive Sampling to specifically choose articles that provided deep, varied, and conceptually pertinent insights in line with the developing theoretical frameworks.

Results

Based on our understanding, creating an effective EWS requires a logical and comprehensive approach that incorporates climate change, SRHR, indigenous knowledge (IK), modern scientific knowledge (MSK), and both quantitative and qualitative data i.e. The foundation of this integration ought to be inclusive coordination involving multiple communities and sectors. The identification of all necessary elements needed to convert a simple early warning system into a comprehensive one that can stop biological hazards, like disease outbreaks, from becoming a disaster is equally crucial. The authors believe that the following essential components are required to guarantee a thorough and effective early warning system as depicted in Fig. 8.1.

There is also a recognition that there is a need to shift attention from the basic as well as single elements of an early warning system to a wholesome one. All the necessary elements of an EWS should be included in the development of an Integrated Early Warning System (IEWS), and emphasis should be on considering their interactions and interrelationships. An effectively integrated EWS that is centered on inclusive coordination, including Climate Change Adaptation and SRHR issues can be achieved if there is integration between forms of knowledge, data, and sectors. No form of knowledge should be treated as a superior or inferior form of data or information in the knowledge corpus. A fusion of scientific and indigenous knowledge ensures overlapping of the weaknesses of the two forms of knowledge. Masipa (2016) aligns with this perspective and reveals that combining the two gives strength to DRR activities to become more effective. While making sure that SRHR concerns are not disregarded, it is necessary to give precedence to workable disaster risk reduction (DRR) strategies that incorporate the two forms of knowledge, climate change adaptation (CCA) and SRHR. The use of information and data that can be used in practical settings should be prioritized guided by a pragmatic approach that ensures

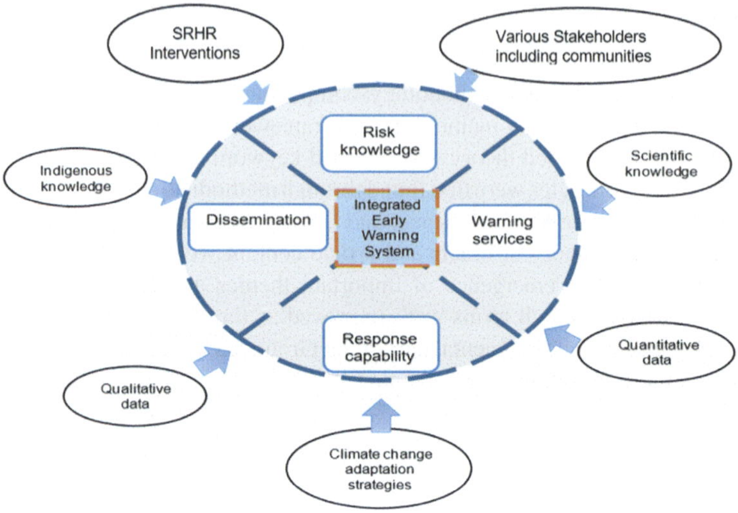

Fig. 8.1 Components of an integrated early warning system. *Source* Rumutsa and Sekgoma (2025)

relevance, usability, and impact. This pragmatic approach in developing an integrated EWS could assist in coming up with an effective EWS and could also be important in establishing a concept that regards different forms of knowledge at par, giving room for each form of data and knowledge to express its distinctive aspects as it is a precondition for integration. The possibility of establishing an Integrated Early Warning System (IEWS) was brought to light by Macherera and Chimbari (2016a, b).

This study emphasizes that bringing out various elements together simultaneously is crucial in the development of an integrated EWS. Not forgetting that DRR is "a conceptual framework of elements well-thought-out with the possibilities to reduce vulnerabilities and disaster risks thus to limit through mitigating and preparing for the adverse impacts of hazards" (Department for International Development 2011). The identification of hazards requires harmonizing information on environmental precursors with timely data and applying models for early detection as well as an early warning (Merkord et al. 2017). The health sector alone can not develop an integrated early warning system, the issue undeniably requires a multi-sectoral approach as buttressed by Macherera and Chimbari (2016a, b).

DRR initiatives should underpin holistic management, involving related and connected issues that altogether influence resilience against disasters. An integrative and target-oriented approach to disasters provides the best basis for preventing societies from threats posed by hazards. In the longer term, adaptation to climate change should be emphasized for hazard prevention and mitigation. An integrative and target-oriented approach works best when informed by specific local conditions. This should include the disaster management continuum divided into phases, pre- and post-exposure. Each of the phases (mitigation, preparedness, response, and recovery) requires different forms of interventions (Twigg 2015).

(a) *Weaving an Integrated biological hazard management continuum web model*

The authors observed that developing an EWS is just like how an orb-weaver spider develops its orb web. The spider develops an orb web to protect its life and eggs, the same way the communities develop EWSs' so that they protect their lives and properties. The first lines of the orb web serve as guidelines and signal lines, likewise, the first phases of an EWS should serve to guide people to work towards disaster mitigation and prevention. Signal lines (prediction measures) should be able to predict hazards and alert communities of an impending disaster in time. In this study, an integrated biological hazards management continuum web was developed embracing the following research concepts, models, and frameworks, which are customarily fragmented and unconnected in their application in hazard management.

1. Evolution of a hazard event
2. Disaster management continuum
3. Epidemiological triad model interventions
4. Components of an effective EWS by ISDR
5. Climate change adaptation
6. Sexual Reproductive Health Rights.

Figure 8.2, shows how the woven orb web, that illustrates how research concepts, models, and frameworks can be brought together in an effort to conceptualize the Integrated EWS that also links climate change and SRHR.

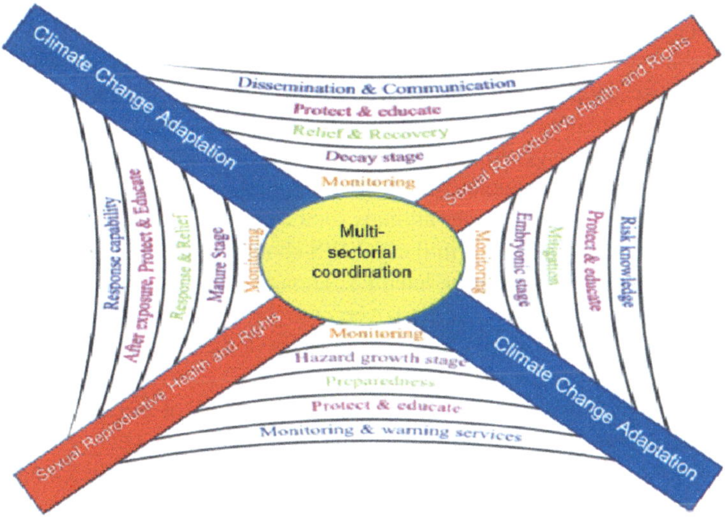

Fig. 8.2 Integrated biological hazard management continuum web model. *Source:* Rumutsa and Sekgoma 2025

Key for different weaving phases.

- Evolution of a hazard event
- Disaster management continuum
- Epidemiological triad model interventions
- Components of an effective EWS by ISDR

Figure 8.2 of the integrated biological hazard management continuum web model developed resembles an orb webs' shape. The characteristic features are, a center portion as well as a series of radiating lines supporting the framework. In this case Climate change Adaptation and SRHR make up the radiating lines supporting the framework while centred on multi-sectorial coordination which is at the centre. As in the orb web created by a spider, the frame is built first and attached to plants or other objects. In the same way, an IEWS should be built on a strong institutional background. If there is no support from various institutions, the whole system can fail to function the way it is intended as attested by Karanath and Dashora (2014). In other words, if the object on which the spider builds its web is not strong, this jeopardizes all the efforts. The center of the IEWS is supposed to be resilient for it to be able to cushion the communities from threatening hazards.

The sticky radii transmit vibrations when something has been caught to alert the spider of danger or threat in its territory. Likewise, both scientific and indigenous knowledge base indicators should assist in giving a warning. Furthermore, such a warning should be noticed as early as possible to alert people when the hazard is still in its embryonic stage. Vibrations which represent warning services should be issued in time to prevent people from getting affected by biological hazards. Some spider webs can even catch bigger animals such as birds. Therefore the strength of the orb web should be durable. Similarly, EWS should be so effective that even the worst anticipated scenarios can be handled. Though investing in an EWS requires costly resources, it is worth having it to overcome the underlying risks of not having it. It is better to invest in an EWS as it is a proactive measure than just waiting to react to disasters as EWS act as safety nets.

It is also prudent to note that, after some time, the silk of the spider web loses its stickiness and becomes inefficient. This normally happens when it is capturing prey. Similarly, in the EWS, some elements and activities need to be continuously improved to maintain their effectiveness and efficiency in protecting communities from hazards. Any hole in the orb web gives a chance for what is meant to be caught by a web to pass through. Loopholes in an EWS web, on any element, cause hazards to pass through the safety net and affect people who are meant to be protected. The spider, after spinning its web, waits on or near the web, in the same way, the communities should always be alert and prepared for action. DRR has to be a full or

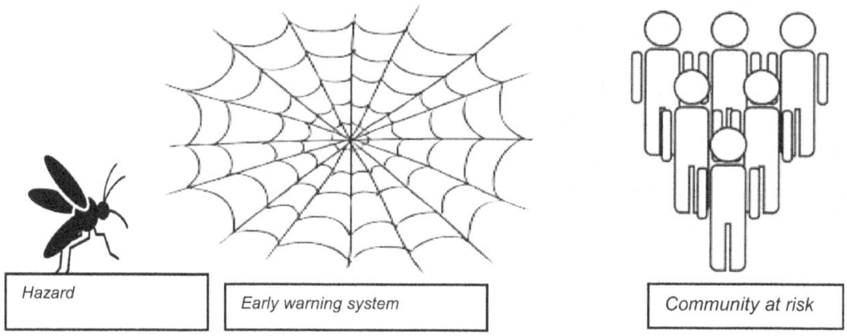

Fig. 8.3 Illustration of an integrated EWS that prevents a hazard from affecting a community. *Source* Rumutsa and Sekgoma (2025)

complete set of elements intricately coming together to form a safety net that should not leave any loophole beyond uncertainty.

Investing in EWS serves resources of managing disasters reactively as the hazards are intercepted by the safety net (web) before it extends to people (as shown in the following Fig. 8.3).

Agreeably, it is imperative to interweave, all salient constituents (threads) and tailor-make an (orb) web of associations which protects (prevents) people from being affected by a hazard. Figure 8.3 has shown an illustration of how the web should protect or shield vulnerable people when finely woven to intercept the biological hazards. There is also a need to bear in mind that fewer resources are required for pro-active measures when it comes to DRR. Similarly an orb web allows a spider to catch prey without having to expend energy by running. Reacting when there is an emergency requires more resources, hence an EWS is worth investing in (Zubir and Amirrol 2011). The orb-web (in this case the EWS) design may affect capture efficiency (Lawania and Mathur 2015). It should be designed in a way such that loopholes are minimized for "*hazard*" capture efficiency. EWS reliability and effectiveness, as well as the precursors for monitoring and predicting events, are automated with the aid of human decisions that influence it. Leaving out an element of an effective EWS e.g. not considering climate change or including SRHR could result in a system that is not effective in intercepting a hazard or vulnerabilities. This leaves communities not fully protected from the hazard as the safety net will have a (loop) hole as shown in Fig. 8.4;

Ostracizing some DRR constituents (threads) and leaving out other priorities in managing and controlling a hazard leaves a gap (ineffectiveness) in the system that will result in exposing the communities to hazard risks and increasing their vulnerabilities as shown on Fig. 8.4.

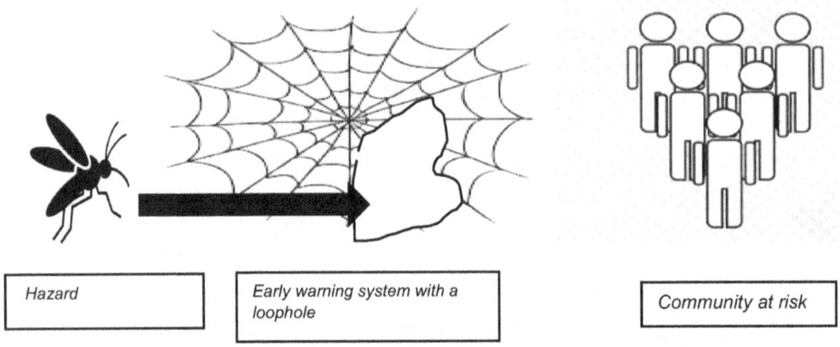

Fig. 8.4 Illustration of integrated EWS with a loophole (elements left out or compromised) that enable the hazard to penetrate through to affect communities.

Discussion

It is crucial to understand what an early warning system is; however, various definitions have been used to describe Early Warning Systems (EWS) by several authors. Firstly, an early warning refers to "a thing that tells you in advance that something serious or dangerous is going to happen" (Cools et al. 2016). International Federation of Red Cross (2012) defines an EWS as "a set of capacities needed to generate and disseminate timely and meaningful warning information to enable individuals, communities and organizations threatened by a hazard to prepare and to act appropriately and in sufficient time to reduce the possibility of harm or loss". Early warning information issuance on its own is not adequate when it is not complemented by an integrated system with different components that enables people to deal with the impending danger. Ideally, an early warning should be augmented by a system that enables people to take action, in consistent with the definition given by the International Federation of Red Cross and Red Crescent Societies (IFCR). IFCR (2012) defines an EWS as "*a set of capacities needed to generate and disseminate timely and meaningful warning information to enable individuals, communities, and organizations threatened by a hazard to prepare and act appropriately in sufficient time to reduce the possibility of harm or loss*".

On the same vein, Macherera et al. (2017) defines a holistic exclusive integrated early warning system (IEWS) as "a series of approaches that refine the understanding of the geographical variation of hazard risk in a dynamic environment and usually comprises forecasting, early warning, and early detection". Cools et al. (2016) define early warning as something that tells you in advance that something serious or dangerous is going to happen. With the definition of Macherera et al. (2017), it can be deduced that such a system as defined is not merely proactive but also reactive, as it includes detection. Detection is the action or process of identifying the presence of something hidden (Cools et al. 2016) hence this is looking for the already affected. Contrarily, an effective Integrated EWS in the first place is supposed to be pro-active, reducing vulnerabilities, preventing damages and losses before they occur. An effective proactive EWS is supposed to prevent at all cost the people at risk from being affected. It is also supposed to be comprised of a strongly linked chain of elements for its effectiveness. Failure of one element or component in the chain or any weak link in the elements of EWS will result in under-performance or failure of the whole system (Macherera and Chimbari 2016a, b; Mukhtar 2018; Twigg 2015). Similarly, while communities may be protected from hazard vulnerabilities, they can still face significant risks related to sexual and reproductive health (SRH). Disasters do not eliminate the human need for intimacy; in some cases, individuals may engage in risky behaviors that increase the likelihood of unintended pregnancies, sexually transmitted infections (STIs), or health complications, particularly if essential SRH commodities, such as contraceptives or antiretroviral (ARV) medication, are inaccessible.

Concentrating on a single element of the components of an effective EWS is not sufficient to cushion communities. It is highly likely that prioritizing some elements of an EWS exposes people to the risks associated with a threatening hazard (Macherera and Chimbari 2016a, b; Mukhtar 2018; Twigg 2015). All the components of an effective EWS should be considered for it to be complete as all the sets of elements come together to form a safety–critical system that intercepts hazards. It has to be comprised of subsystems which are smaller entities that make up the larger system (Simonovic 2015). All the components in the system should have interdependent relationships among them (ibid). The system is supposed to shield people at risk without leaving any loophole beyond uncertainty. It is salient not to leave out any element as each one of them contributes to the increase in the potential of the communities in reducing their vulnerabilities (Twigg 2015). EWS should provide communities with relevant information that enables them to assess levels of risk, for them to make informed decisions to safeguard themselves (Macherera and Chimbari 2016a, b).

The main aim of developing EWSs is to generate information before losses are incurred or damage has been made hence, they act as safety–critical systems. They are a critical element in preventing hazards and vulnerabilities from turning into disasters (Löwe et al. 2013) as it is more than just prediction. They are a cost-effective risk mitigation measure as they provide timely information to prevent or palliate disasters. Effective EWSs must be constructed or developed in an understandable way and should be relevant to the communities that they should serve (Macherera and Chimbari 2016a, b; Baudoin et al. 2016). However, EWS cannot be implemented uniformly across communities. They have to be tailor-made and localized according to the particular needs of each group (Henriksen et al. 2018; Shimaponda-Mataa et al. 2017; Mukhtar 2018). Further more EWS must be driven by the community at risk, facilitating the delivery of services tailored to each community's needs and possibly reducing implementation costs (Baudoin et al. 2016).

IFCR (2012) clarified that "communities and organizations threatened by a hazard should prepare and act appropriately in sufficient time to reduce the possibility of harm or loss" hence communities should not be treated as passive recipients. Instead, an EWS should reduce exposure probability by providing the lead time that enables people to act before an emergency arises. People should have room to make their own decisions when they are notified of a hazard, instead of decisions being imposed on them. Lenton (2013), Baudoin et al. (2016) and Wilkinson et al. (2017), are of the idea that EWS potentially reduces hazard impacts when they provoke the action of those that are at risk. Their designated functions are fulfilled when they can detect hazard events timeously, transfer warning to people at risk to avoid losses (Sättele et al. 2015; Lenton 2013). Their reliability and effectiveness (a relative measure of achieved risk reduction) rely on components and the ability of the system to identify the intended hazard (Sättele et al. 2015). Effectiveness of an EWS is defined as a relative measure of achieved risk reduction. It is not only the technical capacity that guarantees its effectiveness but institutional capacity as well as DRR measures with the efforts to respond at the national level (Fakhruddin and Chivakidakarn 2014). It is essential to constantly assess whether the communities are capacitated to respond to a hazard condition (ibid).

Frameworks, Models, and Theories that Guided the Development of an Integrated EWS

To conceptualize a comprehensive Integrated Early Warning System (IEWS) that addresses both climate change and sexual and reproductive health and rights (SRHR), it is essential to draw upon established frameworks, models, and theories. These conceptual tools provide the foundation for understanding the complexity of hazards and human vulnerabilities, guiding the integration of diverse elements into a cohesive system. This section explores key disaster risk reduction and public health models—including the Disaster Crunch Model, the Epidemiological Triad, and the Disaster Management Continuum—as well as the Systems Theory and relevant early warning frameworks such as the ISDR model. Together, these frameworks offer a multidimensional lens through which the IEWS can be designed to be inclusive, proactive, and responsive to interconnected risks.

(a) *The systems approach to disaster management (Theory)*

This study, adopted the systems approach developed by a German scientist named Ludwig Von Bertalanffy, which was later advanced by Simonovic into the systems approach to disaster management. It was applied in this study as an attempt to integrate elements of the EWS to make it a comprehensive integrated EWS that includes Climate Change Adaptation and SRHR. The ISDR (2006) reveals that the EWS with best practices contains strong inter-linkages as shown in Fig. 8.2 and effective communication channels between elements. The theory assists in clarifying other sub-systems such as individual and societal relationships as well as organisations (Simonovic 2010). It shifts attention from single elements to the whole, permits the inclusion of various factors and interactions to be taken into account as well as helping in dealing with complexities of interacting non-linear systems (Mele et al. 2010; Simonovic 2010).

DRR has to be a full or complete set of elements coming together to form a safety net that should shield people from the risks and not leave any loophole for disastrous mishaps. According to Marczyk et al. (2010), a theory is a conceptualisation or description that attempts to integrate all that is known about a phenomenon into a conscious statement. Merriam (2001) refers to a theoretical framework as "the structure, the scaffolding and the frame of a study". The development of an IEWS in this study is centered on the systems theory. Integration of different essential elements of an EWS improves the identification of problems so that an effective solution can be implemented. The knowledge about a phenomenon is derived from an understanding of a whole system and not just individual elements (Aristotle's Holism). Simonovic (2010) is of the notion that integration is centered in the systems theory as systems are comprised of subsystems with smaller entities that create a larger system. According to Mele et al. (2010), the systemic perspective claims that a phenomenon cannot be fully comprehended by breaking it up into elementary parts and then reforming it, instead there is need for applying a total vision to understand its functioning. Simonovic (2010), converges with Mele et al. (2010) and opinionated that all systems are a collection of diverse elements which are linked by strong interactions, specifically the disaster management system is comprised of four linked subsystems. These include individuals, societies, organisations and the environment.

This theory can lead to a much-needed integration of EWS as each element is affected by at least one other element in the system. According to Simonovic (2010), the systems theory also allows a better analysis of the problem and to identify why the problem exists. Simonovic (2010) also reveals that problems are a sign of malfunctioning processes within a system. The systems theory considers all possible sources of a problem by examining each individual element, including the role it plays in the system. Since most communities proved to fail preventing a disease (biological hazard) from affecting communities in 2019 when COVID19 hit, this is evidence of malfunctioning early warning systems. Henceforth it is vital to examine each component of an early warning system and the processes and measures that are being employed in DRR to establish where the system is failing to prevent disease (biological hazard) outbreaks.

The systems approach is suitable since the study seeks to come up with an IEWS. The systems theory for disaster management is accredited for its integrative character of concerned components and the interrelationships among them, as well as the ability to integrate the knowledge of existing tools in their application to a practical problem. An IEWS can be more effective than unconnected elements operating in an autonomous manner. Several theories of management have demonstrated that their application is inappropriate in the management of disasters involving various stakeholders for risk reduction. None of them has managed to give a comprehensive management solution which embodies all the diverse complexities involved to achieve effective DRR for biological hazards such as COVID19 or Malaria in this case. (Simonovic 2010) discloses that individuals are decision-makers in their own right, with direct roles in disaster management (preparedness, recovery and response). The actions of these individuals drive the behaviour of societies and organisations (systems). As such, individuals cannot be regarded as passive recipients of DRR actioning (Williams and Jones 2004; Twigg 2015). Organisations (systems) are the platform used to produce outcomes that are not possible at the individual level to achieve certain goals.

Organisations or systems have inputs, processes and outputs, and failure in one element or component affects these elements. The systems approach to disaster management allows complex dynamic characters and interdisciplinary needs of management options to be addressed (Simonovic 2010). In this case, the inputs include all resources required in preventing biological hazards, thus in risk awareness, warning services, dissemination and response which are elements of an integrated EWS. Processes include activities carried out by individuals, societies and organizations in prevention, these include monitoring and evaluation processes. The output is supposed to be effective in protecting individuals and societies from the threatening hazard, the focus being on disaster risk reduction. According to the UN (2006), warning, dissemination can be jeopardized if the feedback on the system is ineffective to affect its performance. To continually revolve, a system requires formal feedback processes and these improvements based on feedback and learning from previous experiences (ibid).

(b) *Framework for elements of an effective early warning system*

The framework for elements of an effective EWS by the International Strategy for Disaster Reduction (2006) was found relevant for this study as it determines the components that should be included in developing an integrated EWS. A complete and effective people-centred EWS is comprised of four inter-related elements. The failure of one element could lead to failure of the whole system (Quansah et al. 2010). For the system to succeed, the people within a community have a critical role to assist the government by providing IK, technical and scientific input for the development of an integrated EWS. However, where the capability exists in communities to issue warning timeously, the other three components of the EWS are usually weak, with the main focus being mostly on hazard monitoring and forecasting that neglects community vulnerability assessment (UN 2006). The four elements are risk knowledge, monitoring and warning services, dissemination and communication and response capability (ISDR 2006). According to Karanath and Dashora (2014), all the key elements of an EWS is supposed to be functional hence it is important to anticipate different scenarios including SRHR as well as measuring its performance. These elements of an EWS gave the authors an understanding of the main requirements of an EWS. There are also some models that the authors perceived to play an important role in coming up with an IEWS. According to Nilsen (2015), process models are used for describing or guiding a process to translate research into practice, and this relates to the purpose of this study.

(c) *Disaster crunch model*

To clarify the occurrence of disasters, the crunch model by Wisner is a tool that assists to show how vulnerable communities are caught up by a hazard event if they are unprepared. Wisner et al. (2012) explains how disasters are caused using the disaster crunch model which is also referred to as the "Disaster Pressure and Release Model".

According to Smyth and Hai (2012), the model assists practitioners to understand people's vulnerability to disasters and come up with ways to react. When a hazard affects an unprepared vulnerable community, there is a high probability of a disaster occurring. The people are caught in between a hazard and their vulnerabilities, and as a result, they are crunched in the situation.

If there is no hazard (biological) threatening the community then there can never be a disaster. The root causes of disasters often result from political, economic, social processes or structures which affect the communities, and these are supposed to be addressed at national, regional and international levels (Twigg 2015). Social structures include traditional leaders, religious groups, powerful individuals and government departments. Structures have policies or implementation activities which could be good or bad, and these affect the communities. Structures and processes have deeper roots and are the underlying causes of disasters as they are frequently embedded in the culture and beliefs of a community. Some of the structures can operate from power bases far from the community in which it is influencing. According to St. Cyr (2005), the root causes could be spatial or temporal. The spatial causes include economic and political power for example while temporal causes include the historical pattern of the hazard or it could be bound in cultural assumptions, beliefs, and social relations (ibid). If people have insecure access to resources (e.g. contraceptives), their vulnerability is likely to be increased as they engage in activities that put them at risk (St. Cyr 2005; Twigg 2015).

(d) *Epidemiological triad model*

According to Rockett (2017), the Epidemiological triad model is devised to enhance the understanding of infectious diseases. The authors noted that it is requisite in assisting an understanding of a disease and its components since it enables the establishment of appropriate practical hazard prevention measures. The epidemiological triad model states that biological hazards result from the interaction of the agent, host and the environment (Rockett 2017). The agent is the cause of a biological hazard. The host refers to the organism that is capable of harbouring the disease, in this case, it is the human. The environment refers to the conducive surrounding, thus external to human beings while the social, cultural and physical aspects of the environment are viewed as environmental factors.

For disaster risk reduction experts, it is crucial to identify which side of the epidemiological triad—agent, host, or environment—can be disrupted to prevent or mitigate disease outbreaks. When complemented by an effective early warning system (EWS), this approach can help curb recurring epidemics, particularly in areas where diseases are endemic. As Rockett (2017) notes, outbreaks can be averted by altering, interfering with, or eliminating one element of the triad. The possible interventions between the two (agent and host), is protection, education and altering exposure. An EWS also fits at that phase as it is an essential component and useful tool of DRR (Meissen and Voisard 2010).

(e) *The disaster reduction continuum*

The disaster continuum outlines the chain of events that follow disasters. According to Twigg (2015), disasters are not once-off events to be responded to but are constantly recurring problems that are supposed to be well planned for. It is recognized that disasters in an area, have patterns and sometimes recur after a certain period. The period could be after weeks, months, years or decades and this depends on the nature of the hazard. Taking the disaster continuum into consideration and analysing its phases, contribute immensely to risk reduction. It improves pro-activeness as the continuum illustrates the main phases of disaster reduction, thus the pre- and post-disaster phases. The phases in the disaster continuum are prevention, mitigation, preparedness, response and recovery. This study concentrates on protection initiatives in the pre-disaster phase, thus prevention, mitigation and preparedness, through prediction, early warning and planning of relevant stakeholder coordination.

Communities with their IK can contribute to improved activities in all the four phases of the disaster management continuum as IK is an indispensable empowerment tool that can be used in all the stages of DRR (Masipa 2016). The customary role of disaster management used to focus on taking actions immediately before a disaster, during and shortly after to avoid losses. There is an understanding that disaster reduction is a long-term issue that is not confined in time and risk prevention culture is now being promoted through EWS (Alfieri et al., 2012; IFRC 2012). It is very much possible to avoid biological hazards as they are preventable (Shimaponda-Mataa et al. 2017; WHO 2015). This can be done by taking appropriate precautions. Proactive measures during the pre-disaster phase are more important as they try to prevent people from being exposed to a hazard even at its embryonic stage. Proactiveness continues even when the hazard has reached the decaying stage and at the recovery stage in the disaster continuum phases, so that the disaster does not recur. According to Masipa (2016), ignoring IK in all the phases of the disaster management continuum exacerbates the occurrence and effects of disasters.

An EWS should not be restricted to certain phases of the disaster management continuum or stages of the hazard, since disaster risk management is a continuous process that is not confined to time. A continuous and uninterrupted continuum of a natural disaster provides a basis for improved DRR measures. Systems are supposed to undergo frequent testing, practice, review and warning refinement (Twigg 2015). The same way with the silk of a spider web, after some time, it gets loose in terms of its stickiness and becomes inefficient at capturing prey. The spider recycles the silk to make a new strong web. The mitigation and prevention phases seek to provide a safety-net that intercepts the hazards. This is the phase of the continuum in which communities are shielded from the effects of the hazard. Preparedness towards natural hazards is essential in DRR so as to reduce their impacts on society. Preparedness is important from individual, community and institutional levels. Twigg (2015) points out that the forecasting of events and issuing warnings is also part of preparedness. According to Masipa (2016), the use of the communities' IK contributes immensely in enhanced actions in all phases of the DRR continuum. IK is a precious resource that can facilitate the processes of the disaster continuum sustainably and effectively in terms of cost (Masipa 2016).

In the emergency response phase, activities designed to preserve lives and protect properties should be actioned. In biological hazards context (in this case diseases), it is the phase in which medical care and resources should be readily available. This is the phase that demarcates the end of the embryonic stage of the hazard into the growth stage and failure to handle the emergency can result in morbidity and mortality. Resources and coordination of activities are very important, especially during the period in which outbreaks are expected. It should be noted that emergency relief far exceeds the investment in the resources for DRR capacity development (Zubir and Amirrol 2011). The handling of diseases (biological hazard) outbreaks is still a cause for concern since the practice of relying on customs to deal with emergencies is unable to fit in the disaster management domain. Outbreaks (due to the biological hazards) are usually considered an exclusive mandate of the Department of Health, which by nature has management activities that are overarching and not strictly confined to one domain.

The successful elimination of a hazard requires that the response to emergencies be done through a multi-sectorial approach as opposed to the existing mono-sectorial management system. Integrating various sectors guarantees adequate and combined actions in the recovery phase (De Paula et al. 2017). All disaster recovery-related sectors should have the same view for long term development objectives in the recovery effort (De Paula et al. 2017). It is important to start mitigating both the social and economic impacts of biological hazards on the affected population in the recovery process. Lessons learnt during the recovery stage should be considered for communities to return to normalcy as soon as possible. Plans for re-establishing medical services in preparation for the next season are important in case there is a recrudescence of the epidemic in a short space of time.

Limitations of the Study

This study is subject to several limitations. Firstly, the selection of models, frameworks, and theories was guided by the discretion of the authors as well as influenced by the availability and relevance of literature at the time of writing. As such, the theoretical foundation may not be exhaustive or universally representative, and other researchers might have chosen different conceptual tools based on alternative readings or disciplinary perspectives. Also if the same research process were to be repeated at a later time, it may yield different results due to the continuous emergence of new publications and the dynamic nature of synthesized knowledge in the field of biological hazards. This evolving landscape of information challenges the reproducibility of the study and reflects the fluid and iterative nature of qualitative inquiry. Lastly, while every effort was made to ensure comprehensive coverage of existing literature, the study may have overlooked recent or less accessible publications.

Conclusion

The review has elaborated a prototype Integrated Early Warning System (IEWS) using an orb web. It also revealed that even if it is an early warning system for a biological hazard to be developed, concentrating on a single element of the components of an effective EWS is not sufficient to cushion communities. Prioritizing or ignoring some elements of an effective EWS exposes people to the risks associated with a threatening hazard as well as indirect challenges such as increased SRH. All the components of an effective EWS should be considered for it to be complete as all the sets of elements come together to form a safety net. The system is supposed to shield people at risk without leaving any loophole beyond uncertainty. More so, time and resources that are needed in developing an EWS need a large capital but it will eventually pay off in the long run. Investing in EWS serves resources of managing disasters reactively as the hazards are intercepted by the safety net (web) before extending to people. There is, therefore scope for further work to be done in developing a practical comprehensive integrated early warning system that includes all components of an effective early warning system as well as integrating modern scientific and indigenous knowledge.

References

Alfieri L, Salamon P, Pappenberger F, Wetterhall F, Thielen J (2012) Operational early warning systems for water-related hazards in Europe. Environ Sci Policy 21:35–49

Baudoin MA, Henly-Shepard S, Fernando N, Sitati A, Zommers Z (2016) From top-down to "community-centric" approaches to early warning systems: exploring pathways to improve disaster risk reduction through community participation. Int J Disaster Risk Sci 7(2):163–174

Bhadra S (2022) Exploring dimensions of sexual issues in disasters and conflicts: need to bridge the gaps between policy and practice. Sexologies 15(3):277–290

Charmaz K (2017) The power of constructivist grounded theory for critical enquiry. Qual Inq 23(1):34–45. https://doi.org/10.1177/1077800416657105

Cools J, Innocenti D, O'Brien S (2016) Lessons from flood early warning systems. Environ Sci Policy 58:117–122

Department for International Development (2011) Defining disaster resilience. Department for International Development, London, UK, pp. 1–20. https://www.gov.uk/government/uploads/system/uploads/attachment_data/file/186874/10

Fakhruddin SHM, Chivakidakarn Y (2014) A case study for early warning and disaster management in Thailand. Int J Disaster Risk Reduction 9:159–180

Henriksen HJ, Roberts MJ, van der Keur P, Harjanne A, Egilson D, Alfonso L (2018) Participatory early warning and monitoring systems: a Nordic framework for web-based flood risk management. Int J Disaster Risk Reduction 31:1295–1306

International Federation of Red Cross and Red Crescent Societies (2012) Community early warning systems: guiding principles

Joakim EP, Mortsch L, Oulahen G (2021) Using vulnerability and resilience concepts to advance climate change adaptation. In: Environmental hazards and resilience. Routledge, pp 13–31

Karanath A, Dashora Lalit Kumar RU (2014) Review of multi hazard early warning system of Visakhapatnam (India)

Lawania KK, Mathur P (2015) Study on the pattern and architecture of spider's web with special reference to seasonal abundance in eastern region of Rajasthan, India. IOSR J Environ Sci Ver. I 9(11):2319–2399
Lenton TM (2013) What early warning systems are there for environmental shocks? Environ Sci Policy 27:S60–S75
Löwe P, Wächter J, Hammitzsch M, Lendholt M, Häner R, Moßgraber J, Sabeur Z (2013, June) The evolution of disaster early warning systems in the TRIDEC project. In: The twenty-third international offshore and polar engineering conference. OnePetro
Macherera M, Chimbari MJ (2016a) A review of studies on community based early warning systems, Jàmbá. J Disaster Risk Stud 8:1–11
Macherera M, Chimbari MJ (2016b) Developing a community-centred malaria early warning system based on indigenous knowledge: Gwanda District, Zimbabwe. Jàmbá: J Disaster Risk Stud 8(1):1–10
Macherera M, Chimbari MJ, Mukaratirwa S (2017) Indigenous environmental indicators for malaria: a district study in Zimbabwe. Acta Trop 175:50–59
Marczyk GR, DeMatteo D, Festinger D (2010) Essentials of research design and methodology, biomedical engineering. Wiley. https://doi.org/10.5860/CHOICE.43-5572
Masipa TS (2016) The impact of climate change on food security in South Africa: current realities and challenges ahead. Jàmbá—J Disaster Risk Stud ISSN 9(1):1–7. https://doi.org/10.4102/jamba
Meissen U, Voisard A (2010) Towards a reference architecture for early warning systems. In: International conference on intelligent networking and collaborative systems. pp 513–518. https://doi.org/10.1109/INCOS.2010.81
Mele C, Pels J, Polese F (2010) A brief review of systems theories and their managerial applications. Serv Sci 2(1–2):126–135
Merkord CL et al (2017) Integrating malaria surveillance with climate data for outbreak detection and forecasting: the EPIDEMIA system. Malar J 16(1):89. https://doi.org/10.1186/s12936-017-1735-x
Merriam S (2001) Merriam. In: Qualitative research and case study applications in education, 2nd edn. Jossey-Bass, San Francisco
Mukhtar R (2018) Review of national multi-hazard early warning system plan of Pakistan in context with sendai framework for disaster risk reduction. Procedia Eng 212:206–213
Nilsen P (2015) Making sense of implementation theories, models and frameworks. Implementation Sci 10(1):53
Oktay JS (2012) Grounded theory. Oxford University Press
Quansah JE, Engel B, Rochon GL (2010) Early warning systems: a review. J Terr Obs 2(2)
Rockett I (2017) The gender suicide gap and differential misclassification: a research autobiography. Int Rev Mod Sociol 43:5–32
Sättele M, Bründl M, Straub D (2015) Reliability and effectiveness of early warning systems for natural hazards: concept and application to debris flow warning. Reliab Eng Syst Saf 142:192–202
Segal TR, Giudice LC (2022) Systematic review of climate change effects on reproductive health. Fertil Steril 118(2):215–223
Shimaponda-Mataa NM, Tembo-Mwase E, Gebreslasie M, Mukaratirwa S (2017) Knowledge, attitudes and practices in the control and prevention of malaria in four endemic provinces of Zambia. South Afr J Infect Dis 32(1):29–39
Simonovic SP (2010) Systems approach to management of disasters: methods and applications. John Wiley & Sons
Simonovic SP (2015) Systems approach to management of disasters—a missed opportunity? IDRiM J 5(2):70–81
Smyth I, Hai VM (2012) The disaster crunch model: guidelines for a gendered approach
St. Cyr JF (2005) At risk: natural hazards, people's vulnerability, and disasters. J Homel Secur Emergency Manage 22. https://doi.org/10.2202/1547-7355.1131

Tan S (2024) The effects of climate change on the global meteorological system and trend analysis. Interdisc Human Commun Stud 1(7)
Twigg, J., 2015. Disaster risk reduction. In: Disaster risk reduction, vol 44. Pp 170–175. https://doi.org/10.4324/9781849771016
UNISDR (2006) 2006 UNISDR terminology on disaster risk reduction. International Stratergy for Disaster Reduction (ISDR), pp 1–30. 978-600-6937-11-3
United Nations (2006) Global survey of early warning systems. Matrix, p 56
Wilkinson E, Twigg J, Weingartner L, Peters K, Kirbyshire A, Lovell E, Masson VL, Patel S, Bernard G (2017) Delivering disaster risk reduction by 2030: pathways to progress. Overseas Development Institute, London
Williams HA, Jones CO (2004) A critical review of behavioral issues related to malaria control in sub-Saharan Africa: what contributions have social scientists made? Soc Sci Med 59(3):501–523
Wisner B, Gaillard JC, Kelman I (2012) The Routledge handbook of hazards and disaster risk reduction
World Health Organization (2015) World malaria report 2015. World Health Organization
Zubir SS, Amirrol H (2011) Disaster risk reduction through community participation. WIT Trans Ecol Environ 148:195–206

Brenda Nyeverwai Rumutsa is a Research Officer with the Parliament of Zimbabwe seconded to the SADC Parliamentary Forum's Sexual and Reproductive Health and Rights (SRHR) Sweden Funded Project. She works closely with parliamentary portfolio and thematic committees, integrating research into parliamentary oversight, legislative processes, and advocacy, especially in the areas of climate justice, sexual and reproductive health and rights, among others. Her work highlights the need for integrated and intersectional approaches and she is passionate about using research to influence policy and promote sustainable development across the region. Her dedication to research is evident throughout her career, where she has continuously advanced into more specialized roles across both academic and professional research projects.

Boemo Mmandu Sekgoma Is the Secretary General of the SADC Parliamentary Forum, the main inter-parliamentary organisation in Southern Africa, which regroups 15 Member Parliaments. She has widespread experience in parliamentary initiatives and the management of programmes which hinge on democracy, regional integration, trade, governance, gender, human rights, constitutionalism, and the observance of the rule of law, among others. She has also assisted in spearheading several regional instruments in the form of Model Laws which constitute thematic legal norms and facilitate the domestication of Treaties and Protocols, in the furtherance of the SADC Regional agenda, and serve as gold standard for the benchmarking of national Parliaments, policy makers, and non-state actors including on HIV and AIDS, Elections, eradicating Child Marriages, Gender Based Violence and Public Financial Management. One of the core objectives of the SADC-PF is to facilitate the implementation capacity of SADC, the development of progressive initiatives to address factors that undermine socio-economic empowerment and democratisation, *inter alia* by addressing foreign policy, peace building initiatives, climate justice, and action, as well as issues of gender equality and identity. At the helm of the Forum's Secretariat, Ms Sekgoma has also led initiatives in view of collaborating with SADC countries, and interparliamentary organisations to promote compliance with the Sustainable as well as far-sighted visions such as the SADC Vision 2050 and African Agenda 2063.

Chapter 9
Genealogy of the Concept of Healthy Housing

Parsa Ahmadi Dehrashid, Ahmad Pourahmad, Hossein Mansourian, and Amir Reza Khavarian-Garmsir

Abstract Since ancient times, humans have consistently sought to shield themselves from threats to their health and well-being. This enduring pursuit of safety and comfort—whether explicit or implicit—has played a significant role in shaping the design and organization of human settlements. In the contemporary era, particularly in the wake of the COVID-19 pandemic and the challenges it posed, the notion of "healthy housing" has gained renewed prominence within academic discourse. Despite its widespread use across technical and policy-related domains, the concept is often treated as self-evident and examined primarily through operational or implementation-focused frameworks. As a result, its historical roots and conceptual foundations have received comparatively little attention. Adopting a genealogical approach, this paper traces the evolution of the "healthy housing" concept across various historical periods, illustrating how it has been shaped by shifting social, economic, and cultural forces. It further explores the ways in which the features of healthy housing contribute to the protection and promotion of residents' health—examining elements such as indoor environmental quality, housing design patterns, and the broader external environment. The paper concludes by outlining potential directions for future research on healthy housing, offering a roadmap for the development of housing strategies capable of addressing post-pandemic health challenges.

The original version of the chapter has been revised. A correction to this chapter can be found at https://doi.org/10.1007/978-3-031-99132-5_15

P. Ahmadi Dehrashid (✉) · A. Pourahmad · H. Mansourian
Department of Human Geography and Planning, Faculty of Geography, University of Tehran, Tehran, Iran
e-mail: Ahmadi.parsa@ut.ac.ir

A. Pourahmad
e-mail: apoura@ut.ac.ir

H. Mansourian
e-mail: h.mansourian59@ut.ac.ir

A. R. Khavarian-Garmsir
Department of Geography and Urban Planning, Faculty of Geographical Sciences and Planning, University of Isfahan, Isfahan, Iran
e-mail: a.khavarian@geo.ui.ac.ir

© The Author(s), under exclusive license to Springer Nature Switzerland AG 2025, corrected publication 2025

A. I. Almulhim et al. (eds.), *Urbanization, Climate Change, and Health*, Climate Change Management, https://doi.org/10.1007/978-3-031-99132-5_9

It also proposes actionable solutions to support the role of healthy housing in an era marked by climate change.

Keywords Healthy housing · Housing · Covid-19 · Pandemic · Climate change

Introduction

The concept of "healthy housing" has always been deeply intertwined with humanity's quest for survival and well-being. Since the dawn of human history, housing design and quality have played a pivotal role in maintaining public health, providing protection from environmental hazards, and fostering social cohesion (Krieger and Higgins 2002; Nix et al. 2024). However, the significance of housing as a determinant of health has become even more apparent, especially in the context of recent global crises, including the COVID-19 pandemic and the escalating impacts of climate change. These crises have not only exposed vulnerabilities within housing systems but have also necessitated a re-evaluation of housing models to better address current and future challenges (Afrin et al. 2021; Hernandez et al. 2023).

In pre-modern times, housing served as a fundamental barrier against environmental and health threats, with a strong emphasis on factors such as hygiene, adequate ventilation, and structural integrity in housing design (Bonnefoy 2007; Ilgın 2023). Over time, and particularly during the Industrial Revolution, urbanization brought new challenges like population density and poor living conditions, necessitating broader public health interventions (Reyes et al. 2013). These historical developments paved the way for the integration of health considerations into housing policies and designs, laying the groundwork for more modern approaches that prioritize well-being, sustainability, and resilience as core principles in housing design and planning.

In contemporary discourse, the concept of "healthy housing" has emerged as a multifaceted approach encompassing indoor environmental quality, housing design patterns, and external environmental factors. The COVID-19 pandemic has particularly highlighted the critical link between housing quality and public health, exposing the shortcomings of traditional housing designs in mitigating disease transmission and supporting the physical and mental well-being of occupants during prolonged periods of quarantine (Adediran et al. 2020; Luo et al. 2022; Zarrabi et al. 2021). As a result of these developments, the pandemic has acted as a catalyst for innovation, leading to the adoption of health-promoting design features such as advanced ventilation systems, flexible and multi-functional spaces, and the incorporation of green and open spaces into housing design (Capolongo et al. 2018; Coza 2024; Heydari and Abbasianjahromi 2024; UN-Habitat 2021).

Looking towards the future, the pursuit of "healthy housing" is increasingly driven by two key imperatives: preparedness for future pandemics and adaptation to climate change. Lessons learned from the COVID-19 crisis underscore the need for housing models that can address both immediate and long-term health challenges (Amirzadeh et al. 2023; Cheshmehzangi 2021; de Frutos et al., 2021; Sharifi & Khavarian-Garmsir, 2020). These models must be capable of adapting to unforeseen developments while simultaneously addressing the need for resilient housing solutions in the face of climate change.

Despite the widespread technical and policy-oriented applications of the concept of 'healthy housing,' its historical and conceptual foundations have received little attention. In many contemporary studies, the concept has primarily been examined within practical and technical frameworks, with a focus on physical, health, and technological criteria (Hasselaar and Priemus 2006; Howden-Chapmam 2004; Lawrence 2011; Marco and Burgess 2015; Mood 2005; Ranson 2002; Saegert et al. 2003; Wagner 2020; WHO 1988, 2019, 2021). Furthermore, numerous studies have taken the idea of healthy housing as self-evident, concentrating instead on its operationalization, while overlooking its connection to broader discourses of power, systems of knowledge, and historical developments. Therefore, a significant gap in current research lies in the need for a genealogical analysis of the concept of 'healthy housing,' aimed at identifying its underlying assumptions and tracing its evolving meanings over time. This approach not only offers a deeper understanding of the contemporary relevance of the concept but also strengthens critical thinking and rethinking in the formulation of future housing and urban health policies and programs.

This study employs a genealogical approach to analyze the historical evolution of the concept of 'healthy housing,' investigating how social, economic, and cultural forces have shaped this concept over time. Through this analysis, particularly in the contemporary era, the complex relationships between housing, public health, and sustainable development will be explored in depth. The findings of this research can serve as a valuable resource for urban planners, policymakers, and architects, assisting them in designing adaptable and resilient housing models to address both current and future challenges in an ever-changing world.

Methodology

This research employs a genealogical approach to analyze the concept of "healthy housing" across three historical periods: the past, the present, and the future. The genealogical approach, rooted in historical-analytical methods, is a suitable tool for tracing and analyzing conceptual changes over time (Edgley 2014; Kearins and Hooper 2002; Sylvia 2021). By focusing on identifying the origins, transformations, and reinterpretations of concepts, this method facilitates a deeper understanding of the factors influencing the formation and evolution of ideas (Foucault 1980; Garland 2014; Morrisson 2021).

The application of a genealogical approach to analyze the concept of "healthy housing" is particularly justified, as this method effectively enables the examination of the evolution of concepts across different temporal and spatial contexts (Anaïs 2013). Genealogy, specifically designed to identify and analyze how concepts emerge and change in response to historical, social, and cultural factors, allows researchers to meticulously uncover the semantic complexities and interrelationships between concepts across different historical and cultural periods (Morrison 2016). In the context of "healthy housing", the use of a genealogical approach is valuable because health crises such as pandemics, social changes, and environmental transformations directly influence the definition and expectations associated with healthy housing. Consequently, analyzing the evolution of this concept from the past to the present and forecasting its future trends can significantly enhance our understanding of the impacts of global crises like disease outbreaks and climate change on housing design and policy. By employing genealogy, we can examine the trajectory of this concept not only from a historical perspective but also through the lens of the social and cultural changes of each era.

In this research, data was collected from secondary sources, including historical documents, scholarly articles published in reputable databases such as Scopus, Web of Science, and Google Scholar, policy reports, and previous studies on "healthy housing". These sources were systematically identified and selected to represent the changes and developments associated with the concept of "healthy housing" across the past, present, and future. An initial search yielded 214 articles, which were narrowed down to 73 after a review and screening process that assessed the scholarly credibility, data authenticity, and direct relevance to the research topic. These articles were used to analyze various trends in the concept over different time periods. In addition to scholarly articles, policy-related reports and reliable statistics from international organizations were considered as supplementary sources. To enhance accuracy and comprehensiveness, multiple and diverse sources were utilized to cover various aspects of healthy housing and its evolution over time. This data collection method allows the researcher to obtain a comprehensive and accurate perspective on the conceptual developments of "healthy housing", ensuring that the research benefits fully from the diverse viewpoints of researchers and scholarly sources.

To analyze the collected data, a qualitative content analysis was employed. This method is particularly effective for identifying patterns, themes, and key concepts within each time period. The analysis process involved manually coding the data based on the main themes related to the concept of "healthy housing", followed by a comparative analysis to identify turning points and changes in each period. At this stage, efforts were made to accurately identify the semantic and conceptual transformations of healthy housing in response to social, economic, and health crises such as pandemics.

Results and Discussion

Past

The relationship between humans, the environment, and health is an enduring one. The concept of health has always played a significant role in human settlements. In ancient times, Greek philosophers and physicians such as Hippocrates and Galen emphasized the importance of harmony between humans and the natural environment for maintaining health (Kleisiaris et al. 2014). In Islamic thought, water and green spaces were regarded as essential elements in the design of healthy cities, with water management playing a vital role in improving public health (Hassan 1969; Nawawi 2002). In the modern era, the concept of health has gradually expanded to encompass physical, mental, and social dimensions. Today, health is not merely defined as the absence of disease but also includes well-being and quality of life in urban environments (Bai et al. 2012; Krefis et al. 2018). The development of new theories in the field of urban health is based on the understanding that the built environment, including housing, public spaces, and urban infrastructure, has significant impacts on the health of individuals and communities (Kan et al. 2022; Perdue et al. 2003).

There is ample evidence that in ancient civilizations, housing was not merely a shelter but also a reflection of social and cultural values. For instance, in the classical cities of Greece, the design of houses and public spaces facilitated social interaction and communal life, creating a balance between privacy and public engagement (Dickenson 2017; Hölscher 2007). Spaces such as the agora, theaters, and public squares were integral parts of the city's structure, aimed at enhancing quality of life through cultural and social activities. Similarly, the use of natural materials and architectural techniques that promoted ventilation and access to natural light in traditional Mediterranean housing reflects early efforts to create healthy living conditions, emphasizing the connection between housing design and the physical and mental well-being of residents (Bai et al. 2012; Krefis et al. 2018; Wuerzer 2014).

However, the Relationship Between Humans, the Environment, and Health is not confined solely to the role of the environment in ensuring health; it is also linked to the ongoing endeavor to create healthy and sustainable housing. Since the establishment of the first permanent settlements following the Agricultural Revolution, humans have sought to organize spaces that help them meet their needs, achieve a desirable quality of life, and attain individual and collective goals (Hemathilake and Gunathilake 2022). In other words, by shaping their housing and living environments, people have strived to lay the foundations for factors that contribute to health and well-being. In other words, by shaping their housing and living environments, people have strived to lay the foundations for factors that contribute to health and well-being (Atack et al. 2022; Morris 2013). he deterioration of housing quality and the rise in urban health issues sparked movements advocating for reforms. Utopian urban models, such as Ebenezer Howard's "Garden City", aimed to address the health and social problems associated with industrial cities by combining the benefits of urban living with those of nature. The Garden City concept proposed low-density housing surrounded by green spaces to enhance health, well-being, and social cohesion (Howard 1898, 1902). Furthermore, studies by Graham (1818) and Snow (1849) are examples from the past two centuries that have explored the relationship between housing and health.

The concept of healthy housing has evolved alongside the history of urban development, with its roots in early human settlements where basic needs such as shelter and safety were central to the organization of society. Since ancient times, communities have sought to create living environments that not only provide protection but also promote well-being and a higher quality of life. The concept of a healthy housing model encompasses various dimensions, including physical, environmental, and social aspects that contribute to the well-being of residents. Key factors such as the physical structure of housing, the quality of indoor environments, access to green spaces, and the integration of social spaces are essential components influencing health and the quality of urban life.

Present

The concept of healthy housing has a rich history that has evolved through interactions with social, economic, and environmental changes from past eras to the present. In early societies, housing functioned as a secure space for living and protection against environmental conditions (Anyanwu and Beyer 2024). As civilizations advanced and technology developed, the definition of housing expanded, incorporating new dimensions of quality of life. During the Renaissance, social and cultural transformations heightened attention to aesthetics and design in housing (de Arce 2014; Divitiis and Christian 2018). In the nineteenth century, with the rise of public health and industrial movements, the quality and hygiene of housing, especially in urban areas, became a fundamental challenge (Krieger and Higgins 2002). In recent decades, scientific advancements and growing social awareness of health and well-being issues have increasingly brought the concept of healthy housing to the forefront, establishing it as a key focus in urban planning and social policies.

The concept of Healthy City, as one of the essential pillars of a Healthy City, has been shaped by social, economic, and health-related transformations throughout history. This concept formally emerged in the 1940s, following World War II, with the adoption of the Universal Declaration of Human Rights (1948), which emphasized the right to adequate housing. The declaration not only acknowledged human rights but also highlighted social health, serving as a foundation for future housing policy. In the 1970s, the first United Nations Conference on Human Settlements in Vancouver (1976) addressed the necessity of housing and human environmental considerations, followed by the Ottawa Charter (1986), which introduced principles for health promotion in housing. The charter underscored the impact of housing conditions on social well-being and public health, highlighting the critical role of the living environment in disease prevention and quality of life enhancement. Additionally, the concept of Sick Building Syndrome in the 1980s responded to increasing reports of health issues associated with buildings, emphasizing the need for improvements in residential conditions to ensure occupant health (Burge et al. 1987; Finnegan et al. 1984; Jafari et al. 2015). Ultimately, the World Health Organization's introduction of the healthy housing concept added a new dimension to the literature in public health and urban planning, reinforcing the importance of designing and providing suitable living conditions to promote individual and social health.

Healthy housing, as a fundamental concept, refers to residential conditions that can meet the physical, mental, and social needs of its occupants (WHO 1988, 2018). This concept is based on the premise that housing quality can profoundly influence residents' health outcomes and is considered one of the indicators of a healthy city. A healthy housing environment should be designed and constructed to provide favorable living conditions, addressing not only physical health but also the psychological and social well-being of its residents. The significance of this concept has increased in response to the growing health and environmental challenges in contemporary societies, positioning it as a key solution for enhancing quality of life and social welfare (Iddi et al. 2022; Saegert et al. 2003). Consequently, attention to the characteristics of healthy housing not only contributes to individual quality of life but also promotes sustainable development and improves overall community conditions.

To achieve a comprehensive understanding of the components contributing to a healthy housing model, it is essential to examine the complex pathways that link housing characteristics with residents' health. This exploration reveals several key dimensions that significantly impact the maintenance and promotion of residents' well-being. Below, we provide an in-depth analysis of three critical dimensions: indoor environmental quality, housing design patterns, and the quality of the external environment. Each of these dimensions plays a vital role in shaping residents' lived experiences and their overall well-being.

Indoor Environmental Quality

The quality of the indoor environment within a residential unit is fundamental for meeting the essential needs of residents for comfort, safety, and security. Key elements such as spatial adequacy, ventilation, natural light, and thermal comfort have been shown to profoundly impact the mental and emotional health of residents (NIOSH 2013; Samet and Spengler 2003; Zalejska-Jonsson and Wilhelmsson 2013). For instance, research indicates that well-designed indoor environments that effectively harness natural light and optimize ventilation can lead to significant improvements in life satisfaction, reduced stress levels, and enhanced psychological well-being. Moreover, the adequacy of interior spaces—designed to meet the diverse needs of various demographic groups—facilitates better functionality and accessibility. This adaptability not only enhances everyday living experiences but also fosters a deeper sense of belonging and stability within the community. Additionally, interior design elements that promote aesthetic appeal and emotional comfort can contribute to positive psychological outcomes. The integration of personal expression through spatial organization can further empower residents and support their overall mental health and satisfaction with the living environment.

The quality of the indoor environment in a residential unit is fundamental for addressing the essential needs of residents for comfort, safety, and security. Key factors such as spatial adequacy, ventilation, natural light, and thermal comfort have been shown to significantly impact the mental and emotional health of occupants. For instance, studies reveal that well-designed indoor environments that effectively utilize natural light and optimize ventilation can lead to considerable improvements in life satisfaction, reductions in stress levels, and enhanced psychological well-being. Furthermore, the adequacy of interior spaces—tailored to accommodate the diverse needs of various demographic groups—facilitates better functionality and accessibility. This adaptability not only enriches everyday living experiences but also fosters a deeper sense of belonging and stability within the community. Additionally, interior design elements that enhance aesthetic appeal and emotional comfort can result in positive psychological outcomes. The incorporation of personal expression through spatial organization can further empower residents, contributing to their overall mental health and satisfaction with their living environment.

The quality of the indoor environment within a residential unit is essential for meeting the fundamental needs of residents for comfort, safety, and security. Key elements such as spatial adequacy, ventilation, natural light, and thermal comfort have been shown to have a profound impact on the psychological and emotional health of occupants. For instance, research indicates that well-designed indoor environments, which effectively harness natural light and optimize ventilation, can lead to significant improvements in life satisfaction, reductions in stress levels, and enhancements in psychological well-being (Edwards and Torcellini 2002; Kang et al. 2014; Reinikainen and Jaakkola 2001). Moreover, the adequacy of interior spaces—designed to cater to the diverse needs of various demographic groups—facilitates improved functionality and accessibility (Bonnefoy 2007; Klepeis et al. 2001; Rosa-Jimenez and Jaime-Segura 2021). This adaptability not only enhances daily living experiences but also fosters a deeper sense of belonging and stability within the community (Femenias and Geromel 2020). Additionally, interior design elements that enhance aesthetic appeal and emotional comfort can yield positive psychological outcomes (Kearns et al. 2000; Zhang 2016). The integration of personal expression through spatial organization can further empower residents, contributing to their overall mental health and satisfaction with their living environment (Gurstein 2001; Hasselaar and Priemus 2006; Magee 2000).

Housing Design Patterns

Housing design is a significant determinant of social interactions and community cohesion, which influences both individual and collective well-being. Attributes such as aesthetic quality, architectural diversity, and human-scale design not only enhance the visual appeal of residential areas but also evoke positive emotional responses among residents (Bettaieb and Alsabban 2020; Peters and Halleran 2021). Thoughtful housing design that promotes natural surveillance—such as the strategic placement of windows and shared spaces—encourages positive social interactions, thereby strengthening neighborhood relationships and enhancing residential well-being (Akbari et al. 2021; Veitch and Galasiu 2012). Moreover, inclusive design considerations that address the diverse needs of various demographic groups—such as families, elderly residents, and individuals with disabilities—are essential for creating spaces that support a range of social dynamics (Usher et al. 2020). By fostering environments that promote inclusivity and interaction, housing design can facilitate stronger social bonds and a greater sense of belonging, contributing to overall mental well-being.

Quality of the External Environment

The quality of the external environment surrounding residential units plays a fundamental role in influencing residents' well-being. This dimension encompasses access to green spaces and communal areas beyond the immediate residential unit, extending to the building, site, and block levels. Access to such spaces provides health benefits, including stress reduction and increased opportunities for physical activity (Podrekar et al. 2020; Sallis et al. 2016; Yuen and Nyuk Hien 2005). Moreover, well-maintained public spaces are crucial for facilitating social interactions and enhancing social participation, both of which are essential for fostering a sense of belonging and building social capital (Aydin and Sayar 2020; Williams et al. 2019). The availability of safe and appealing outdoor environments encourages outdoor activities and social engagement, directly contributing to overall life satisfaction (Bowlby and Jupp 2021; Prochorskaite et al. 2016; Wågø et al. 2016). Access to amenities such as playgrounds, walking paths, and public gathering areas enhances residents' quality of life, promotes active lifestyles, and strengthens social relationships (Can and Heath 2016; Fezi 2020). Additionally, attention to the aesthetic quality of the external environment—through landscaping, public art, and thoughtful urban design—can improve emotional well-being and create a more vibrant and attractive social atmosphere.

In the 2020s, the COVID-19 pandemic underscored the importance of healthy housing, especially in combating pandemics. This global health crisis had profound and widespread impacts on individuals' living patterns and residential conditions, emphasizing the need for increased attention to housing quality (Albreht 2023; Peters and Halleran 2021; Pollard et al. 2020; Rogers and Power 2020). During the crisis, residents spent extended periods in their residential environments, directly affecting their physical and mental health (Awada et al. 2021; Megahed and Ghoneim 2020; Zarrabi et al. 2021).

The dimensions and features of healthy housing play a crucial role in reducing the risk of contamination within residential settings during future pandemics, such as COVID-19. Materials that are easy to clean help prevent the accumulation of pathogens. Additionally, acoustic comfort reduces the need to open windows, thereby minimizing the infiltration of airborne pollutants (Akbari et al. 2021; D'alessandro et al. 2020). Thermal comfort and proper ventilation systems decrease reliance on window ventilation, reducing the airborne transmission of viruses. Spaces with adequate lighting facilitate regular cleaning and disinfection while ensuring safety measures that prevent accidents and lessen the burden on healthcare systems (Aydin and Sayar 2020; Kim and Kim 2023). High indoor air quality, achieved through ventilation and filtration, aids in reducing viral transmission and contributes to the overall well-being of residents, helping to alleviate stress related to the illness (Molaei et al. 2022).

In this context, the design of supportive social spaces helps counteract isolation, which can adversely affect health, and optimal spatial arrangements facilitate infection control measures. Access to indoor and outdoor green spaces also supports mental health. By integrating these features into the design and management of housing, policymakers can create resilient environments that reduce contamination risks during pandemics. Thus, it can be asserted that the application of healthy housing principles is essential for developing residential models to address future pandemics. This approach ensures that housing developments effectively protect and enhance the health and well-being of residents, providing a robust framework for future housing policies and practices. Ultimately, recognizing the significance of healthy housing as a key factor in preventing viral transmission and mitigating associated health complications underscores the need for quality and healthy housing, serving as a fundamental lesson for future housing policy and urban design.

Future

The future of healthy housing, as a critical issue in urban and housing planning—particularly in the post-COVID era and in light of climate change—requires a thorough re-evaluation and deep reflection. In this context, two main areas demand serious attention: preparedness for future pandemics and the impact of climate change on the design and management of Healthy Housing.

Post-COVID Housing and Preparedness for Future Pandemics

The COVID-19 pandemic underscored the crucial role that housing quality and the design of living spaces play in safeguarding public health. The pandemic highlighted vulnerabilities within current housing models, revealing how inadequate design could exacerbate the spread of infectious diseases and affect the overall well-being of residents. Consequently, it has become evident that the future of housing design must evolve beyond traditional paradigms to address both immediate health crises and long-term societal needs.

In preparing for future pandemics, housing must not only focus on preventing disease transmission but also prioritize the enhancement of residents' physical and mental health. The design of flexible, adaptable spaces will be essential to accommodate the shifting needs of residents during times of crisis. This adaptability can range from the ability to reconfigure spaces for quarantine or home office usage, to the integration of technologies that facilitate remote work and learning, ensuring that housing remains functional even during emergencies.

Future housing models should incorporate health-promoting features, including easily accessible medical facilities, advanced ventilation systems, and the use of non-toxic, easily cleanable materials. These elements will play a vital role in reducing the risk of airborne diseases and supporting general well-being. In addition, the integration of open and green spaces within housing designs will be essential, as such spaces can foster social interaction while promoting mental health. By allowing for safe outdoor activities, these spaces can mitigate the psychological strain that often accompanies isolation during health crises.

Furthermore, the design of multi-functional spaces will become paramount in enabling residents to perform daily tasks, engage in social interactions, and maintain healthy lifestyles without needing to leave their homes. The importance of these spaces is amplified during periods of crisis, as they provide a refuge that reduces stress and anxiety. These spaces could be designed to cater to a variety of functions, including home offices, gyms, or entertainment areas, allowing residents to carry on with their daily lives in a comfortable and safe environment.

Research in the post-pandemic era should delve deeper into how housing can be designed to optimize quality of life during crises, not only from a physical health perspective but also from a psychological and social standpoint. Understanding how social and cultural factors influence mental health will be critical to developing housing solutions that foster resilience and well-being. Future studies must explore innovative design strategies that promote community cohesion, provide psychological support, and cater to diverse social needs. Ultimately, the focus should be on creating adaptable, resilient, and holistic housing solutions that serve as both protective and supportive environments in times of crisis..

Climate Change and Healthy Housing

In today's world, climate change has emerged as a formidable and complex challenge to achieving healthy housing. This necessitates a shift from traditional paradigms towards innovative, adaptive, and preventive approaches. Rising temperatures, erratic rainfall patterns, and increasing frequency of natural disasters have profound implications for housing quality, occupant health, and overall urban resilience. Addressing these challenges requires a comprehensive integration of environmental, social, and technological factors in housing design and management. Additionally, the links between climate change and public health crises, such as respiratory diseases, heat-related illnesses, and the increased spread of infectious diseases during extreme weather events, must be clearly acknowledged. The climate-related health risks significantly heighten the urgency of developing housing models that not only support environmental sustainability but also safeguard public health (Ansah et al. 2024; Or and Seppänen 2024; WHO 2020Younger et al. 2008a).

To confront these challenges, policymakers, urban planners, and architects must conduct comprehensive climate-related risk assessments and prioritize the creation of resilient and adaptable housing solutions. Key strategies in this regard include the use of renewable and sustainable materials, the incorporation of green infrastructure such as green roofs and urban gardens, and the optimization of energy consumption through passive design and renewable energy technologies (Jo et al., 2009; Karimi et al., 2023; Lowe, 2003; MacGregor et al., 2018; Mihalakakou et al., 2023; Olabi et al., 2025; Younger et al. 2008). Moreover, special attention should be given to stormwater management systems and the mitigation of urban heat island effects to reduce the negative impacts of global warming. The design of climate-responsive buildings, for instance, can contribute significantly to reducing the impact of extreme heat waves, which are becoming increasingly common due to climate change (Liu et al. 2023; Liyanage et al. 2024; McNeilly Smith et al. 2024; Peri et al. 2024; Saifudeen and Mani 2024).

Future research should focus on interdisciplinary and context-sensitive approaches, emphasizing the integration of environmental and social dimensions in healthy housing design. For example, culturally appropriate and climate-responsive housing solutions can enhance occupant quality of life while achieving broader sustainability goals. Promoting community-based initiatives that encourage sustainable lifestyles, such as waste reduction, water conservation, and energy efficiency, can contribute to climate change adaptation at the grassroots level. Future research could explore local knowledge and traditional building practices that have historically been effective in mitigating the effects of harsh climates, adapting them to modern needs.

Furthermore, healthy housing has the transformative potential to extend beyond its role as mere shelter (WHO 1988, 2018). It can serve as a catalyst for improving public health, strengthening social equity, and enhancing urban resilience. Given the intersection of climate change with global health crises such as pandemics, housing strategies must evolve to address both immediate needs and long-term challenges. In this context, forward-looking housing models that incorporate smart technologies and nature-based solutions can offer innovative pathways to strengthen adaptive capacity and mitigate risks. Future research must address how housing can integrate cutting-edge technologies such as smart climate control systems, real-time environmental monitoring, and health surveillance systems to enhance resilience and adaptability in the face of future health crises.

Ultimately, emphasizing healthy housing in the context of climate change is not only a response to emerging challenges but also an opportunity to redefine urban environments and the way we live. Policymaking must adopt a forward-looking perspective and ensure that housing serves as a foundation for resilience, sustainability, and equity, contributing to an improved quality of life for all inhabitants. By embracing evidence-based strategies and strengthening intersectoral collaborations, the future of healthy housing can provide a robust defense against the uncertainties of climate change and other global crises. Future research should concentrate

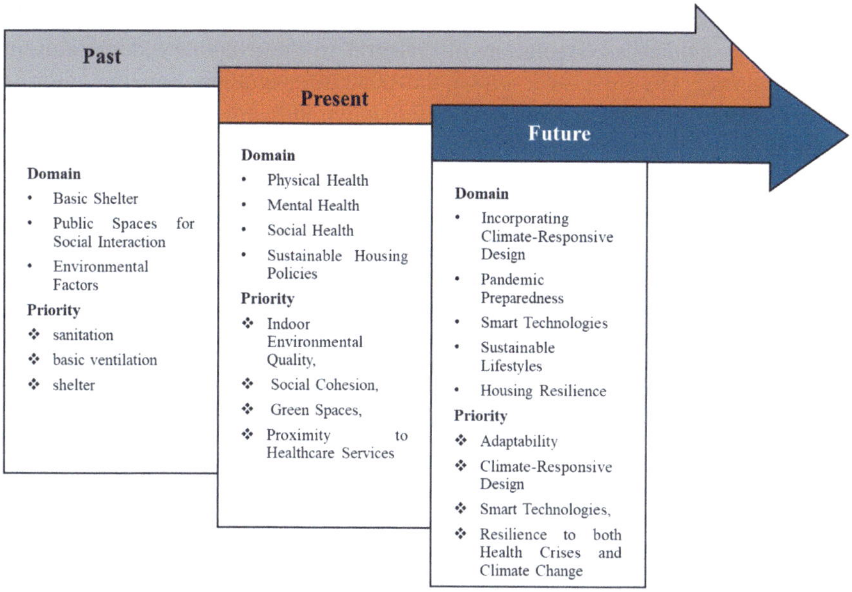

Fig. 9.1 A historical overview of the domain and priorities of healthy housing

on refining climate risk modeling for housing, establishing clear guidelines for the adaptation of housing to future climate scenarios, and developing global standards for climate-resilient housing (Fig. 9.1).

Conclusion and Recommendations

This study has critically re-examined the concept of "healthy housing" through a genealogical and discourse-informed lens, revealing the limitations of dominant paradigms that reduce housing to a set of technical or hygienic specifications. Instead, it argues for a reconceptualization of healthy housing as a historically situated and socially constructed phenomenon, shaped by power relations, urban governance practices, environmental contexts, and cultural norms. This perspective allows for a more nuanced understanding of how health, space, and dwelling interact under conditions of rapid ecological change and increasing urban vulnerability.

The analysis highlights the pressing need to address the complex interdependencies between climate change, public health, and housing. As climate-related risks—such as heatwaves, air pollution, infectious disease outbreaks, and flooding—intensify, healthy housing must evolve into a forward-looking model of resilience and adaptability. This involves not only mitigating environmental risks but also advancing equity and social inclusion in housing policy and design.

One of the key contributions of this research is the articulation of smart technologies and nature-based solutions as integral to contemporary definitions of healthy housing. The study underscores the value of integrating digital monitoring systems, energy-efficient automation, and adaptive infrastructure with ecological interventions such as green roofs, urban vegetation, and passive cooling systems. These approaches can significantly enhance housing resilience, reduce environmental burdens, and improve both physical and mental well-being. However, their implementation must be context-sensitive, incorporating local knowledge systems and socio-cultural practices to ensure relevance and sustainability.

From a policy and planning perspective, the findings call for a shift away from technocratic and standardized frameworks toward holistic, multisectoral approaches that foreground health, equity, and climate resilience. Policymakers and urban designers must adopt integrative strategies that align environmental sustainability with public health priorities and cultural specificity. This includes revising building regulations, incentivizing green retrofitting, and supporting community-based design initiatives.

Future research should deepen this agenda by exploring interdisciplinary methodologies that link urban Planning, urban climatology, public health, and critical housing studies. Special attention should be given to refining climate risk modeling in housing contexts, evaluating the long-term health impacts of smart and green technologies, and developing inclusive guidelines for climate-resilient housing across diverse geographic and socio-economic settings.

Ultimately, by reframing healthy housing as a dynamic and relational construct—technological, ecological, and cultural—this study contributes to a broader redefinition of urban resilience. It positions housing not merely as a shelter, but as a vital infrastructure for sustaining life, fostering social justice, and shaping a more equitable and sustainable urban future.

Acknowledgements We used the Gemini AI tool to enhance the translation from Persian to English and ChatGPT to improve the readability and clarity of the text. The purpose of using these tools was to enhance the quality of the translation and make the text more fluent, facilitating the presentation of scientific content. The use of AI was solely as an auxiliary tool in the writing process, while all ideas, analyses, and findings presented in this chapter are entirely the result of the authors' research and scientific expertise.

References

Adediran A, Oladejo SO, Akinwande TO, Ajibade S-SM, Moveh S (2020) Housing quality standard and Covid-19 pandemic: a call for attention in Nigeria. J Sci Eng Technol Manage 2(2). ISSN: 2795-3955

Afrin S, Chowdhury FJ, Rahman MM (2021) COVID-19 pandemic: rethinking strategies for resilient urban design, perceptions, and planning. Front Sustain Cities 3.https://doi.org/10.3389/frsc.2021.668263

Akbari P, Yazdanfar S-A, Hosseini S-B, Norouzian-Maleki S (2021) Housing and mental health during outbreak of COVID-19. J Build Eng 43:102919. https://doi.org/10.1016/j.jobe.2021.102919

Albreht T (2023) Challenges to global health emerging from the COVID-19 pandemic. Sustainability 15(9):7633. https://doi.org/10.3390/su15097633
Amirzadeh M, Sobhaninia S, Buckman ST, Sharifi A (2023) Towards building resilient cities to pandemics: a review of COVID-19 literature. Sustain Cities Soc 89:104326. https://doi.org/10.1016/j.scs.2022.104326
Anaïs S (2013) Genealogy and critical discourse analysis in conversation: texts, discourse, critique. Crit Discourse Stud 10(2):123–135. https://doi.org/10.1080/17405904.2012.744321
Ansah EW, Amoadu M, Obeng P, Sarfo JO (2024) Health systems response to climate change adaptation: a scoping review of global evidence. BMC Public Health 24(1):2015. https://doi.org/10.1186/s12889-024-19459-w
Anyanwu C, Beyer KMM (2024) Intersections among housing, environmental conditions, and health equity: a conceptual model for environmental justice policy. Soc Sci Human Open 9:100845. https://doi.org/10.1016/J.SSAHO.2024.100845
Atack J, Margo RA, Rhode PW (2022) Industrialization and urbanization in nineteenth century America. Reg Sci Urban Econ 94:103678. https://doi.org/10.1016/j.regsciurbeco.2021.103678
Awada M, Becerik-Gerber B, Hoque S, O'Neill Z, Pedrielli G, Wen J, Wu T (2021) Ten questions concerning occupant health in buildings during normal operations and extreme events including the COVID-19 pandemic. Build Environ 188.https://doi.org/10.1016/j.buildenv.2020.107480
Aydin D, Sayar G (2020) Questioning the use of the balcony in apartments during the COVID-19 pandemic process. Archnet-IJAR: Int J Architect Res 15(1):51–63. https://doi.org/10.1108/ARCH-09-2020-0202
Bai X, Nath I, Capon A, Hasan N, Jaron D (2012) Health and wellbeing in the changing urban environment: complex challenges, scientific responses, and the way forward. Curr Opin Environ Sustain 4(4):465–472. https://doi.org/10.1016/J.COSUST.2012.09.009
Bettaieb DM, Alsabban R (2020) Emerging living styles post-COVID-19: housing flexibility as a fundamental requirement for apartments in Jeddah. Archnet-IJAR: Int J Architect Res 15(1):28–50. https://doi.org/10.1108/ARCH-07-2020-0144
Bonnefoy X (2007) Inadequate housing and health: an overview. Int J Environ Pollut 30(3–4). https://doi.org/10.1504/IJEP.2007.014819
Bowlby S, Jupp E (2021) Home, inequalities and care: perspectives from within a pandemic. Int J Hous Policy 21(3):423–432. https://doi.org/10.1080/19491247.2020.1840901
Burge S, Hedge A, Wilson S, Bass JH, Robertson A (1987) Sick building syndrome: a study of 4373 office workers. Ann Occup Hyg 31(4A):493–504
Can I, Heath T (2016) In-between spaces and social interaction: a morphological analysis of Izmir using space syntax. J Housing Built Environ 31(1):31–49. https://doi.org/10.1007/s10901-015-9442-9
Capolongo S, Rebecchi A, Dettori M, Appolloni L, Azara A, Buffoli M, Capasso L, Casuccio A, Oliveri Conti G, D'Amico A (2018) Healthy design and urban planning strategies, actions, and policy to achieve salutogenic cities. Int J Environ Res Public Health 15(12):2698
Cheshmehzangi A (2021) Housing and health evaluation related to general comfort and indoor thermal comfort satisfaction during the COVID-19 lockdown. J Hum Behav Soc Environ 31(1–4):184–209. https://doi.org/10.1080/10911359.2020.1817225
Coza H (2024) The impact of the COVID-19 pandemic on architectural space needs. Kent Akademisi 17(1):60–71. https://doi.org/10.35674/kent.1365028
D'alessandro D, Gola M, Appolloni L, Dettori M, Fara GM, Rebecchi A, Settimo G, Capolongo S (2020) COVID-19 and living space challenge. Well-being and public health recommendations for a healthy, safe, and sustainable housing. Acta Biomed 91. https://doi.org/10.23750/abm.v91i9-S.10115
de Arce RP (2014) Urban transformations and the architecture of additions. Routledge
de Frutos F, Cuerdo-Vilches T, Alonso C, Martin-Consuegra F, Frutos B, Oteiza I, Navas-Martin MA (2021) Indoor environmental quality and consumption patterns before and during the COVID-19 lockdown in twelve social dwellings in Madrid, Spain. Sustainability 13(14). https://doi.org/10.3390/su13147700

Dickenson CP (2017) On the agora: the evolution of a public space in hellenistic and roman Greece (c. 323 BC–267 AD), vol 398. Brill

Divitiis B, Christian K (2018) Local antiquities, local identities: art, literature and antiquarianism in Europe, c. 1400–1700

Edgley C (2014) A genealogy of accounting materiality. Crit Perspect Account 25(3):255–271. https://doi.org/10.1016/j.cpa.2013.06.001

Edwards L, Torcellini P (2002) Literature review of the effects of natural light on building occupants

Femenias P, Geromel F (2020) Adaptable housing? A quantitative study of contemporary apartment layouts that have been rearranged by end-users. J Housing Built Environ 35(2):481–505. https://doi.org/10.1007/s10901-019-09693-9

Fezi BA (2020) Health engaged architecture in the context of Covid-19. J Green Building 15(2):185–212. https://doi.org/10.3992/1943-4618.15.2.185

Finnegan MJ, Pickering CA, Burge PS (1984) The sick building syndrome: prevalence studies. BMJ 289(6458):1573–1575. https://doi.org/10.1136/bmj.289.6458.1573

Foucault M (1980) Power/knowledge, brighton. Harvester Press 206:381–393

Garland D (2014) What is a "history of the present"? On Foucault's genealogies and their critical preconditions. Punishment Soc 16(4):365–384. https://doi.org/10.1177/1462474514541711

Graham R (1818) Practical observations on continued fever, especially that form existing at present as an epidemic

Gurstein P (2001) Wired to the world, chained to the home: telework in daily life. UBC Press.

Hassan R (1969) The nature of islamic urbanization—a historical perspective. J Pak Hist Soc 17(2):77

Hasselaar E, Priemus H (2006) Health performance of housing: indicators and tools. In: Sustainable urban areas, vol 10: Vol. phd

Hemathilake DMKS, Gunathilake DMCC (2022) Agricultural productivity and food supply to meet increased demands. In: Future foods: global trends, opportunities, and sustainability challenges. pp 539–553.https://doi.org/10.1016/B978-0-323-91001-9.00016-5

Hernandez JRE, Torabi Moghadam S, Lombardi P (2023) Sustainability assessment in social housing environments: an inclusive indicators selection in Colombian post-pandemic cities. Sustainability 15(3). https://doi.org/10.3390/su15032830

Heydari A, Abbasianjahromi H (2024) Evaluating the resilience of residential buildings during a pandemic with a sustainable construction approach. Heliyon 10(10):e31006. https://doi.org/10.1016/j.heliyon.2024.e31006

Hölscher T (2007) Urban spaces and central places. The Greek world

Howard E (1898) To-morrow: a peaceful path to real reform. Cambridge University Press

Howard E (1902) Garden city of tomorrow. Passim, London

Howden-Chapmam P (2004) Housing standards: a glossary of housing and health. J Epidemiol Community Health 58(3). https://doi.org/10.1136/jech.2003.011569

Iddi S, Muindi K, Gitau H, Mberu B (2022) Characterization of healthy housing in africa: method, profiles, and determinants. J Urban Health 99(1):146–163. https://doi.org/10.1007/s11524-021-00603-5

Ilgın H (2023) High-rise residential timber buildings: emerging architectural and structural design trends. Buildings 14(1):25. https://doi.org/10.3390/buildings14010025

Jafari MJ, Khajevandi AA, Najarkola SAM, Yekaninejad MS, Pourhoseingholi MA, Omidi L, Kalantary S (2015) Association of sick building syndrome with indoor air parameters. Tanaffos 14(1):55

Jo JH, Golden JS, Shin SW (2009) Incorporating built environment factors into climate change mitigation strategies for Seoul, South Korea: a sustainable urban systems framework. Habitat Int 33(3):267–275. https://doi.org/10.1016/j.habitatint.2008.10.020

Kan Z, Kwan M-P, Ng MK, Tieben H (2022) The impacts of housing characteristics and built-environment features on mental health. Int J Environ Res Public Health 19(9):5143. https://doi.org/10.3390/ijerph19095143

Kang NN, Kim JT, Lee TK (2014) A study on the healthy housing quality of multi-family attached house according to dwelling unit age. Energy Procedia 62:595–602. https://doi.org/10.1016/j.egypro.2014.12.422

Karimi H, Adibhesami MA, Bazazzadeh H, Movafagh S (2023) Green buildings: human-centered and energy efficiency optimization strategies. Energies 16(9):3681. https://doi.org/10.3390/en16093681

Kearins K, Hooper K (2002) Genealogical method and analysis. Acc Auditing Accountability J 15(5):733–757. https://doi.org/10.1108/09513570210448984

Kearns A, Hiscock R, Ellaway A, MaCintyre S (2000) "Beyond Four Walls". The psycho-social benefits of home: evidence from West Central Scotland. Housing Stud 15(3):387–410. https://doi.org/10.1080/02673030050009249

Kim M-K, Kim E-J (2023) Apartment space planning directions for infectious disease prevention and management: insights based on residents' experiences. Buildings 13(9). https://doi.org/10.3390/buildings13092203

Kleisiaris CF, Sfakianakis C, Papathanasiou IV (2014) Health care practices in ancient Greece: the hippocratic ideal. J Med Ethics History Med 7:6. http://www.ncbi.nlm.nih.gov/pubmed/25512827

Klepeis NE, Nelson WC, Ott WR, Robinson JP, Tsang AM, Switzer P, Behar JV, Hern SC, Engelmann WH (2001) The national human activity pattern survey (NHAPS): a resource for assessing exposure to environmental pollutants. J Expo Sci Environ Epidemiol 11(3):231–252.https://doi.org/10.1038/sj.jea.7500165

Krefis A, Augustin M, Schlünzen K, Oßenbrügge J, Augustin J (2018) How does the urban environment affect health and well-being? A systematic review. Urban Sci 2(1):21. https://doi.org/10.3390/urbansci2010021

Krieger J, Higgins DL (2002) Housing and health: time again for public health action. Am J Public Health 92(5):758–768. https://doi.org/10.2105/AJPH.92.5.758

Lawrence RJ (2011) Housing for health promotion. Int J Public Health 56(6):577–578. https://doi.org/10.1007/s00038-011-0307-z

Liu S, Yin S, Hua J, Ren C (2023) Climate-responsive architectural and urban design strategies for adapting to extreme hot events. In: Adapting the built environment for climate change. Elsevier, pp 253–272. https://doi.org/10.1016/B978-0-323-95336-8.00002-0

Liyanage DR, Hewage K, Hussain SA, Razi F, Sadiq R (2024) Climate adaptation of existing buildings: a critical review on planning energy retrofit strategies for future climate. Renew Sustain Energy Rev 199:114476. https://doi.org/10.1016/j.rser.2024.114476

Lowe RJ (2003) Preparing the built environment for climate change. Build Res Inform 31(3–4):195–199

Luo W, Baldwin E, Jiang AY, Li S, Yang B, Li H (2022) Effects of housing environments on COVID-19 transmission and mental health revealed by COVID-19 participant experience data from the all of us research program in the USA: a case-control study. BMJ OPEN 12(12). https://doi.org/10.1136/bmjopen-2022-063714

MacGregor C, Dowdell DC, Jaques RA, Bint L, Berg BL (2018) The built environment and climate change: a review of research, challenges and the future. BRANZ Jugendford, New Zealand

Magee JL (2000) Home as an alternative workplace: negotiating the spatial and behavioral boundaries between home and work. J Inter des 26(2):35–47. https://doi.org/10.1111/j.1939-1668.2000.tb00357.x

Marco E, Burgess S (2015) Healthy housing. In The Routledge handbook of planning for health and well-being. Routledge, pp 403–416

McNeilly Smith R, Tavares S, Stevens N (2024) Urban design and planning for extreme heat: an empirical study of built environment professionals' perceptions in South East Queensland, Australia. Cities Health 8(4):653–665. https://doi.org/10.1080/23748834.2023.2290901

Megahed NA, Ghoneim EM (2020) Antivirus-built environment: lessons learned from Covid-19 pandemic. Sustain Cities Soc 61.https://doi.org/10.1016/j.scs.2020.102350

Mihalakakou G, Souliotis M, Papadaki M, Menounou P, Dimopoulos P, Kolokotsa D, Paravantis JA, Tsangrassoulis A, Panaras G, Giannakopoulos E, Papaefthimiou S (2023) Green roofs as a nature-based solution for improving urban sustainability: progress and perspectives. Renew Sustain Energy Rev 180:113306. https://doi.org/10.1016/j.rser.2023.113306

Molaei P, Hashempour P, Tang LM (2022) Semi-open spaces of apartments considering COVID-19 pandemic: general expectations of balcony design in the post-pandemic world. Architect Eng des Manage 18(5):705–722. https://doi.org/10.1080/17452007.2021.2021385

Mood EW (2005) Fundamentals of healthful housing: their application in the 21st century. In: Unhealthy housing. Taylor & Francis, pp 303–337

Morris AEJ (2013) History of urban form before the industrial revolution. Routledge. https://doi.org/10.4324/9781315841199

Morrison DA (2016) Genealogies: pedigrees and phylogenies are reticulating networks not just divergent trees. Evol Biol 43(4):456–473. https://doi.org/10.1007/s11692-016-9376-5

Morrisson I (2021) Nietzsche, the anthropologists, and the genealogy of Trauma. Genealogy 5(1):23. https://doi.org/10.3390/genealogy5010023

Nawawi NM (2002) Islamic perspective to healthcare architecture—an overview of the medieval islamic world with case study of contemporary healthcare architecture in Malaysia. Academic Press

NIOSH (2013) Indoor environmental quality. https://www.cdc.gov/niosh/topics/indoorenv

Nix E, Ibbetson A, Zhou K, Davies M, Wilkinson P, Ludolph R, Pineo H (2024) Getting to effective housing policy for health: a thematic synthesis of policy development and implementation. Cities Health 8(3):486–503. https://doi.org/10.1080/23748834.2024.2328951

Olabi AG, Shehata N, Issa UH, Mohamed OA, Mahmoud M, Abdelkareem MA, Abdelzaher MA (2025) The role of green buildings in achieving the sustainable development goals. Int J Thermofluids 25:101002. https://doi.org/10.1016/j.ijft.2024.101002

Or Z, Seppänen A-V (2024) The role of the health sector in tackling climate change: a narrative review. Health Policy 143:105053. https://doi.org/10.1016/j.healthpol.2024.105053

Perdue WC, Stone LA, Gostin LO (2003) The built environment and its relationship to the public's health: the legal framework. Am J Public Health 93(9):1390–1394. https://doi.org/10.2105/ajph.93.9.1390

Peri G, Cirrincione L, Mazzeo D, Matera N, Scaccianoce G (2024) Building resilience to a warming world: a contribution toward a definition of "Integrated Climate Resilience" specific for buildings—literature review and proposals. Energy Build 315:114319. https://doi.org/10.1016/j.enbuild.2024.114319

Peters T, Halleran A (2021) How our homes impact our health: using a COVID-19 informed approach to examine urban apartment housing. ARCHNET-IJAR Int J Architect Res 15(1, SI):10–27. https://doi.org/10.1108/ARCH-08-2020-0159

Podrekar N, Kozinc Ž, Šarabon N (2020) The effects of cycle and treadmill desks on work performance and cognitive function in sedentary workers: a review and meta-analysis. Work 65(3):537–545. https://doi.org/10.3233/WOR-203108

Pollard CA, Morran MP, Nestor-Kalinoski AL (2020) The COVID-19 pandemic: a global health crisis. Physiol Genomics 52(11):549–557. https://doi.org/10.1152/physiolgenomics.00089.2020

Prochorskaite A, Couch C, Malys N, Maliene V (2016) Housing stakeholder preferences for the "Soft" features of sustainable and healthy housing design in the UK. Int J Environ Res Public Health 13(1). https://doi.org/10.3390/ijerph13010111

Ranson R (2002) Healthy housing: a practical guide. Taylor & Francis

Reinikainen LM, Jaakkola JJK (2001) Effects of temperature and humidification in the office environment. Arch Environ Health: Int J 56(4):365–368. https://doi.org/10.1080/00039890109604469

Reyes R, Ahn R, Thurber K, Burke TF (2013) Urbanization and infectious diseases: general principles, historical perspectives, and contemporary challenges. In: Challenges in infectious diseases. Springer New York, pp 123–146. https://doi.org/10.1007/978-1-4614-4496-1_4

Rights C (1948) Universal declaration of human rights. In: General assembly resolution 217A III
Rogers D, Power E (2020) Housing policy and the COVID-19 pandemic: the importance of housing research during this health emergency. Int J Hous Policy 20(2):177–183. https://doi.org/10.1080/19491247.2020.1756599
Rosa-Jimenez C, Jaime-Segura C (2021) Living space needs of small housing in the post-pandemic era: Malaga as a case study. J Contemp Urban Aff 6(1):51–58. https://doi.org/10.25034/ijcua.2022.v6n1-5
Saegert SC, Klitzman S, Freudenberg N, Cooperman-Mroczek J, Nassar S (2003) Healthy housing: a structured review of published evaluations of US interventions to improve health by modifying housing in the United States, 1990–2001. Am J Public Health 93(9):1471–1477
Saifudeen A, Mani M (2024) Adaptation of buildings to climate change: an overview. Front Built Environ 10.https://doi.org/10.3389/fbuil.2024.1327747
Sallis JF, Bull F, Guthold R, Heath GW, Inoue S, Kelly P, Oyeyemi AL, Perez LG, Richards J, Hallal PC (2016) Progress in physical activity over the Olympic quadrennium. The Lancet 388(10051):1325–1336. https://doi.org/10.1016/S0140-6736(16)30581-5
Samet JM, Spengler JD (2003) Indoor environments and health: moving into the 21st century. Am J Public Health 93(9):1489–1493
Sharifi A, Khavarian-Garmsir AR (2020) The COVID-19 pandemic: impacts on cities and major lessons for urban planning, design, and management. Sci Total Environ 749:142391. https://doi.org/10.1016/j.scitotenv.2020.142391
Snow J (1849) On the mode of communication of cholera. John Churchill
Sylvia JJ IV (2021) A genealogical analysis of information and technics. Information 12(3):123. https://doi.org/10.3390/info12030123
UN-Habitat (2021) Cities and pandemics: towards a more just, green and healthy future. https://unhabitat.org/sites/default/files/2021/03/cities_and_pandemics-towards_a_more_just_green_and_healthy_future_un-habitat_2021.pdf
Usher K, Bhullar N, Jackson D (2020) Life in the pandemic: social isolation and mental health. J Clin Nurs 29(15–16):2756–2757. https://doi.org/10.1111/jocn.15290
Veitch JA, Galasiu AD (2012) The physiological and psychological effects of windows, daylight, and view at home: review and research agenda. PsycEXTRA Dataset. https://doi.org/10.1037/e554552013-001
Wagner JK (2020) Health, housing, and 'direct threats' during a pandemic. J Law Biosci 7(1). https://doi.org/10.1093/jlb/lsaa022
Wågø S, Hauge B, Støa E (2016) Between indoor and outdoor: Norwegian perceptions of well-being in energy-efficient housing. J Architect Plann Res 326–346
WHO (1986) Ottawa charter for health promotion, 1986. World Health Organization, Regional Office for Europe
WHO (1988) Guidelines for healthy housing
WHO (2018) WHO housing and health guidelines
WHO (2019) Healthy housing: raising standards, reducing inequalities. Available on http://www.who.int/sustainable-development/factsheet-housing-190517.pdf
WHO (2020) WHO guidance for climate-resilient and environmentally sustainable health care facilities. World Health Organization. https://iris.who.int/bitstream/handle/10665/335909/9789240012226-eng.pdf
WHO (2021) Policies, regulations and legislation promoting healthy housing: a review
Williams KJH, Lee KE, Sargent L, Johnson KA, Rayner J, Farrell C, Miller RE, Williams NSG (2019) Appraising the psychological benefits of green roofs for city residents and workers. Urban for Urban Greening 44:126399. https://doi.org/10.1016/j.ufug.2019.126399
Wuerzer T (2014) Urban health. In: Encyclopedia of quality of life and well-being research. Springer Netherlands, pp 6835–6837. https://doi.org/10.1007/978-94-007-0753-5_3127
Younger M, Morrow-Almeida HR, Vindigni SM, Dannenberg AL (2008) The built environment, climate change, and health. Am J Prev Med 35(5):517–526. https://doi.org/10.1016/j.amepre.2008.08.017

Yuen B, Nyuk Hien W (2005) Resident perceptions and expectations of rooftop gardens in Singapore. Landsc Urban Plan 73(4):263–276. https://doi.org/10.1016/j.landurbplan.2004.08.001

Zalejska-Jonsson A, Wilhelmsson M (2013) Impact of perceived indoor environment quality on overall satisfaction in Swedish dwellings. Build Environ 63:134–144. https://doi.org/10.1016/j.buildenv.2013.02.005

Zarrabi M, Yazdanfar S-A, Hosseini S-B (2021) COVID-19 and healthy home preferences: the case of apartment residents in Tehran. J Build Eng 35:102021. https://doi.org/10.1016/j.jobe.2020.102021

Zhang D (2016) Courtyard housing in North America: Chinese design for health and happiness. Urban des Int 21(4):281–297. https://doi.org/10.1057/udi.2016.6

Chapter 10
Green and Blue Spaces: Enhancing Urban Health and Resilience

Sardar Al Imran

Abstract As urbanization intensifies, cities face mounting environmental and health challenges, exacerbated by climate change. This chapter, titled *Green and Blue Spaces: Enhancing Urban Health and Resilience*, explores how integrating green and blue spaces into urban planning can address these challenges, fostering sustainable and resilient urban environments. Green spaces, such as parks, urban forests, and green roofs, contribute to cooler temperatures, cleaner air, and increased opportunities for physical activity, while blue spaces, including rivers, lakes, and constructed ponds, play a critical role in water management, habitat support, and mental health enhancement. Emphasizing three dimensions—public well-being, climate resilience—the chapter explored the many advantages of urban green and blue areas. Methodologically, a literature review conducted, drawing on recent studies, meta-analyses, global reports data, and reputed journal articles to provide a well-rounded analysis of the topic. Studies selected based on relevance to urban health, climate resilience with additional emphasis on socioeconomically diverse urban settings. The chapter synthesized those findings, highlighting green and blue spaces as essential "natural infrastructure" for sustainable urban planning, focusing on urban climate and health. Also, this chapter sheds light on different cities' utilizing the UGBS concept around the world. By presenting green and blue spaces as foundational elements of resilient cities, this chapter contributes to the broader discourse on sustainable urban development, aligning with the book's theme of addressing urbanization, climate change, and health to build sustainable, resilient cities.

Keywords Urban climate · Health and resilience · Green–blue infrastructure · Literature review approach

---

S. Al Imran (✉)
BRAC University, Dhaka, Bangladesh
e-mail: imran.sardar@link.cuhk.edu.hk

Department of Geography and Resource Management, The Chinese University of Hong Kong, Hong Kong, China

Sustainable Development Management, Faculty of Society and Economics, Rhine-Waal University of Applied Science, Kleve, Germany

Development Studies Discipline, Khulna University, Khulna, Bangladesh

© The Author(s), under exclusive license to Springer Nature Switzerland AG 2025

A. I. Almulhim et al. (eds.), *Urbanization, Climate Change, and Health*, Climate Change Management, https://doi.org/10.1007/978-3-031-99132-5_10

Introduction

Nowadays, most people on the planet reside in urban areas. Globally, it is predicted that about 7 billion people would reside in cities by 2050. It is crucial to change urban infrastructure to provide the expanding urban population with a better standard of living and improved health. Initiatives in line with maintaining the health of the planet should be given top priority in the challenging surroundings of climate change. Climate change and fast urbanization are aggravating issues for metropolitan environments (Yang et al. 2020). Urbanization has rendered natural surfaces impermeable, altering the materials, energy, radiation, and composition of the near-surface layer (Yang et al. 2020). It modifies land surfaces, building materials, and emits anthropogenic heat, creating the urban heat island (UHI) effect. Urban temperatures are higher than rural ones (Yang et al. 2020; Yu et al. 2019). It is anticipated that the urban heat island effect, cold waves, and heat waves will be exacerbated by global climate change (Coumou and Robinson 2013; Montazeri et al. 2017). Indirect and direct negative impacts abound from urban heat islands (UHI). It contaminates metropolitan air, therefore increasing energy and water use and compromising the health and sustainability of urban people (Gunawardena et al. 2017; Yang et al. 2020). Urban green infrastructure, sometimes called urban green and blue spaces, includes water bodies and vegetation in urban areas (Wang et al. 2014; Yang et al. 2020). Other research suggests that urban blue-green space development could replace cool materials due to its low cost, low environmental impact, and political acceptability (Carvalho et al. 2017; Yang et al. 2020). Urban green space is cheaper than cool roofs at lowering outdoor temperatures, according to a global analysis of 245 major cities (Hopkins et al. 2022). Larger rising cities inevitably stress the surrounding areas and present huge sustainable development issues (Nautiyal et al. 2023). Modern technologies like public transit, green buildings, and clean lakes with water management systems work best in sustainable, green, blue cities. In these places, changes to the way people live include clean, filtered water, bikes, walking, and equipment that uses less energy (Beatley Washington and Aru 2000). Water features, open spaces, and verdant vegetation that enhance urban environments are together known as urban green and blue areas (UGBSs) (Nautiyal et al. 2023).

Distribution, location, accessibility, and proximity also influence UGBS performance (Nautiyal et al. 2023). Urban cooling systems and pollution reduction suffered from green and blue space density and connectedness (Vargas-Hernández et al. 2018). Among the many ecological services UGBS offers are recharging groundwater, controlling microclimates, noise reduction, air purification, and recreational possibilities (Bolund and Hunhammar 1999). UGBSs improve access to residential and commercial property, self-esteem, water, temperature control, beauty, wildlife food and habitat (Elmqvist et al. 2016). Urban green areas (UGS) withstand natural calamities and disturbances, therefore adjusting to urbanism and climate change (Vargas-Hernández et al. 2018). The local climatic energy budgets are affected by its higher albedo through evapotranspiration and shadow (Nautiyal et al. 2023).

Vegetation's cooling effects can lower temperatures by 2–8 °C compared to developed environs (Nautiyal et al. 2023). Areas with many trees receive less diffuse and direct sunlight than those with herbaceous plants, asphalt, or other man-made surfaces (Nautiyal et al. 2023). Figure 10.1, A modified paradigm from Hunter et al. (2023) outlines the health outcomes and co-benefits of UGBS. To fill the huge gaps in information on UGBS contribution's types, determinants, and magnitude, the framework was created. In this modified paradigm, the author integrated health and environmental outcomes and showed a loop of how UDBS affects urban health and environment. Parks, gardens, rivers, and lakes are the cornerstone of urban ecosystems, delivering essential ecosystem services that improve urban living. These places filter pollutants, reduce the urban heat island effect with shade and evapotranspiration, and manage stormwater. Urban trees can cool surrounding temperatures by 5 °F, safeguarding vulnerable populations during extreme heat events. These green and blue spaces are crucial to climate/environmental resilience, especially as cities confront more climate change-related flooding and heatwaves. Permeable pavements and rain gardens absorb and filter runoff, minimizing flooding concerns. Public health outcomes are linked to urban space access. Residents near parks are more likely to exercise, reducing the risk of obesity and related chronic diseases. Access to green places reduces melancholy and anxiety, improving mental health. Nature's tranquility improves health and quality of life. Healthy communities are more likely to participate in climate change activities like tree planting and conservation, which improves urban green and blue areas. However, urban development that prioritizes these natural places improves public health and climate resilience. Urban green and blue spaces, climate resilience, and health are interconnected, therefore integrating them into sustainable urban development methods benefits the environment and urban inhabitants.

Aim and Objectives

With an emphasis on public health, climate change and resilience, and cities perspective of UGBS, this chapter seeks to emphasize the many advantages of urban green and blue areas. It also elucidates a revised framework that demonstrates the interconnection of green and blue spaces along additional dimensions and offers policy suggestions for a developing nation to solve the UGBS.

Materials and Methods

Data required for evaluating green and blue spaces concerning health, resilience, and urban environments were sourced from electronic databases, including Google Scholar, PubMed, Web of Science, and ScienceDirect. The Boolean keywords selected for their relevance to urban health and environmental stability, with a focus

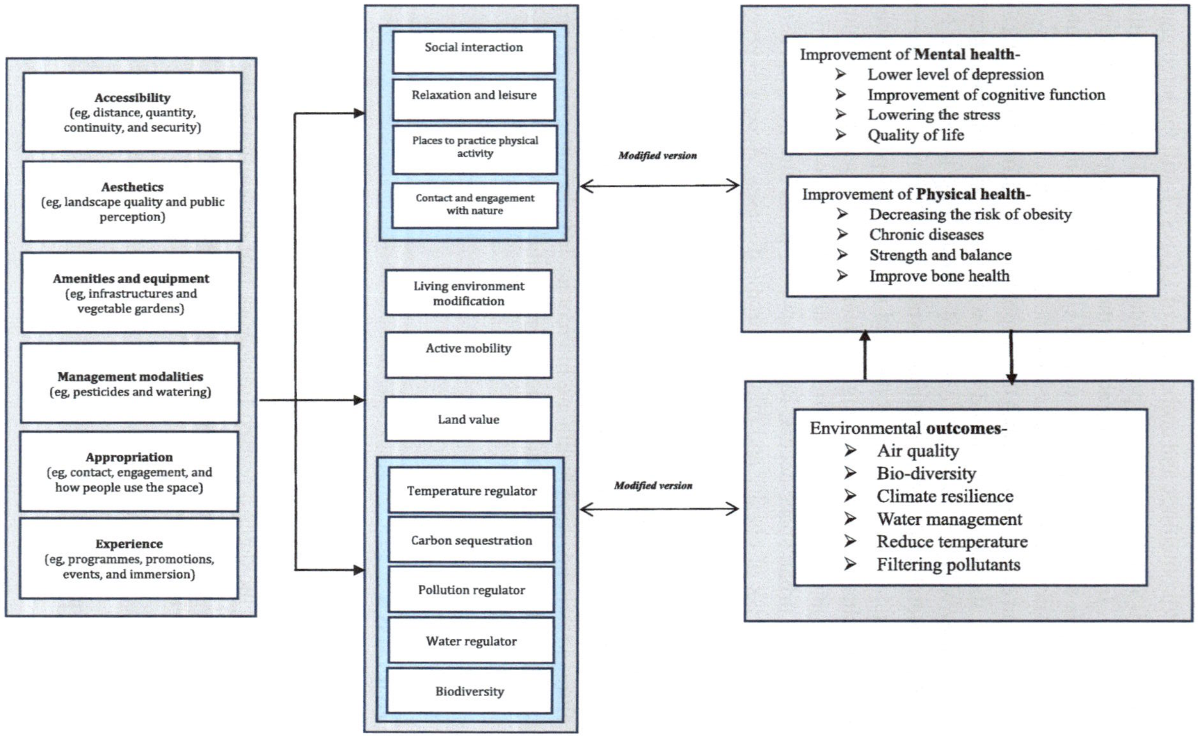

Fig. 10.1 Modified framework on UGBS characteristics, and health outcomes with other co-benefits. *Source* Modified framework, derived from Hunter et al. (2023)

on socioeconomically diverse urban settings, including health and green space, air quality and blue spaces, health benefits, green and blue infrastructure, and environmental and ecological benefits. This chapter utilized published journal articles in English on related topics. The items examined were sourced from published journals on the subject, without any restrictions on publication year. A total of 282 articles and abstracts were generated by the search. The published sources were filtered for relevance, ensuring that only the most appropriate were selected for this chapter. Consequently, the chapter focused on re-evaluating established claims and reports instead of gathering and analyzing new field data.

Defining Green and Blue Spaces in Urban Contexts

Green–blue infrastructures or spaces are the most important of the many elements that impact urban expansion, which leads to a constant evolution of the wide geographic dispersion of landscape areas and the broader urbanized structure (Sunita et al. 2023). Modern urban landscapes depend on UGBS for environmental sustainability, human well-being, and socio-economic development. Green–blue spaces in cities include vegetated regions and waterways. Blue-green space improves biodiversity, climate change adaptation, and human well-being in cities (Ebi and Bowen 2023). The whole pattern of greenery and blueness that makes up a city's landscape, including things like water and trees, is collectively known as green–blue regions (Eggimann 2022). Examples of "green space" include mountains, forests, farms, grasslands, biological corridors, expansive green areas, protective green belts, public open spaces, and green roofs in cities (Wang et al. 2022). Both flat and linear bodies of water, such rivers and marshes, are included in the "blue space" category. A regulating component strongly linked to urban sustainable development and residents' lives; urban green–blue space is an inseparable "organic aggregation" of the urban ecological system (Rana and Ilina 2021). These spaces provide aesthetic value, relaxation and wellbeing, climate regulation and reduction of urban heat island effect, disturbance prevention, purification, especially of water and air, and habitat for urban biodiversity (Ebi and Bowen 2023). If we want ecological stability, robust systems, improved urban functions, and high-quality urban development, we must increase the percentage of green and blue areas in cities (Plieninger et al. 2022).

Typologies Urban Green and Blue Spaces and Their Roles

Public Parks and Gardens

Public parks, such as Central Park in New York, serve as multifunctional spaces providing ecological and social benefits. They regulate urban temperatures through evapotranspiration and provide psychological relief to residents. Parks contribute to urban biodiversity by serving as habitats for birds and pollinators (Elmqvist et al. 2015). Public parks have long been a vital component of urban landscapes, providing recreational spaces and enhancing the aesthetic value of cities. Historically, parks have undergone numerous planning and design phases, evolving from the industrial revolution through to the mid-twentieth century. The formalized approach to park planning, particularly through standardized methods such as those outlined in the Stephenson–Hepburn Metropolitan Regional Plan in Perth, has aimed to ensure adequate green space by dedicating at least 10% of new subdivisions to public parks (Middle et al. 2014). Although this standardization is effective in ensuring that there is enough parkland, it frequently results in areas that are boring and dominated by structured leisure pursuits, such sports fields. This architectural approach limits the parks' capacity to sustain biological processes and satisfy the community's varied recreational requirements (Middle et al. 2014). Because they lack the diversity and natural elements that would improve their usefulness, parkland in suburbs, particularly those designed according to the "Garden Cities" concept, might feel practical and uninviting (Grose 2009; Middle et al. 2014). Parks' ability to satisfy local ecological and social demands is limited when the community is not included in the planning process since choices are frequently made by organizations that are not directly involved with the park's everyday users (Middle et al. 2014). But community gardens serve purposes beyond just food production; they also provide environmental education, civic involvement, and neighborhood revitalization (Hou et al. 2009). A citizen-driven movement to remedy city planners' perceived shortcomings in developing acceptable public spaces usually resulted in these gardens. Community gardens benefit from fostering social relationships and community. They encourage communities to engage, building social links and bridges (Kaźmierczak 2013). This is particularly significant in urban environments, where public spaces are frequently constrained, and social isolation may pose a challenge. In contrast to formal parks that often reinforce existing relationships, community gardens promote interactions among diverse social groups and strengthen the social fabric of neighborhoods (Cattell et al. 2008). Many community gardens also organize social events that further promote community building and create opportunities for cross-cultural exchange (Glover 2003).

Urban Forests and Woodlands

Urban forests like Rio de Janeiro's Tijuca Forest help to reduce carbon emissions, improve air quality, and lower risk of climate change. They help to reduce noise levels and provide cultural activities. Moreover, their ecological role in preventing soil erosion and preserving many species is vital for urban resilience (McPherson et al. 2016). Urban forests and woodlands offer a wide range of benefits, extending from intangible psychological and aesthetic values to tangible environmental improvements (Tyrväinen et al. 2005). Historically, urban forests have primarily contributed to health, aesthetics, and recreation in industrialized cities, while also offering subsistence benefits such as food, fodder, and timber (Tyrväinen et al. 2005). Today, urban woodlands symbolize personal, local, community, and cultural meanings, providing environments that help in stress recovery, memory revival, and children's environmental education in otherwise artificial urban settings (Tyrväinen et al. 2005). Aesthetically, trees play a significant role in urban planning by shaping open spaces, integrating built structures with the environment, and enhancing the visual quality of landscapes through diversity in color, form, and texture (Kaplan and Kaplan 1989; Tyrväinen et al. 2005). By reducing air temperatures, improving air quality, and therefore lessening the urban heat island effect, urban woods also significantly help to enhance urban climates (Eliasson 2000; Tyrväinen et al. 2005). Research indicates that parks may lower urban air temperatures by 2–3 °C, especially when the park area surpasses 10 ha (Sukopp 1998). Economically, while urban forests offer limited consumptive goods like timber and mushrooms, their non-consumptive values such as recreational use, property value enhancement, and environmental regulation—are substantial and increasingly recognized (Tyrväinen and Tutkimuskeskus 1999). Hedonic pricing studies demonstrate that proximity to forests can significantly raise property values, reflecting people's willingness to pay for green amenities (Tyrväinen and Miettinen 2000). Despite these extensive benefits, urban forests are often undervalued in city planning, leading to conflicts between development pressures and green-space preservation (Tyrväinen et al. 2005). As cities densify, the challenge remains to strategically integrate urban woodlands into urban policies that balance ecological, social, aesthetic, and economic needs (Germann-Chiari and Seeland 2002).

Green Roofs and Walls

Reducing surface and ambient temperatures of green roofs and walls helps to cool cities. Incorporating biodiversity into densely populated urban areas while effectively managing runoff is possible, as demonstrated by projects like Milan's Bosco Verticale (Dunnett and Kingsbury 2008). Another, primary advantages of green roofs are stormwater management, as they capture precipitation in the growing media and vegetation, preventing excess runoff. Green roofs may decrease stormwater runoff by 60–100%, contingent upon the design and plant species used (Denardo et al. 2005). This substantial decrease in runoff alleviates the adverse effects of impermeable surfaces, including floods and combined sewer overflows (Getter and Rowe 2006). Green roofs provide delayed runoff by absorbing water, which is gradually released through evapotranspiration, thereby reducing the burden on municipal stormwater systems (Moran et al. 2005). In addition to water management, green roofs can provide energy conservation benefits by improving the thermal performance of buildings. Green roofs diminish solar heat input via insulation and shade, which lowers interior temperatures and, in turn, lessens the demand for air conditioning (Peck et al. 1999; Hien Wong et al. 2003). Research has demonstrated that green roofs can reduce the temperature of buildings by 3–4 °C, resulting in an annual energy consumption reduction of up to 15% (Hien Wong et al. 2003). Green roofs also protect standard roof materials from UV rays and weather changes, which makes them last longer by making roofing coverings last longer (Getter and Rowe 2006).

Urban Lawns and Medows

Urban lawns are vital to urban green infrastructure, delivering ecosystem services that improve the environment and human health. Urban lawns cool cities by evapotranspiration and shade, reducing the urban heat island effect (Armson et al. 2013). By encouraging infiltration and lowering surface runoff, lawns help replenish groundwater reserves and regulate urban floods (Amani-Beni et al. 2018). Furthermore, lawns offer home for a variety of tiny species, including insects like bees and butterflies, which perform critical roles in pollination (Gaston et al. 2005; Ignatieva and Hedblom 2018). Lawns are also cultural and recreational venues in many cities that host events like picnics, sports, and community meetings (Ignatieva 2017). In addition to improving mental and physical well-being, these green areas also provide visual value to metropolitan landscapes, fostering a feeling of place and social cohesiveness in urban areas (Johnson 2013). However, urban meadows are crucial to urban green infrastructure, delivering ecosystem services that can improve environmental quality and human well-being. Unmown meadows with grass and blossoming plants can improve stormwater management, biodiversity, beauty, and social value. Mown grasses lack biodiversity compared to urban meadows. Studies demonstrate that urban park visitors prefer meadows with higher structural complexity and species richness, making them valuable (Southon et al. 2017; Brose 2003). More visually appealing meadows with denser vegetation or a diversity of plant species were consistently evaluated higher. In addition to increasing the area's aesthetics, structural diversity feeds more insects and other species, which helps pollinate and control pests (Gaston et al. 2005; Ignatieva and Hedblom 2018). Furthermore, urban meadows may promote human health. Meadows and other urban landscapes improve mental health, reduce stress, and stimulate exercise (Fuller et al. 2007).

Urban River

Like the Seine in Paris, urban rivers provide both visual and recreational value in addition to important stormwater control purposes. Projects for restoration, like the Cheonggyecheon Stream in Seoul, show how blue spaces may be restored to enhance urban quality of living and environmental health (Bae 2011). As cities expand, rivers change in a number of ways including culverting, channel alterations, and increasing impermeable surfaces. These alterations disturb the natural hydrological cycles, and urban rivers' ability to provide essential ecosystem services like flood control, water filtration, and recreational areas is degraded (Grimm et al. 2008; Baron et al. 2002; Groffman et al. 2003). Urbanization has also undermined rivers' capacity to preserve natural processes, leading to declining water quality from higher concentrations of contaminants like metals, fertilisers, and hydrocarbons (House et al. 1993; Landscape et al. 2001). Urban river restoration may also restore their ecological functions. Water, food, flood control, recreation, and biodiversity and nutrient cycling services are included. These ecological services become progressively more important for urban resilience, health, and quality of life as cities increase (Everard and Moggridge 2012). Restoration initiatives have demonstrated that urban rivers may regenerate the ecosystem and improve public health. The restoration of Mayesbrook Park in East London has increased biodiversity and created green areas for leisure and community interaction (Everard 2009). Restorations may also reduce urban floods and bring aesthetic, cultural, and educational advantages to locals (Everard and Moggridge 2012). These initiatives demonstrate how river restoration may change urban environments and provide ecological and social benefits.

Wetlands

By acting as organic sponges, wetlands reduce the danger of flooding and filter pollutants. The East Kolkata Wetlands, which promote biodiversity and provide employment possibilities, are one example (Ghosh 2005). Urban green and blue space ratios vary widely, reflecting local environmental legislation, geography, and socioeconomic interests. Urban wetlands can improve community well-being via socio-environmental advantages. As recreational, relaxing, and social settings, these ecosystems promote mental and physical wellness. Interacting with natural habitats, especially water-based ones, has been shown to alleviate stress and improve mental health (Hartig et al. 2003; Korpela and Hartig 1996). People often like settings with water elements because they provide chances for both passive and active leisure, which increases emotions of wellbeing and connection (Carter 2015; Ulrich 1993). Wetlands' sense of place is especially significant in cities because they provide a haven from the constructed world, fostering psychological healing and emotional fulfilment (Carter 2009; Carter 2015). Urban wetlands are also important places for culture and spirituality. Wetlands are culturally significant to many people, providing

Table 10.1 Percentage of urban green and blue spaces across cities

City	Green space (%)	Blue space (%)	Source
Amsterdam, Netherlands	21	11	Kabisch et al. (2015)
Singapore	47	3	Tan et al. (2013)
Dhaka, Bangladesh	19	5	Shammi et al. (2023)
New York City, USA	27	9	Gómez-Baggethun and Barton (2013)
Tokyo, Japan	20	6	Otsuka et al. (2021)

a link to history and identity. Indigenous people have deep spiritual connections to wetland habitats, which support traditional customs and give them a feeling of continuity and inclusion (Mooney and Tan 2012). Wetlands improve health by fostering social cohesiveness and community resilience via spiritual and cultural connections (Carter 2015). In addition to its obvious cultural and health advantages, urban wetlands promote community participation in conservation initiatives. In addition to enhancing the biological health of these areas, taking part in wetland management and restoration projects builds community links and fosters a feeling of civic duty and pride (Moore et al. 2006; Miles et al. 1998). It has been shown that when people actively participate, their mental health improves because they feel empowered by their role in preserving the environment and promoting communal well-being (Miles et al. 2000; Wood an Giles-Corti 2008). Urban wetlands support recreational activities, enhance environmental quality, and cultivate a sense of place and cultural identity, rendering them essential assets for urban populations. Table 10.1 presents data on the percentage of urban green and blue space coverage in selected cities.

Converging Influences on Health, Climate Change, and Cities with UGBS

Enhanced Relaxation, Recovery and Physical Activity

Contact with nature and green spaces is associated with numerous well-documented health benefits, including mental restoration and relaxation (Hartig et al. 2014; Kabisch et al. 2015). Some research suggests that those experiencing high levels of stress may benefit from exposure to nature, such as views of green space, by entering a more positive mental state mentally fatigued people feel better when exposed to natural stimuli (Kabisch et al. 2015). "Significant evidence speaks to the potential benefits of contact with nature for avoiding health problems traceable to chronic stress and attentional fatigue," according to Hartig et al. (2014). However, they also noted that most of the earlier research only showed the short-term restorative benefits of one-time nature experiences. A UK study employed wearable sensors to show how a brief walk in a green space and sitting near the canals affected brain activity, which may improve relaxation and repair (Aspinall et al. 2015). Additionally, it has been demonstrated that walking in natural settings, such as parks and lakes, yields greater immediate cognitive advantages than walking in an urban residential setting (Gidlow et al. 2016). Gardening reduces acute stress faster than reading, according to cortisol tests (Kabisch et al. 2015). Evidence also suggests that spending time in nature can help ease depressive symptoms and lower activity in the subgenual prefrontal cortex (Bratman et al. 2015).

Physical activity improves cardiovascular, mental, neurocognitive, and general well-being and prevents obesity, cancer, and osteoporosis (Owen et al. 2010). Creating visually appealing and easily accessible urban spaces could promote outside time and physical activity (Bedimo-Rung et al. 2005). Physical engagement in green–blue spaces, such as parks, gardens, or lakes, may be very helpful for city dwellers with mental health issues (Kabisch et al. 2015). Green–blue spaces that make outdoor activity fun and easy may also help other populations or subgroups. Using satellite imagery, GPS, and accelerometer data collected from kids in multiple Californian towns, Almanza et al. (2012) showed that higher levels of residential GBS were positively correlated with more strenuous forms of physical exercise. Researchers in Sweden (de Jong et al. 2012) discovered that people were more active when they had access to large, clean water sheds, high-quality open spaces, and other similar amenities. Figure 10.2 illustrates how proximity to green and blue spaces correlates with health benefits, emphasizing the role of accessible green and blue spaces in promoting physical activity and reducing mental stress.

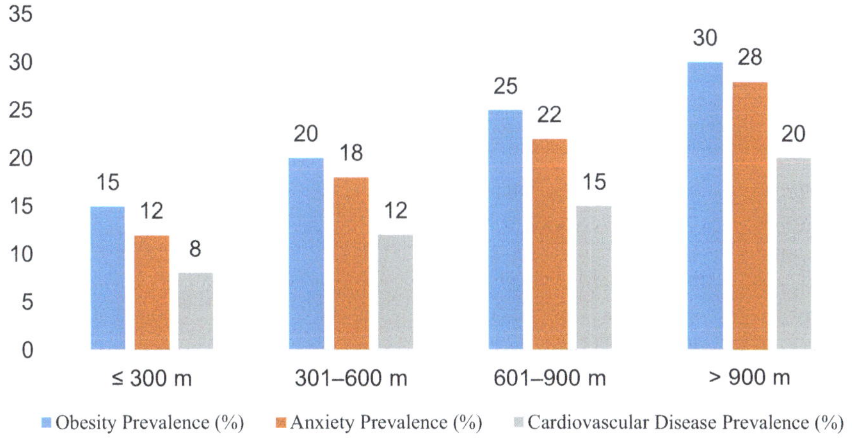

Fig. 10.2 Relationship between proximity to UGBS and percentage of health benefit. *Source* WHO (2016)

Climate Regulation and Urban Cooling

Urban regions frequently have hotter temperatures than their rural counterparts because of their dense development and sparse vegetation. Through mechanisms like evapotranspiration and the creation of shade, vegetation in green areas lowers ambient temperatures and cools the air (Bowler et al. 2010). For urban development, Blue-Green Infrastructure (BGI) has a lot of potential. Its implementation in cities not only tackles upcoming issues but also complies with the requirements of improving the urban atmosphere (Krishnan and Shanthi Priya 2022). Temperature reductions of up to 2 °C are possible with the BGI solutions (Przestrzelska et al. 2024). Establishing urban gardens for vegetable production and managing the city's temperature with plants helps to maintain a steady flow of energy and water (Przestrzelska et al. 2024). Designed wetlands can generate nutrient-dense effluent water and organic matter to irrigate energy crops, therefore lowering energy use. Urban wetlands close by could lower ambient temperatures by more than 2.65 °C (Xue et al. 2019). BGI may coexist with other energy sources, such photovoltaic panels on large green rooftops (Pochodyła et al. 2021). BGI could cut cooling power consumption. In buildings with Blue Infrastructure (BI), Green Infrastructure (GI), and Blue-Green Infrastructure (BGI), BI saved the most cooling energy (8%), followed by BGI (6.71%) and GI (4.73%) (Ayad et al. 2019). Moreover, enhanced water coverage results in a higher potential for energy savings. According to the findings, it is advisable to prioritize the use of BI for sustainable and resilient urban planning focused on minimizing energy consumption (Przestrzelska et al. 2024). If BI fails, GI is recommended. The combination of BI and GI does not save energy but may be more attractive (Przestrzelska et al. 2024). In Table 10.2, different countries show the significant reduction of temperature and the energy savings for cooling by using the BGI.

Table 10.2 Percentage of temperature reduction and cooling in terms of BGI usage in different countries

Country	Infrastructure type	Temperature reduction (°C)	Energy savings for cooling (%)	Sources
Singapore	Green infrastructure	1.5–2.0	7–12	Teo et al. (2022)
China	Blue-green infrastructure	2.0–3.5	12–18	Sun et al. (2019)
India	Green infrastructure	0.8–2.5	5–10	Menon and Sharma (2021)
Australia	Blue infrastructure	1.0–3.0	8–14	Coutts et al. (2013)
United States	Blue-green infrastructure	1.8–2.8	10–16	Klompmaker et al. (2023)

Improvement of Urban Air Quality and UGBS

Though they don't look as good as trees on a local or city scale, studies have shown that urban green walls and roofs significantly reduce pollution (Speak et al. 2012). Baik et al. (2012) demonstrated that green roofs, characterized by the presence of vegetation, have led to a notable reduction of more than 30% in air pollution concentration by enabling a cooling intensity of 2 °C at breathing levels within a canyon. As cooler air descends through the canyon, the cooling gradient has exacerbated the canyon vortices and created a greater capacity for dispersion. Green roofs have been demonstrated to significantly reduce pollutants (by about 24%) when placed close to traffic corridors and vehicle emissions (Speak et al. 2012). According to data gathered from a trial experiment on a Singaporean green rooftop, SO_2 was significantly reduced by 37%, but nitrogen dioxide showed mixed effects (Tan and Sia 2005). Additionally, it has been demonstrated that urban blue spaces, such as ponds, lakes, canals, rivers, ports, and fountains, improve the local microclimate, particularly by lowering the ambient temperature by 2–6 °C by evaporative cooling (Dhyani et al. 2022). Blue spaces encourage adjacent plants and foliage transpiration, cooling the atmosphere. Urban studies benefit from understanding the dynamic interactions between the lake surface and urban canopy layer, which affect latent heat flux and air mass movement (Dhyani et al. 2022). When wind strikes the lake's surface, water droplets are carried by the wind and dispersed throughout the urban environment downstream, lowering the temperature and raising the absolute humidity (Masiero and Souza 2015). Computational experiments show that the downwind bank of a 230-m-wide river has 0.4 g/m^3 more absolute humidity and 0.5 °C lower temperature than the upwind bank (Masiero and Souza 2015).

Enhancement of UGBS in Climate Resilience

Urban green–blue spaces (GBSs) can facilitate climate change adaptation in cities through risk reduction, adaptation strategies, and recovery efforts. Urban resilience refers to the capacity of cities to anticipate, absorb, recover from, and adapt to climate-related disasters while maintaining essential services (IPCC 2015). Ecosystem services and cost-effective, nature-based climate vulnerability solutions from GBSs increase these capabilities. Urban microclimate management is key to climate change resistance. Urban environments may grow hotter than rural places because to less foliage and more heat-absorbing surfaces like asphalt and concrete; GBSs minimize this (Bowler et al. 2010). During periods of intense heat, evapotranspiration and shade from trees and plants in green areas reduce ambient temperatures and guard against heat-related diseases (Gill et al. 2017). Having blue places like rivers and lakes around makes it less likely that towns will flood when it rains. Soak up rain through plants and water sources in the wild. This slows flow and makes it easier for city sewage systems to work (Nesshöver et al. 2017). Wetlands, for example, act as a natural

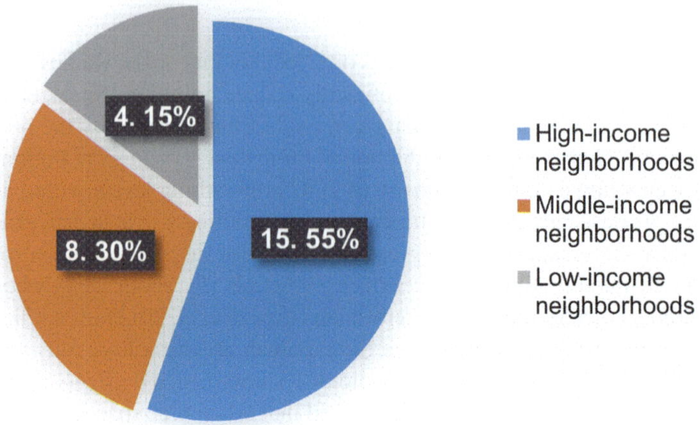

Fig. 10.3 Average access to green/blue space (square meters per capita). *Source* Rigolon et al. (2018)

sponge, absorbing rainwater and releasing it gradually to prevent flooding. Furthermore, plants and soil act as organic filters, removing contaminants from runoff and improving water quality (Markogianni et al. 2014). An additional significant factor is the potential of GBSs to improve carbon sequestration and air quality. Vegetation and urban forests play a significant role in mitigating climate change through the sequestration of carbon dioxide, thereby decreasing greenhouse gas concentrations (Nowak et al. 2006). These things clean the air and make cities better places to live, even if they make the air quality worse (Haaland and van den Bosch 2015). In Fig. 10.3, it shows significant differences in access based on neighborhood income levels.

Successful Cities with Urban Green and Blue Spaces

China's Sponge Cities Program

By participating in the "Sponge Cities" program, which seeks to enhance urban infrastructure's absorption capacity, Chinese cities have become known for their green–blue projects. Twenty percent of city lots must be filled up with features shaped like sponges. Achieving efficient absorption or utilization of approximately 80% of precipitation by 2030 is the principal objective of this program (Stadler 2021). The main goal of the sponge city program is to incorporate natural processes into urban infrastructure to retain, filter, penetrate, and absorb rainfall, much like a sponge.

Initiated by the Chinese government in 2015, this project has as its major objectives improved urban flood control, water quality, water conservation, and environmental preservation (Li et al. 2017). Starting with the choice of first pilot towns in 2015 and then adding additional cities in 2016, China's sponge city initiative is being carried out in stages. These cities get significant annual financial assistance from the central government—between 400 and 600 million Chinese Yuan—along with matching money acquired via public–private partnerships (PPP) (Li et al. 2017). This money helps to put creative water management techniques—permeable pavements, green roofs, rain gardens, and infiltration swales—which taken together improve the sustainability of metropolitan environments—into use (Li et al. 2017).

Singapore's Gardens by the Bay

One of the most ambitious urban greening initiatives worldwide is turning a reclaimed site into a famous green park that strikes a balance between public involvement and environmental needs Gardens by the Bay in Singapore. Through the integration of native plants with giant tree-like structures known as Super trees, this program seeks to enhance biodiversity and reduce the impact of the urban heat island (Friess 2017). The lake system shows the parks and green areas play in stormwater control as a runoff filter. The project greatly enhances the urban scene of the city by including several aspects of environmental, cultural, and leisure sustainability. One significant advantage of Gardens by the Bay is its contribution to improving environmental sustainability. The inclusion of features such as Dragonfly and Kingfisher Lakes enhances water quality and facilitates natural filtration, thereby aiding in the management of stormwater runoff and supporting the ecological health of the Marina Reservoir (Kenneth et al. 2011). The lakes function as eco-filters, capturing runoffs and employing aquatic plants to purify the water prior to their return to the reservoir, thus improving the local ecosystem (Kenneth et al. 2011). Furthermore, Gardens by the Bay incorporates creative solutions that provide both practical and aesthetic advantages to adapt to Singapore's tropical environment. The Conservatories' design provides an educational and sensory experience by simulating various climatic conditions, such as Mediterranean and tropical montane climates (Kenneth et al. 2011). These regulated habitats showcase many plant species that would not thrive in Singapore's natural temperature, drawing local and international tourists. Sustainable cooling methods reduce energy use and meet Singapore's energy efficiency targets (Kenneth et al. 2011).

New York City's High Line

An elevated linear park built on an abandoned train track; the High Line in New York City provides a rare chance to see nature among the busy metropolitan scene. Benefits of this park's regeneration of the surroundings include improved air quality, a refuge for wildlife, and a boon to the local economy (Littke et al. 2016). By selecting native plant species, The High Line supports urban biodiversity; community events unite people and get them active. The restoration of the High Line has boosted economic activity, which may be its most significant impact, even if it has effectively linked many New Yorkers with their city (Ashley Hassell 2023). Estimates made when the High Line first proposed reusing the building put the city's tax revenue at about $250 million more over 20 years from "increased local real estate values"; however, those numbers grossly understated the project's eventual success (Kate Ascher et al. 2015). The High Line has generated an extra $900 million in tax income for New York City and created almost $2 billion in new economic activity in the city's "High Line district" (Kate Ascher et al. 2015). Through distinctive city views and access to green space that may be hard to locate amid the city's concrete walkways, the High Line has let New Yorkers escape the fast-paced New York lifestyle (Ashley Hassell 2023). The High Line is mainly to blame for the revitalization of Chelsea and the huge rise in economic activity, which has brought in over $900 million in tax money for the city (Ashley Hassell 2023).

France ÉcoQuartiers Programme

Established in France under the ÉcoQuartiers Program are urban eco-districts for instance, this initiative has turned Paris's 17th district, Clichy-Batignolles (Jóźwik 2020). The eco-district consists of public and private sectors; in the former, rainfall flows down green rooftops and into the sewer system, while in the latter some of it is used for flushing toilets. France employs permeable surfaces like bike lanes and walkways to promote environmentally friendly urban practices (Langeveld et al. 2022). For instance, the Clichy-Batignolles eco-district runs 83% of its heating need on geothermal energy and generates a significant amount of its power demand using solar rooftops. These energy plans complement the green and blue agendas by helping the neighborhood to have less carbon impact (Amy et al. 2019). Furthermore, included in the eco-district are stormwater management techniques and rainfall collecting systems including biotope basins and built wetland ditches, therefore enhancing water quality and biodiversity in addition to helping to reduce floods. These qualities demonstrate how well the program may integrate urban resilience with ecological purpose, therefore enabling other communities striving to include climate adaptation strategies into their urban growth (Amy et al. 2019). Moreover, the project pays for social sustainability of high importance such that the eco-districts foster community engagement and provide affordably priced homes. Combining houses, social housing, and capped-rent buildings ensures that a broad range of individuals might benefit from the ecologically friendly, robust neighborhoods. By encouraging socioeconomic and generational diversity, the ÉcoQuartier Program allows communities to be long-term resilient, therefore increasing solidarity and raising resident quality of life (Amy et al. 2019).

Conclusion

Green and blue spaces are vital infrastructure for social justice, climate change and resilience, and health; they are more than simply beautiful embellishments to cityscapes. This chapter reviews the benefits to mental and physical health, resilience to climate change, and various cities, using evidence from scientific literature and real-world case studies. Legislators and urban planners should prioritize parks and other green spaces in future municipal planning systems with an eye on equity and accessibility. With well-thought-out rules and community input, cities can design green and blue spaces that serve a variety of purposes, mitigate the effects of climate change, and construct resilient urban ecosystems. Also, another concerning area of investigation could be the interconnection of green–blue infrastructure into the food-water-energy nexus. Moreover, future research should also shed light on the trade-offs and synergies between the green–blue infrastructure and its function. As urban issues evolve, it is imperative that future research and planning refine these tactics to guarantee that parks and other green spaces are adaptable and easily accessible.

Acknowledgements Artificial Intelligence (AI) was utilized in various stages of developing this book chapter, primarily to enhance writing quality, ensure clarity. Firstly, the author manually paraphrased the writing and based on his own writing author used various AI Tools such as Grammarly and Quillbot in refining grammar, sentence structure, and readability. The use of AI was justified by its ability to optimize time efficiency, improve writing precision, and ensure consistency throughout the chapter. Grammarly and Quillbot were particularly beneficial in maintaining an academic tone while eliminating grammatical errors and redundancy. Furthermore, AI-assisted content refinement helped in improving the overall readability, ensuring that complex ideas were effectively communicated to a diverse audience. By integrating AI into this writing process, the research remained focused, well-structured, and aligned with academic standards. The use of AI was not intended to replace critical thinking or original research but rather to augment the efficiency and quality of scholarly work. This responsible application of AI contributed to producing a well-articulated and impactful write up especially this chapter.

References

Almanza E, Jerrett M, Dunton G, Seto E, Ann Pentz M (2012) A study of community design, greenness, and physical activity in children using satellite, GPS and accelerometer data. Health Place 18(1):46–54. https://doi.org/10.1016/j.healthplace.2011.09.003

Amani-Beni M, Zhang B, di Xie G, Xu J (2018) Impact of urban park's tree, grass and waterbody on microclimate in hot summer days: a case study of Olympic Park in Beijing, China. Urban For Urban Greening 32:1–6. https://doi.org/10.1016/j.ufug.2018.03.016

Amy O, Emilie G, Isabelle T (2019) Paris

Armson D, Stringer P, Ennos AR (2013) The effect of street trees and amenity grass on urban surface water runoff in Manchester, UK. Urban For Urban Greening 12(3):282–286. https://doi.org/10.1016/j.ufug.2013.04.001

Ashley Hassell (2023) How the high line reshaped New York City. www.bdcnetwork.com/blog/high-line-effect-placemaking-economic-

Aspinall P, Mavros P, Coyne R, Roe J (2015) The urban brain: analysing outdoor physical activity with mobile EEG. Br J Sports Med 49(4):272–276. https://doi.org/10.1136/bjsports-2012-091877

Ayad A, Fahmy M, Kamel W (2019) Urban green and blue infrastructure simulation in a changing climate from microclimate to energy consumption: a case study in Alexandria, Egypt. Build Simul Conf Proc 5:3499–3507. https://doi.org/10.26868/25222708.2019.210676

Bae H (2011) Urban stream restoration in Korea: design considerations and residents' willingness to pay. Urban For Urban Greening 10(2):119–126. https://doi.org/10.1016/j.ufug.2011.02.001

Baik JJ, Kwak KH, Park SB, Ryu YH (2012) Effects of building roof greening on air quality in street canyons. Atmos Environ 61:48–55. https://doi.org/10.1016/j.atmosenv.2012.06.076

Baron JS, LeRoy Poff N, Angermeier PL, Dahm CN, Gleick PH, Hairston NG, Jackson RB, Johnston CA, Richter BD, Steinman AD (2002) Meeting ecological and societal needs for freshwater. Ecol Appl 12(5):1247–1260. https://doi.org/10.1890/1051-0761(2002)012[1247:MEASNF]2.0.CO;2

Beatley Washington T, Aru C (2000) Learning from European cities

Bedimo-Rung AL, Mowen AJ, Cohen DA (2005) The significance of parks to physical activity and public health a conceptual model.https://doi.org/10.1016/j.ampre.2004.10.024

Bolund P, Hunhammar S (1999) Ecosystem services in urban areas. In: Ecological economics, vol 29

Bowler DE, Buyung-Ali L, Knight TM, Pullin AS (2010) Urban greening to cool towns and cities: a systematic review of the empirical evidence. Landscape Urban Planning 97(3):147–155. https://doi.org/10.1016/j.landurbplan.2010.05.006

Bratman GN, Hamilton JP, Hahn KS, Daily GC, Gross JJ (2015) Nature experience reduces rumination and subgenual prefrontal cortex activation. Proc Natl Acad Sci USA 112(28):8567–8572. https://doi.org/10.1073/pnas.1510459112

Brose U (2003) Bottom-up control of carabid beetle communities in early successional wetlands: mediated by vegetation structure or plant diversity? Oecologia 135(3):407–413. https://doi.org/10.1007/s00442-003-1222-7

Carter ME (2009) Health and the nature of urban green spaces. https://ro.ecu.edu.au/theses

Carter M (2015) Wetlands and health: how do urban wetlands contribute to community well-being? In: Wetlands and human health. Springer Netherlands, pp 149–167. https://doi.org/10.1007/978-94-017-9609-5_8

Carvalho D, Martins H, Marta-Almeida M, Rocha A, Borrego C (2017) Urban resilience to future urban heat waves under a climate change scenario: a case study for Porto urban area (Portugal). Urban Climate 19:1–27. https://doi.org/10.1016/j.uclim.2016.11.005

Cattell V, Dines N, Gesler W, Curtis S (2008) Mingling, observing, and lingering: everyday public spaces and their implications for well-being and social relations. Health Place 14(3):544–561. https://doi.org/10.1016/j.healthplace.2007.10.007

Coumou D, Robinson A (2013) Historic and future increase in the global land area affected by monthly heat extremes. Environ Res Lett 8(3). https://doi.org/10.1088/1748-9326/8/3/034018

Coutts AM, Tapper NJ, Beringer J, Loughnan M, Demuzere M (2013) Watering our cities: the capacity for Water Sensitive Urban Design to support urban cooling and improve human thermal comfort in the Australian context. Prog Phys Geogr 37(1):2–28. https://doi.org/10.1177/0309133312461032

Denardo JC, Jarrett AR, Manbeck HB, Beattie DJ, Berghage RD (2005) Stormwater mitigation and surface temperature reduction by green roofs. Trans ASAE 48(4):1491–1496

De Jong K, Albin M, Skärbäck E, Grahn P, Björk J (2012) Perceived green qualities were associated with neighborhood satisfaction, physical activity, and general health: results from a cross-sectional study in suburban and rural Scania, southern Sweden. Health & place 18(6):1374–1380. https://doi.org/10.1016/j.healthplace.2012.07.001

Dhyani S, Basu M, Santhanam H, Dasgupta R (2022) Blue-green infrastructure across Asian countries: improving urban resilience and sustainability. In: Blue-green infrastructure across Asian

countries: improving urban resilience and sustainability. Springer Nature. https://doi.org/10.1007/978-981-16-7128-9
Dunnett N, Kingsbury N (2008) Planting green roofs and living walls: vol. II, 2004th edn
Ebi KL, Bowen K (2023) Green and blue spaces: crucial for healthy, sustainable urban futures. The Lancet 401(10376):529–530. https://doi.org/10.1016/S0140-6736(23)00096-X
Eggimann S (2022) Expanding urban green space with superblocks. Land Use Policy 117.https://doi.org/10.1016/j.landusepol.2022.106111
Eliasson IÈ (2000) The use of climate knowledge in urban planning
Elmqvist T, Setälä H, Handel SN, van der Ploeg S, Aronson J, Blignaut JN, Gómez-Baggethun E, Nowak DJ, Kronenberg J, de Groot R (2015) Benefits of restoring ecosystem services in urban areas. Curr Opin Environ Sustain 14:101–108. https://doi.org/10.1016/j.cosust.2015.05.001
Elmqvist T, Zipperer WC, Giineralp B (2016) 54-I 66. Rbanization: Evid 2(1)
Everard M (2009) The business of biodiversity
Everard M, Moggridge HL (2012) Rediscovering the value of urban rivers. Urban Ecosyst 15(2):293–314. https://doi.org/10.1007/s11252-011-0174-7
Friess DA (2017) Singapore as a long-term case study for tropical urban ecosystem services. Urban Ecosyst 20(2):277–291. https://doi.org/10.1007/s11252-016-0592-7
Fuller RA, Irvine KN, Devine-Wright P, Warren PH, Gaston KJ (2007) Psychological benefits of greenspace increase with biodiversity. Biol Let 3(4):390–394. https://doi.org/10.1098/rsbl.2007.0149
Gaston KJ, Warren PH, Thompson K, Smith RM (2005) Urban domestic gardens (IV): the extent of the resource and its associated features. Biodivers Conserv 14(14):3327–3349. https://doi.org/10.1007/s10531-004-9513-9
Germann-Chiari C, Seeland K (2002) 1-s2.0-S1389934102000679-main
Getter KL, Rowe DB (2006) The role of extensive green roofs in sustainable development. HortSci 41(5)
Ghosh D (2005) Ecology and traditional wetland practice: lessons from wastewater utilization in the East Calcutta Wetlands. https://prod-qt-images.s3.amazonaws.com/indiawaterportal/import/sites/default/files/iwp2/ecology_1.pdf
Gidlow CJ, Jones MV, Hurst G, Masterson D, Clark-Carter D, Tarvainen MP, Smith G, Nieuwenhuijsen M (2016) Where to put your best foot forward: psycho-physiological responses to walking in natural and urban environments. J Environ Psychol 45:22–29. https://doi.org/10.1016/j.jenvp.2015.11.003
Gill SE, Handley JF, Ennos AR, Pauleit S (2017) Adapting cities for climate change: the role of the green infrastructure. https://doi.org/10.2148/benv.33.1.115
Glover TD (2003) The story of the Queen Anne Memorial Garden: resisting a dominant cultural narrative. J Leis Res 35(2):190–212. https://doi.org/10.1080/00222216.2003.11949990
Gómez-Baggethun E, Barton DN (2013) Classifying and valuing ecosystem services for urban planning. Ecol Econ 86:235–245. https://doi.org/10.1016/j.ecolecon.2012.08.019
Grimm NB, Faeth SH, Golubiewski NE, Redman CL, Wu J, Bai X, Briggs JM (2008) Global change and the ecology of cities. Science 319(5864):756–760. https://doi.org/10.1126/science.1150195
Grose MJ (2009) Changing relationships in public open space and private open space in suburbs in south-western Australia. Landsc Urban Plan 92(1):53–63. https://doi.org/10.1016/j.landurbplan.2009.02.006
Groffman PM, Bain DJ, Band LE, Belt KT, Brush GS, Grove JM, Pouyat RV, Yesilonis IC, Zipperer WC (2003) Down by the riverside: urban riparian ecology. Front Ecol Environ 1(6):315–321. https://doi.org/10.1890/1540-9295(2003)001[0315:DBTRUR]2.0.CO;2
Gunawardena KR, Wells MJ, Kershaw T (2017) Utilising green and bluespace to mitigate urban heat island intensity. Sci Total Environ 584–585:1040–1055. https://doi.org/10.1016/j.scitotenv.2017.01.158
Haaland C, van den Bosch CK (2015) Challenges and strategies for urban green-space planning in cities undergoing densification: a review. Urban For Urban Greening 14(4):760–771. https://doi.org/10.1016/j.ufug.2015.07.009

Hartig T, Evans GW, Jamner LD, Davis DS, Gärling T (2003) Tracking restoration in natural and urban field settings. J Environ Psychol 23(2):109–123. https://doi.org/10.1016/S0272-4944(02)00109-3

Hartig T, Mitchell R, De Vries S, Frumkin H (2014) Nature and health. Annu Rev Public Health 35:207–228. https://doi.org/10.1146/annurev-publhealth-032013-182443

Hien Wong N, Chen Y, Leng Ong C, Sia A (2003) Investigation of thermal beneÿts of rooftop garden in the tropical environment. Build Environ 38. www.elsevier.com/locate/buildenv

Hopkins LP, January-Bevers DJ, Caton EK, Campos LA (2022) A simple tree planting framework to improve climate, air pollution, health, and urban heat in vulnerable locations using non-traditional partners. Plants People Planet 4(3):243–257. https://doi.org/10.1002/ppp3.10245

Hou J, Johnson JM, Lawson LJ (2009) Greening cities, growing communities learning from Seattle's urban community gardens. www.washington.edu/uwpress

House MA, Ellis JB, Herricks EE, Hvitved-Jacobsen T, Seager J, Lijklema L, Aalderink H, Clifforde IT (1993) Urban drainage—impacts on receiving water quality. Water Sci Technol 27(12):117–158. https://doi.org/10.2166/wst.1993.0293

Hunter RF, Nieuwenhuijsen M, Fabian C, Murphy N, O'Hara K, Rappe E, Sallis JF, Lambert EV, Duenas OLS, Sugiyama T, Kahlmeier S (2023) Advancing urban green and blue space contributions to public health. The Lancet Public Health 8(9):e735–e742. https://doi.org/10.1016/S2468-2667(23)00156-1

Ignatieva M (2017) Manual lawn alternatives in Sweden from theory to practice

Ignatieva M, Hedblom M (2018) An alternative urban green carpet: how can we move to sustainable lawns in a time of climate change? Science 362(6411):148–149. https://doi.org/10.1126/science.aau6974

IPCC (2015) Climate change 2014 : synthesis report : longer report. Intergovernmental Panel on Climate Change

Johnson PG (2013) Priorities for Turfgrass management and education to enhance urban sustainability worldwide

Jóźwik R (2020) Rainwater management solutions and their impact on shaping inner city areas undergoing transformation (case study of the ZAC clichy-batignolles area in Paris). J Ecol Eng 21(3):209–219. https://doi.org/10.12911/22998993/118277

Kabisch N, Qureshi S, Haase D (2015) Human-environment interactions in urban green spaces—a systematic review of contemporary issues and prospects for future research. Environ Impact Assess Rev 50:25–34. https://doi.org/10.1016/j.eiar.2014.08.007

Kaplan R, Kaplan S (1989) The experience of nature a psychological perspective the right of the Cambridge University Press Cambridge New York Port Chester Melbourne Sydney

Kate Ascher P, BuroHappold Engineering, Sabina Uffer, & Head of Research (2015) The high line effect

Kaźmierczak A (2013) The contribution of local parks to neighbourhood social ties. Landsc Urban Plan 109(1):31–44. https://doi.org/10.1016/j.landurbplan.2012.05.007

Kenneth E, Michelle L, Andrew G (2011) CITYGREEN #1 A Centre for Urban Greenery and Ecology Publication

Klompmaker JO, Hart JE, Bailey CR, Browning MHEM, Casey JA, Hanley JR, Minson CT, Scott Ogletree S, Rigolon A, Laden F, James P (2023) Racial, ethnic, and socioeconomic disparities in multiple measures of blue and green spaces in the United States. Environ Health Perspect 131(1). https://doi.org/10.1289/EHP11164

Korpela K, Hartig T (1996) Restorative qualities of favorite places. J Environ Psychol 16

Krishnan G, Shanthi Priya R (2022) Blue Green infrastructure: a panacea for urban environmental challenges. case study: Thiruvananthapuram city, Kerala, India. Mater Today: Proc 68:2646–2652. https://doi.org/10.1016/j.matpr.2022.09.562

Landscape U, Paul MJ, Meyer JL (2001) Streams in the landscape urban. Source: Ann Rev Ecol Syst 32

Langeveld JG, Cherqui F, Tscheikner-Gratl F, Muthanna TM, Juarez MFD, Leitão JP, Roghani B, Kerres K, do Céu Almeida M, Werey C, Rulleau B (2022) Asset management for blue-green

infrastructures: a scoping review. Blue-Green Syst 4(2):272–290. https://doi.org/10.2166/bgs.2022.019
Li H, Ding L, Ren M, Li C, Wang H (2017) Sponge city construction in China: a survey of the challenges and opportunities. Water (Switzerland) 9(9). https://doi.org/10.3390/w9090594
Littke H, Locke R, Haas T (2016) Taking the high line: elevated parks, transforming neighbourhoods, and the ever-changing relationship between the urban and nature. J Urban 9(4):353–371. https://doi.org/10.1080/17549175.2015.1063532
Markogianni V, Dimitriou E, Karaouzas I (2014) Water quality monitoring and assessment of an urban Mediterranean lake facilitated by remote sensing applications. Environ Monit Assess 186(8):5009–5026. https://doi.org/10.1007/s10661-014-3755-0
Masiero É, de Souza LCL (2015) Among winds, water bodies and urban elements. ICUC9, (1973). http://www.meteo.fr/icuc9/presentations/UDC/UDC3-3.pdfpdf
McPherson EG, van Doorn NS, Peper PJ (2016) Urban tree database. Forest Service Research Data Archive. https://doi.org/10.2737/RDS-2016-0005
Menon JS, Sharma R (2021) Nature-based solutions for co-mitigation of air pollution and urban heat in Indian cities. Front Sustain Cities 3. https://doi.org/10.3389/frsc.2021.705185
Middle I, Dzidic P, Buckley A, Bennett D, Tye M, Jones R (2014) Integrating community gardens into public parks: an innovative approach for providing ecosystem services in urban areas. Urban For Urban Greening 13(4):638–645. https://doi.org/10.1016/j.ufug.2014.09.001
Miles I, Sullivan WC, Kuo FE (1998) Ecological restoration volunteers: the benefits of participation, vol 2
Miles I, Sullivan WC, Kuo FE (2000) Psychological benefits of volunteering for restoration projects 18(4)
Montazeri H, Toparlar Y, Blocken B, Hensen JLM (2017) Simulating the cooling effects of water spray systems in urban landscapes: a computational fluid dynamics study in Rotterdam, The Netherlands. Landsc Urban Plan 159:85–100. https://doi.org/10.1016/j.landurbplan.2016.10.001
Mooney C, Tan PL (2012) South Australia's river murray: social and cultural values in water planning. J Hydrol 474:29–37. https://doi.org/10.1016/j.jhydrol.2012.04.010
Moore M, Townsend M, Oldroyd J (2006) Linking human and ecosystem health: the benefits of community involvement in conservation groups. EcoHealth 3(4):255–261. https://doi.org/10.1007/s10393-006-0070-4
Moran A, Hunt B, Jennings G (2005) A North Carolina field study to evaluate Greenroof runoff quantity, runoff quality, and plant growth
Nautiyal S, Gupta AK, Goswami M, Imran Khan YD (2023) The Palgrave handbook of socio-ecological resilience in the face of climate change: contexts from a developing country. Springer Nature. https://doi.org/10.1007/978-981-99-2206-2
Nesshöver C, Assmuth T, Irvine KN, Rusch GM, Waylen KA, Delbaere B, Haase D, Jones-Walters L, Keune H, Kovacs E, Krauze K, Külvik M, Rey F, van Dijk J, Vistad OI, Wilkinson ME, Wittmer H (2017) The science, policy and practice of nature-based solutions: an interdisciplinary perspective. In: Science of the total environment, vol 579. Elsevier B.V., , pp 1215–1227. https://doi.org/10.1016/j.scitotenv.2016.11.106
Nowak DJ, Crane DE, Stevens JC (2006) Air pollution removal by urban trees and shrubs in the United States. Urban For Urban Greening 4(3–4):115–123. https://doi.org/10.1016/j.ufug.2006.01.007
Otsuka N, Abe H, Isehara Y, Miyagawa T (2021) The potential use of green infrastructure in the regeneration of brownfield sites: three case studies from Japan's Osaka Bay Area. Local Environ 26(11):1346–1363. https://doi.org/10.1080/13549839.2021.1983791
Owen N, Ve G, Healy N, Matthews CE, Dunstan DW, Owen N, Healy GN, Matthews CE, Dunstan DW (2010) Too much sitting: the population health science of sedentary behavior. Exerc Sport Sci Rev 38(3). www.acsm-essr.org

Peck SW, Callaghan C, Kuhn ME, Arch B, Bass B (1999) Greenbacks from green roofs: forging a new industry in canada status report on benefits, barriers and opportunities for green roof and vertical garden technology diffusion. Canada Mortgage and Housing Corporation

Plieninger T, Thapa P, Bhaskar D, Nagendra H, Torralba M, Zoderer BM (2022) Disentangling ecosystem services perceptions from blue infrastructure around a rapidly expanding megacity. Landsc Urban Plan 222.https://doi.org/10.1016/j.landurbplan.2022.104399

Pochodyła E, Glińska-Lewczuk K, Jaszczak A (2021) Blue-green infrastructure as a new trend and an effective tool for water management in urban areas. Landscape Online 92:1–20. https://doi.org/10.3097/LO.202192

Przestrzelska K, Wartalska K, Rosińska W, Jurasz J, Kaźmierczak B (2024) Climate resilient cities: a review of blue-green solutions worldwide. Water Resour Manage. https://doi.org/10.1007/s11269-024-03950-5

Rana MMP, Ilina IN (2021) Climate change and migration impacts on cities: lessons from Bangladesh. Environ Challs 5:100242. https://doi.org/10.1016/j.envc.2021.100242

Rigolon A, Browning M, Jennings V (2018) Inequities in the quality of urban park systems: an environmental justice investigation of cities in the United States. Landsc Urban Plan 178:156–169. https://doi.org/10.1016/j.landurbplan.2018.05.026

Shammi M, Reza F, Sarker AC, Sakib AA (2023) Prospects and challenges of achieving sustainable urban green-spaces: a case study of urban greening in Dhaka North City Corporation (DNCC), Bangladesh. PLOS Sustain Transform 2(5):e0000061. https://doi.org/10.1371/journal.pstr.0000061

Southon GE, Jorgensen A, Dunnett N, Hoyle H, Evans KL (2017) Biodiverse perennial meadows have aesthetic value and increase residents' perceptions of site quality in urban green-space. Landsc Urban Plan 158:105–118. https://doi.org/10.1016/j.landurbplan.2016.08.003

Speak AF, Rothwell JJ, Lindley SJ, Smith CL (2012) Urban particulate pollution reduction by four species of green roof vegetation in a UK city. Atmos Environ 61:283–293. https://doi.org/10.1016/j.atmosenv.2012.07.043

Stadler O (2021) Blue-green infrastructure: an opportunity for Indian cities. https://www.researchgate.net/publication/353616196

Sukopp H (1998) Urban ecology-scientific and practical aspects

Sun Y, Xie S, Zhao S (2019) Valuing urban green spaces in mitigating climate change: a city-wide estimate of aboveground carbon stored in urban green spaces of China's Capital. Glob Change Biol 25(5):1717–1732. https://doi.org/10.1111/gcb.14566

Sunita K, Shahnawaz D, Shekhar S (2023) Evaluating urban green and blue spaces with space-based multi-sensor datasets for sustainable development. Comput Urban Sci 3(1). https://doi.org/10.1007/s43762-023-00091-0

Tan PY, Sia A (2005, May) A pilot green roof research project in Singapore. In: Proceedings of third annual greening rooftops for sustainable communities conference, Awards and Trade Show, Washington, DC. https://www.osti.gov/etdeweb/biblio/20861923

Tan PY, Sia A (n.d.) A pilot green roof research project in Singapore

Tan PY, Wang J, Sia A (2013) Perspectives on five decades of the urban greening of Singapore. Cities 32:24–32. https://doi.org/10.1016/j.cities.2013.02.001

Teo YH, Makani MABH, Wang W, Liu L, Yap JH, Cheong KH (2022) Urban heat island mitigation: GIS-based analysis for a tropical city Singapore. Int J Environ Res Public Health 19(19). https://doi.org/10.3390/ijerph191911917

Tyrväinen L, Miettinen A (2000) Property prices and urban forest amenities. J Environ Econ Manag 39(2):205–223. https://doi.org/10.1006/jeem.1999.1097

Tyrväinen L, Pauleit S, Seeland K, de Vriesc S (2005) Chapter 4 part I

Tyrväinen L, Tutkimuskeskus V (1999) Monetary valuation of urban forest amenities in Finland. In: Metsäntutkimuslaitoksen Tiedonantoja, 739

Ulrich RS (1993) Biophilia, biophobia, and natural landscapes design guidelines view project healthcare architecture view project. https://www.researchgate.net/publication/284655696

Vargas-Hernández JG, Pallagst K, Zdunek-Wielgołaska J (2018) Urban green spaces as a component of an ecosystem. In: Handbook of engaged sustainability. Springer International Publishing, pp 1–32. https://doi.org/10.1007/978-3-319-53121-2_49-1

Wang Y, Bakker F, de Groot R, Wörtche H (2014) Effect of ecosystem services provided by urban green infrastructure on indoor environment: a literature review. Build Environ 77:88–100. https://doi.org/10.1016/j.buildenv.2014.03.021

Wang J, Liu J, Mei C, Wang H, Lu J (2022) A multi-objective optimization model for synergistic effect analysis of integrated green-gray-blue drainage system in urban inundation control. J Hydrol 609.https://doi.org/10.1016/j.jhydrol.2022.127725

WHO (2016) Urban green spaces and health

Wood L, Giles-Corti B (2008) Is there a place for social capital in the psychology of health and place? J Environ Psychol 28(2):154–163. https://doi.org/10.1016/j.jenvp.2007.11.003

Xue Z, Hou G, Zhang Z, Lyu X, Jiang M, Zou Y, Shen X, Wang J, Liu X (2019) Quantifying the cooling-effects of urban and peri-urban wetlands using remote sensing data: case study of cities of Northeast China. Landsc Urban Plan 182:92–100. https://doi.org/10.1016/j.landurbplan.2018.10.015

Yang G, Yu Z, Jørgensen G, Vejre H (2020) How can urban blue-green space be planned for climate adaption in high-latitude cities? A seasonal perspective. Sustain Cities Soc 53.https://doi.org/10.1016/j.scs.2019.101932

Yu Z, Yao Y, Yang G, Wang X, Vejre H (2019) Spatiotemporal patterns and characteristics of remotely sensed region heat islands during the rapid urbanization (1995–2015) of Southern China. Sci Total Environ 674:242–254. https://doi.org/10.1016/j.scitotenv.2019.04.088

Chapter 11
Sustainable Urban Mobility: Bridging Health, Climate, and Mobility for a Better Future

Khalid Mohammed Almatar

Abstract Sustainable urban mobility is among the most important components of improving quality of life in cities, issues of environmental concern, and public health. The book chapter reflects on emerging challenges and opportunities for introducing sustainable urban mobility strategies in developed and developing economies and will consider integrating innovative technologies that could propel sustainable transport systems, such as digital tools for planning and monitoring. It focuses on the health benefits of active modes, like walking and cycling, and local authority capacity building to make sustainable urban mobility effective. It pays special attention to some primary barriers to sustainable urban mobility implementation in terms of institutional fragmentation, financial constraints, and the dependence of society on private vehicles. Case studies from cities such as Durres, Nairobi, and New York offer insights into how successful models of sustainable urban mobility could be established while offering lessons learned and strategies for overcoming common challenges. The future directions include the role of digitalization, increased policy support, and multi-stakeholder collaboration in scaling sustainable urban mobility globally. The book chapter concludes with actionable recommendations for cities seeking long-term sustainability, equity, and improved public health through urban mobility reforms.

Keywords Sustainability · Sustainable urban mobility plan · Public health · Urban mobility

K. M. Almatar (✉)
Department of Urban and Regional Planning, College of Architecture and Planning, Imam Abdulrahman Bin Faisal University, P.O. Box 1982, Dammam 31451, Saudi Arabia
e-mail: kmalmatar@iau.edu.sa

© The Author(s), under exclusive license to Springer Nature Switzerland AG 2025

A. I. Almulhim et al. (eds.), *Urbanization, Climate Change, and Health*, Climate Change Management, https://doi.org/10.1007/978-3-031-99132-5_11

Introduction

The twenty-first century has been shaped by urbanization, in which cities have become central to economic activity, cultural exchange, and social development (Scott 2017). More than 55% of the world's population lives in cities; this figure is expected to rise to 68% by 2050 (Angel et al. 2018). However, rapid growth in cities has posed challenges, especially in mobility systems that cannot meet the complex needs of modern urban populations (Angel et al. 2018). The legacy of the Second Industrial Revolution has brought an extremely heavy reliance on private vehicles, which, though convenient, has resulted in traffic congestion, air pollution, and major public health issues (Sperling 2018). Among the most important reasons for urgent innovation in the approach to urban mobility is the increase in respiratory diseases, such as asthma, and the spread of sedentary lifestyles brought about by car-centric urban designs (Lewis 2024; Sperling 2018).

Sustainability Urban Mobility Plans or SUMPs are gaining prominence as the strategic framework to address challenges (Torrisi et al. 2020). The plans focus on sustainable practice as they allow for environmentally friendly modes of transport such as public transport, cycling, and walking besides providing stakeholder engagement through a participatory approach to planning. Besides environmental issues, SUMPs focus much on public health outcomes. For instance, active transportation modes, such as cycling and walking, have resulted in low prevalence rates of obesity, chronic diseases, and respiratory conditions because of air pollution (Kahlmeier et al. 2024; Torrisi et al. 2020).

The importance of SUMPs is also recognized at the international level. For example, the European Commission has already demanded that all Trans-European Transport Network (TEN-T) urban nodes adapt to SUMPs by 2027 (Rutka et al. 2024). Measures of this kind highlight the central role that such plans must play in achieving more integrative sustainability goals, such as reducing climate change impacts and making a city livable. New studies underpin the transformative potential of SUMPs by estimating a saving of health and economic resources because of active transportation and a reduced car dependency (Rutka et al. 2024).

This book chapter explores the potential role SUMPs can play in mitigating the challenges that arise in addressing the issues of urban mobility and improving public health outcomes. This includes embedding health-focused policies into urban mobility planning for the reduction of emissions, support for active lifestyles, and the general enhancement of the quality of life of residents. This chapter seeks to present action-oriented recommendations and strategies through close examination of global case studies and best practices toward creating a sustainable urban environment. This chapter explores sustainable urban mobility's role in enhancing public health, environmental sustainability, and urban equity. Objectives include assessing the health benefits of active transport, barriers to SUM implementation, and international best practices in promoting health-oriented, sustainable mobility solutions.

Methodology

Electronic databases used to search for the information required to find the development and implementation of SUMPs and their impact on health and urbanization include Google Scholar, PubMed, ScienceDirect, and SpringerLink. Keywords or phrases used include sustainable urban mobility, traffic congestion, active mobility, climate change, public health, and car dependency. This study relied on articles published in English and comprehensive review of the existing literature, policy documents, and case studies instead of collecting primary data. Only the sources that best answered the research questions were analyzed in this study. Through analyzing how SUMPs encourage emission cuts, enhance air quality, and support active mobility, the study is directly aligned with the larger research goal of connecting sustainable urban mobility with climate change adaptation and health improvement.

Results and Discussion

Sustainable Urban Mobility Plan

In recent decades, unprecedented urbanization has been witnessed, with the world's population expected to reach 60% urbanization by 2030 (Soliman 2021). This rapid expansion brings several problems: increasing traffic congestion, air pollution, and inadequate housing. This tends to lead to an over-reliance on automobiles, which again fuels environmental degradation and enhances the deterioration of the city environment. Therefore, a need for efficient yet sustainable urban mobility comes while the cities grow (Soliman 2021).

In this context, the transport system not only help reduce traffic congestion but should also reduce air pollution and carbon emissions. Such mobility models based mainly on private vehicles are unsustainable. Hence, there is an urgent need to rethink both the design of the cities and transportation strategies. Sustainable urban mobility as shown in Fig. 11.1 seeks to develop integrated transport systems prioritizing public transportation, cycling, and walking to promote fair access and reduce environmental impacts (Lah 2019). This change needs to be done for better economic opportunity and social equity in rapidly growing cities. SUMP is the strategic framework that addresses the problem of urban mobility in the context of sustainable development goals (Hristoski et al. 2020; Jordová and Brůhová-Foltýnová 2021). They advocate for cooperation among city planners, policymakers, and residents to design efficient and environmentally responsible modes of transportation. SUMPs promote an integrated system of modes of transport while emphasizing a sustainable approach. This approach plays a prime role in making resilient urban environments capable of adapting to the demands of growing populations (Monardo and Ravagnan 2022). SUMPs aim to integrate environmental, economic, and social considerations into urban mobility planning, addressing the full spectrum of urban mobility needs, ranging from daily trips to freight transport (Kasraian et al. 2024). SUMPs support a transition away from car-dependent transport systems to more sustainable modes of transport, such as public transport, walking, cycling, and shared mobility. By including health-focused policies, SUMPs directly link urban mobility with improved public health, such as reduced air pollution, fewer traffic injuries, and more physical activity (Márquez et al. 2021; Zhang et al. 2018).

In the following sections, the chapter will proceed with additional discussion on the principles and components of SUMPs, their use in various urban contexts, and their broader implications for public health and environmental sustainability. This initial knowledge on SUMPs is the backdrop to analyze the specific drivers towards adopting sustainable urban mobility culture, the challenges encountered, and the lessons learned from international case studies.

Key Drivers for Adopting Sustainable Urban Mobility Practices

There are several interlinked factors supporting sustainable adoption of urban mobility practice. These factors also form a growing realization that transport systems need to be made more efficient, equitable, and environmentally friendly (Finger et al. 2024). Environmental considerations, public health, economic efficiency, and social equity are the most important critical factors shaping the future of urban mobility (Fig. 11.2) (Butler et al. 2020). The following section describes key drivers and critically discusses implications for public health of sustainable mobility and its benefits to society.

(a) ***Environmental Concerns***

Climate change urgency has placed it at the top of being the primary motivating factor leading to the use of best practices in sustainable urban mobility by cities. Carbon footprints are generated through greenhouse gas emissions by 70% of total emissions (Atmaca 2024). This implies that cities have been involved in the reduction of these emissions through better modes of transport. Public transport promotion, cycling, and walking are also essential sustainable practices in mobility as cities aim to achieve climate goals. Studies have shown that all these modes of travel considerably decrease emission (Arsenio et al. 2016; Tiboni et al. 2021). For instance, cities that have invested in upgrading their public transit infrastructure have reduced per capita emissions by as much as 30% (Buehler and Pucher 2011). Environmental benefits directly align with public health because less pollution and cleaner air lead to healthier urban populations.

(b) ***Economic Efficiency***

Economic factors also spur the transition to sustainable urban mobility. Sustainable modes of transport can save municipalities the infrastructure maintenance cost of extensive road networks. For instance, it has been demonstrated that investment in public transport will return four dollars into the economy for every dollar spent

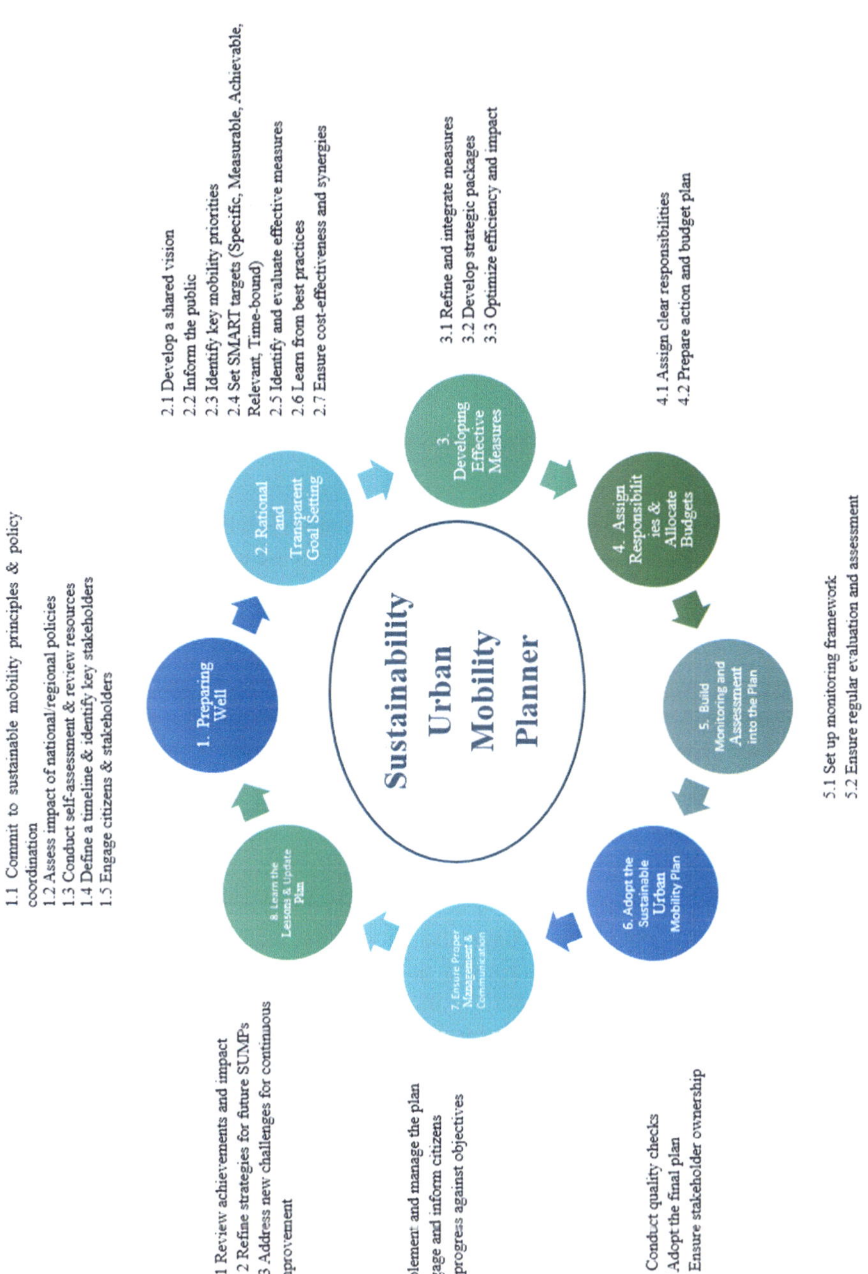

Fig. 11.1 Sustainability urban mobility plan. *Source* Author

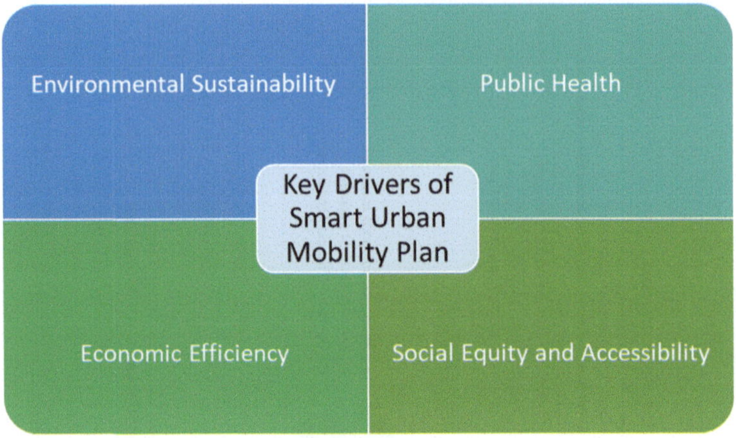

Fig. 11.2 Key drivers of sustainable urban mobility plan

because of access to employment and service opportunities (Mallett 2020). Cities that have adopted sustainable mobility have managed to increase economic productivity due to the access to markets and reduced travel time.

(c) ***Social Equity***

Social equity is increasingly being considered as an integral element of the urban policy. Sustainable practices for mobility are now expected to render transport accessible to all the sections of society (Meerow et al. 2019). Research indicates that the most vulnerable communities are restricted from access to safe means of transport, and it further contributes to the already existing inequalities (Oviedo 2021). This calls for accessible transportation planning, such as increasing more public transit routes and enhancing pedestrian infrastructure, toward helping cities overcome these inequities and toward more balanced access to employment opportunities and other necessary services. Other studies have also shown that increases in public transit access yield positive implications for low-income people (Pathak et al. 2017). For instance, in Helsinki, it is seen that the increased public transport has helped to improve the job opportunities available to the residents in the deprived area (Broberg and Sarjala 2015). In this regard, social equity should be embedded into the framework of sustainable urban mobility so that the citizens are given a chance to enjoy the changeover to more sustainable and equitable means of transportation. Besides, sustainable transport systems provide social cohesion through more inclusive public spaces. The more people walk and cycle, the more vibrant and accessible their cities will become in terms of community interaction and better social well-being.

(d) ***Public Health***

Public health considerations are increasingly becoming integral parts of sustainable urban mobility. Moving away from auto-dependent transport is no longer an environmental or economic issue alone; it has intrinsic links with the health outcomes that affect populations in such cities. The implications of the mobility of a city for health cannot be restricted to air quality and thus translate into overall well-being and the efficiency of health facilities. Car dependency has been associated with all kinds of negative health impacts.

Motor vehicle use ranks among the leading causes of air pollution, and has been linked to respiratory diseases, cardiovascular conditions, and every type of cancer (Maji et al. 2023). The WHO stated that exposure to motor vehicle emissions is one of the main risk factors for NCDs, including type 2 diabetes and coronary heart disease (World Health Organization 2019). The car dependency lifestyle further brings about the increase in the rates of obesity and related chronic diseases. Many studies show that cities with an automobile-orientated design have increased rates of obesity due to less time available for active physical life (Rojas-Rueda et al. 2021).

Encouraging active modes of travel like walking and cycling yields a lot of health benefits. General practice of these activities promotes improvements in bodily fitness and mental health; therefore, the possibility of chronic diseases decreases. One systematic review shows that environments for walking enhance the level of physical activity and psychological well-being among individuals dwelling in those environments (Sundling and Jakobsson 2023). The WHO states that active mobility not only reduces the risks of NCDs but also saves a great deal for health systems because it reduces the prevalence rates of lifestyle-related diseases (WHO 2019). For instance, cities that invest in cycling and walking infrastructure have documented low rates of obesity as well as better mental health. A research conducted in Barcelona found that the Superblock Model, intended to reduce motor vehicle traffic, may save as many as 667 early deaths each year due to its impact on reducing pollution and enhancing physical activity among the residents (Mueller et al. 2020). This model thus shows how urban design how urban design can have direct effects on public health through designing spaces to promote active transportation.

The COVID-19 pandemic added further emphasis to public health in city planning. As the lockdowns and social distancing measures hit cities, noticeable shifts towards outdoor spaces and alternatives to motorized transport came about. It was thus evident that cities implementing pandemic-time measures of pedestrianization could thus witness an improvement in air quality and an increase in general community well-being (Kakderi et al. 2021). These changes have helped to make safer movement during the pandemic and revitalized an appreciation for active mobility as a way to promote public health. The pandemic has made the need for more resilient, health-conscious urban mobility systems with the prioritization of public health in the face of global challenges (Kakderi et al. 2021).

The Health Impacts of Urban Mobility

The health impacts of urban mobility systems range from air quality to physical activity, mental well-being, and traffic safety. This section deals with the health implications of urban mobility and the different transport modes with their respective impacts on health outcomes.

(a) **Air Quality and Respiratory Health**

The relationship between urban mobility and air quality is well established. Emissions from vehicles have emerged as one of the main sources of air pollution, leading to respiratory diseases. Air quality issues in cities with high traffic volumes and the use of conventional combustion engine vehicles are enormous. For instance, in Beijing and New Delhi cities, air pollution has so intensely affected the population resulting in respiratory conditions such as asthma, bronchitis, and even lung cancer (Joshi et al. 2023).

One of the best ways to improve urban air quality and reduce health risks is by transitioning to electric vehicles. Cities like Copenhagen that have fewer vehicle emissions tend to show improvements in air quality and related health outcomes. A study done in Copenhagen showed that the elaborative cycling infrastructure has really improved respiratory health since people rely less on cars and thus less emissions and clean air (Glazener and Khreis 2019).

(b) **Active Transportation and Physical Health**

Active transport modes like walking and cycling have been related to several health benefits. The modes improve physical activity, which is essential in preventing and managing chronic conditions like obesity, heart disease, and diabetes. A WHO study revealed that cities where cycling and walking are quite high experience lower healthcare costs because there are fewer cases of obesity, diabetes, and cardiovascular diseases (WHO 2019). It would also engage people in being more active at work, avoiding any health problem-related absenteeism.

Healthy populations also contribute positively to gains in economic productivity, with people using active transportation to travel to work. Research indicates that cities with good cycling infrastructure like Copenhagen and Bogotá have very low cases of obesity, and the people rarely have heart disease cases (Albrecher et al. 2023). Also, green spaces and pedestrian-friendly infrastructure allow people to walk and cycle more often, promoting more physical activity and enhanced health. Active transportation is, therefore, essential in urban areas because campaigns for public health encourage people to walk or cycle to avoid using their cars. For instance, Bogotá has developed the "Ciclovía" program by converting main roads into a car-free area every Sunday. The increase in physical activity amongst its inhabitants has been attributed to cardiovascular health and fewer cases of obesity (Montero 2019).

Furthermore, studies indicate a high correlation between cycling rates and obesity prevalence among countries. The Netherlands and Denmark have a 36% and 23% cycling rate, respectively. They also have low percentages of obesity at 20.4% and 19.7%. Meanwhile, the United States and Canada have low cycling rates of 1% and 1.2%, respectively, which corresponds to higher obesity rates of 36.2% and 29.4%. Germany (11%) and Sweden (12%) also had moderate cycling rates accompanied by relatively lower obesity rates of 22.3% and 20.6%, respectively, further supporting the notion that even small increases in cycling may benefit public health. Similarly,

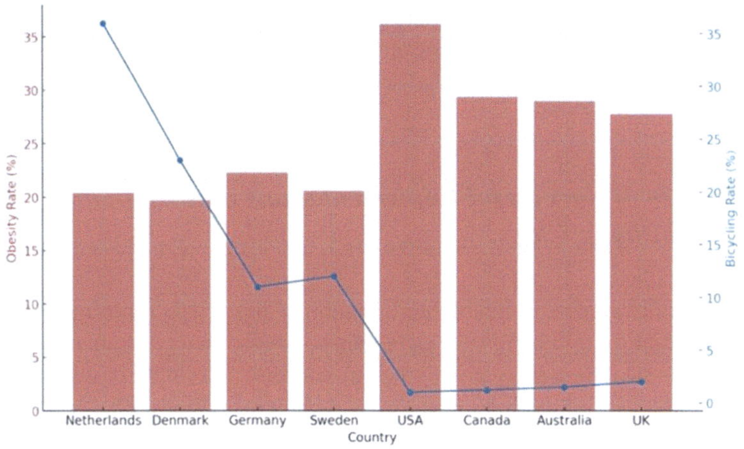

Fig. 11.3 Correlation between cycling rates and obesity prevalence (Cyclist, July 25, 2024)

Australia has the lowest cycling rate at 1.5%, while the UK has a rate of 2%, with obesity rates at 29% and 27.8%, respectively (Cyclist, July 25, 2024). The data shows an inverse relationship between cycling rates and obesity prevalence (Fig. 11.3). Thus, developing active transportation infrastructure will be part of public health strategies to reduce obesity and improve community health outcomes.

(c) **Mental Health and Urban Mobility**

Urban mobility systems also affect mental health. Stress from transportation, such as congestion and long traveling times, has been linked to anxiety, depression, and other mental disorders. Such reduction in traffic congestion and easy access to public transport would reduce such stressors and improve mental well-being. Other studies have shown that cities with proper public transport systems have reduced complaints of transportation stress, like Berlin and New York (Glazener and Khreis 2019). Active transport means, such as cycling and walking, enhance mental well-being through increased outdoor activities and reduced social isolation. Cities that support infrastructure for active transportation tend to increase social interaction, leading to improved community cohesion and overall mental well-being.

(d) **Traffic-Related Injuries and Safety**

Traffic-related injuries have been one of the important challenges to public health in urban communities. Such injury and death become critical safety concerns in cities that have heavy traffic accidents, including São Paulo and New York. Introducing pedestrian zones, improving road safety, and taking traffic-calming measures will most likely minimize traffic-related injuries and deaths to a significant extent. New York's Vision Zero has successfully reduced the number of injuries and deaths through street redesign to enhance safety for all road users (Kang 2019) Health impact assessments are increasingly being applied to determine transportation projects' potential risks and benefits. Such analyses will help identify ways urban mobility interventions could mitigate or worsen health risks, guiding cities toward safer, healthier mobility solutions.

Case Studies of Successful Implementations of Sustainable Urban Mobility in Cities

This transition towards sustainable urban mobility has led to innovative practices across cities, demonstrating how the right transportation strategies can make a difference in public health, reduce environmental impacts, and improve quality of life. In addition, this section outlines best practice implementations, elicits key lessons learned from them, and discusses how that same best practice can be applied to other urban environments, with a main focus on health outcomes.

(a) ***Barcelona, Spain: Superblock Initiative***

An innovation which reduces automobile traffic in an urban city is the Superblock project in Barcelona. Barcelona has transformed its public space into a vibrant activity place for community engagement by restricting automobiles from the roads and emphasizing walkers and cyclists (López et al. 2020). This model has cut down the level of air pollution by as much as 25%. This has done that by limiting the cases of respiratory diseases and improving the health of the public. This has further increased the rate of physical activities amongst the residents; walking and cycling rates have more than doubled since the introduction of the program, as studies shows. The superblock model shows how urban design immediately impacts health outcomes in the creation of environments that encourage active transportation (López et al. 2020).

(b) ***Copenhagen, Denmark***

Copenhagen is said to be the world leader on sustainable urban mobility as large infrastructure and policies in that city support active transportation. Bike lanes, bike-sharing programs, and a well-managed traffic system aid cycling. The interventions taken on cycling in the city as demonstrated by Rojas-Rueda et al. had reduced car traffic by 40% at peak hours with increased cycling rates. This has also improved air quality, reduced emissions of greenhouse gases, and health benefits since it has been shown that cycling daily reduces the risks of chronic diseases such as heart disease and diabetes by up to 50%. In the case of Copenhagen, cycling has been proved to be very crucial in this area of urban transport pertaining to public health (Rojas-Rueda 2021).

(c) ***Durres, Albania: Designing Sustainable Urban Mobility Plans***

Durres is one typical example of how cities in developing countries can implement sustainable urban mobility practices. Responding to traffic congestion worsened by seasonal tourism, the Municipality of Durres took the first step toward creating its first SUMP. It has more bicycle lanes, public transport access, and pedestrian pathways.

Early reviews suggest that such initiatives promote mobility and influence healthy lifestyles among residents. For instance, it is reported that the residents walk or cycle for a shorter distance, which improves the level of physical activity of the residents (Miltiadou et al. 2021).

(d) ***Valencia, Spain: Integrated Bike-Sharing System***

Valencia's bike-sharing system reflects the advantages of sustainable mobility on public health (Fernández-Aguilar et al. 2023). Since 2016, the system has gained popularity rapidly, allowing residents affordable bicycle access throughout the city. A study in Sustainability found that introducing bike-sharing has resulted in a significant increase in cycling among residents. That increase corresponds to better metrics of air quality and a decrease in respiratory illnesses in the population. Valencia is an example of how this bike-sharing system would work in the best interest of public health and active transportation of the community (Qiu and He 2018).

(e) ***Amsterdam: Health-oriented mobility policy***

Amsterdam is a well-known example of how a health-oriented mobility policy changes the urban environment. In its progressive approach, Amsterdam has introduced several initiatives that promote health and Sustainability. Extensive cycling infrastructure has been developed, with over 500 km of dedicated bike lanes (Cavill et al. 2008). As a result, an active mobility culture has been promoted; today, 63% of the population cycles daily. Air quality management measures include strict regulation of vehicle emissions, banning diesel cars, and reducing air pollution in the city. Amsterdam promotes active mobility through campaigns and programs encouraging walking and cycling, supported by pedestrian-friendly zones and accessible bike-sharing schemes (Des Enffans d'Avernas and Paque). Lessons learnt from the city of Amsterdam's experience include investments in active mobility infrastructure and cultural changes to attain sustainable practices. Success stories of such initiatives demonstrate value in health integration within SUMPs and become a model for other cities to seek improvement in their pursuit of enhancing urban Sustainability and improving health outcomes.

Barriers to the Implementation of Sustainable Urban Mobility

The difficulties of the application of Sustainable Urban Mobility Plans (SUMPs) are not universal but rooted firmly in the specific socio-economic, political, and infrastructural conditions of the cities. Although the previous sections have outlined the overall benefits of SUMPs, such as their positive impact on the health of the population, air quality, and environmental sustainability, the actual application of the principles is a complex set of issues that have to be understood. Transition to sustainable urban mobility often faces institutional, policy, and community-level barriers. Recognizing the main challenges helps in developing strategies to counter

Institutional and Policy-Level Challenges	Resistance to Change from Car-Dependent Populations	Infrastructure and Funding Constraints
• Lack of coordination among government agencies • Inconsistent or outdated regulations • Political influence and short-term policy focus • Limited stakeholder engagement in decision-making • Insufficient funding allocation for sustainable mobility projects	• Cultural attachment to private vehicles • Public perception of inconvenience in alternative transport options • Lack of awareness about sustainable mobility benefits • Mistrust in the efficiency and reliability of public transport • Economic concerns (e.g., car loans, insurance costs) preventing change	• Inadequate public transport infrastructure • Poor integration between different transport modes • High costs of upgrading or expanding mobility infrastructure • Limited government and private sector investment • Challenges in maintaining and operating sustainable transport networks

Fig. 11.4 Barriers to the implementation of sustainable urban mobility

these issues. This section describes a few barriers that arise from current scholarly research (Fig. 11.4).

(a) ***Institutional and Policy-Level Challenges***

By synthesizing information from multiple secondary sources, this study provides a more holistic understanding challenges in implementing SUMP. One of the major barriers to implementing Sustainable Urban Mobility Plans is institutional structures and policy integration. As discussed in earlier sections, cities face different barriers from fragmented governance systems involving many agencies responsible for planning and implementing transportation (Rosan 2016). The outcome is inconsistencies in the goals of the policies as well as fragmentation that does not favor concerted action. For instance, Copenhagen faces a lack of coordination in horizontal and vertical levels of government, which complicates the process of implementing SUMPs.

Long decision-making processes due to political differences and lack of funds delay sustainable mobility initiatives (Wefering et al. 2013). This will overcome the institutional challenges of transport planning. Coordination between the agencies needs to be increased. Inter-agency committees can be established for communication and coordination among various departments of the government in transport planning (Wefering et al. 2013). This will also contribute to the better alignment of objectives and streamlined decision-making. Training programs for local government officials in sustainable transport planning can also be developed to enhance the skills and effectiveness of SUMPs. Collaborations with academic institutions or international organizations can serve as an excellent source of resources and knowledge-sharing.

The political landscape determines the emphasis of sustainable mobility initiatives. Short-term political interests mostly override long-term sustainability goals, meaning that policy support is usually inconsistent. Research findings have shown that political will greatly influence the implementation of SUMPs; without it, initiatives may lack the necessary funding and public backing. Advocating to stakeholders, such as community organizations, businesses, and health advocates, can help in establishing a coalition that supports initiatives for sustainable mobility. Most local governments also lack the knowledge and resources to develop and implement an effective SUMP. Poor capacity-building and training measures further weaken the effectiveness of urban mobility policies, especially in smaller municipalities. Such a lack of institutional capacity leads to poorly formulated policies and are, therefore, unable to effectively respond to the needs of local mobility. This lack of guidelines on implementing SUMP makes the above problems worse since local governments might have challenges aligning with other broader sustainability goal. The national or regional governments should offer detailed guidelines on developing SUMP, including best practices, case studies, and tools to the local governments.

(b) ***Resistance to Change from Car-Dependent Populations***

The other major constraint is the resistance from car-dependent populations. Most residents view their cars as symbols of freedom and convenience, hence resisting the need to accept other modes of transport as alternatives. A survey of several cities showed that those with high car ownership rates were not likely to support policies geared towards reducing car usage, such as congestion pricing or restrictions on car access in city centers (Nieuwenhuijsen and Khreis 2016). Cultural attitudes towards transport are very critical in setting public opinion. In certain regions, car culture is so deep-seated that an attempt to popularize public or active transport will immediately be met with skepticism or outright opposition (Berger et al. 2014). Studies have shown that resistance in the public can delay or stop proposed cycling infrastructure projects outright due to safety or even perceived inconvenience fears. Public engagement campaigns that target this resistance with education on the advantages of sustainable mobility options for the residents and addressing their fears will help overcome it.

(c) ***Infrastructure and Funding Constraints***

Infrastructure is the other most significant hurdle in implementing sustainable urban mobility plans. A lack of appropriate facilities for walking, cycling, and public transport makes it very hard for the people of cities to shift from their dependency on cars. It has been observed that low investment in non-motorized transport infrastructure leads to reduced rates of walking and cycling. For example, in developing countries, the cities mainly have poor road infrastructure and fewer public transport facilities. This further enhances the usage of private vehicles (Attias and Mira-Bonnardel 2017).

Funding constraints is another huge challenge. Most municipalities do not have enough money to afford the improvements in the needed infrastructure or innovative mobility alternatives. According to an article on urban mobility planning, municipalities will not achieve their sustainability vision without a stable funding stream, such as government grants or public–private partnerships (Attias and Mira-Bonnardel 2017). Competitive priorities within the municipal budget then end up underfunding projects for transportation sustainability. Other funding sources that cities will have to pursue are far more diverse than municipal budgets alone. Grants can come from national governments and international climate change organizations. Partnership development with private sector stakeholders who may be interested in investing in sustainable transport solutions as part of their corporate social responsibility initiative can add to funding opportunities.

Lessons Learned from Policies Promoting Sustainable Urban Mobility

The future of urban mobility is changing radically nowadays with the innovative role of technologies, the deepening influence of digitalization, and the scaling up of sustainable urban mobility plans worldwide. Improvement must be focused not just on efficiency but also on public health outcomes so that these urban environments become healthier and more sustainable. Building upon the analysis of urban mobility policies and their effectiveness in previous sections, this section highlights several lessons that can be learned with regard to sustainable urban mobility. These lessons offer key insights that not only inform future policy-making but also provide actionable recommendations for enhancing sustainable urban mobility systems.

(a) **Community Engagement**

Capacity building is a crucial first step because most cities lack the capability to develop and implement SUMPs. Training programs for government officials and stakeholders, facilitated by international organizations, can raise technical capacity. Active community participation is a critical element of effective, sustainable urban mobility. For instance, in the transition of Bogotá toward sustainable transport, especially with its TransMilenio BRT system, there is evidence of the role of communities in shaping transport policies (Morales 2010). The city has maintained a strong vision for equitable and sustainable urban mobility across different political administrations by actively involving citizens in discussions about transport needs and priorities. This makes the residents feel a sense of ownership over the city, thus making them more likely to support and utilize new mobility solutions.

(b) **Integrated Planning**

Holistic approaches that integrate different modes of transport, like cycling, walking, and public transit, are more effective than isolated initiatives (Bakker et al. 2014). The innovative technologies such as EVs, shared mobility solutions, and autonomous transport systems are core in reshaping urban mobility (Paiva et al. 2021). Increasingly, EVs are viewed as the solution which will reduce urban air pollution and greenhouse gas emissions. The studies reveal that universal adoption of EVs is going to greatly reduce emissions from urban transport, most notably in cities where internal combustion engine vehicles dominate the roads (Paiva et al. 2021). The reduction in emissions leads to quality air directly, which is important for respiratory health and general public well-being. The city of Copenhagen has made itself a haven for cyclists, thus reducing traffic congestion and, more importantly, promoting healthier lifestyles. In that regard, investments in cycling infrastructure within the city have enhanced air quality, promoted physical activity, and enhanced benefits on health (Conticelli et al. 2021; Rojas-Rueda et al. 2021).

Shared mobility services, amongst other car-sharing and ride-hailing services, remain popular alternatives to private automobile ownership (Tirachini 2020). One study reported that half of the respondents said they would replace their personal cars with shared mobility services over the next decade (Ho et al. 2020). This reduces the number of cars on the roads, therefore allowing for easier parking in such tight urban settings and easing the congestion problem. Additionally, reduced levels of anxiety and depression within populations in cities have been linked to lower traffic congestion. Autonomous vehicles have emerged as a transformative force in urban mobility. Pilot projects with autonomous shuttles are already in place in several cities and have proven to be very important in enhancing public transport systems by ensuring first-mile and last-mile connectivity (Jaroudi 2024). These technologies can be merged into current transportation frameworks for more efficient and sustainable urban mobility ecosystems while improving access for mobility-challenged individuals.

(c) **Health-Oriented Policies**

Robust policy frameworks is another critical component; governments must put in place regulations that promote sustainable transport and discourage car dependency. Policies for active transport improve mobility and positively influence public health outcomes. Such cities with priorities given to walking and cycling infrastructure tend to have reduced levels of obesity and more favorable mental health conditions amongst their residents. For example, researchers have pointed out that cities with enhanced infrastructures for cycling have significantly reduced cases of lifestyle diseases, such as diabetes and heart conditions (Rojas-Rueda 2021). Besides that, the COVID-19 pandemic highlighted the interconnection between urban mobility and public health (Kakderi et al. 2021). Cities that could respond quickly to their transport policies to promote outdoor spaces and nonmotorized modes of transport experienced better air quality and community well-being. The results are indicative of how sustainable urban mobility plays an important role in creating a healthier urban environment.

(d) **Flexibility and Adaptation**

In scaling up sustainable urban mobility plans (SUMPs), adaptability and flexibility are essential to allowing these plans to change according to evolving conditions, innovative technology, and unique local requirements. One of the key strategies to improve flexibility and adaptability is through Public–Private Partnerships (PPPs). Governments should actively seek Public–Private Partnerships (PPP) to co-finance and implement SUMPs. PPPs can leverage private sector expertise, innovation, and efficiency without compromising public accountability and consistency with long-term urban development goals. Such collaborations also offer flexibility in risk-sharing, allowing both the public and private sectors to modify their strategies based on real-time information and current trends in transportation (Kakderi et al. 2021; Velez 2018).

(e) **Role of Digitalization in Enhancing Urban Mobility Systems**

Digitalization significantly develops the urban mobility system by improving data collection, analysis, and user experience. Cities can optimize their transportation networks through smart mobility solutions like real-time traffic management systems and integrated mobility platforms. It has been researched that cities using live traffic analytics can respond dynamically to the conditions, thus reducing congestion and improving overall efficiency. These will increase the activity level among residents as active modes of transport become easier to access (Alsaawy et al. 2022). Digital platforms aggregating different modes of transport help users better plan their journeys. For instance, multimodal apps combine public transport, cycling, and ride-sharing in a single journey plan.

Through integration, it becomes easier for people to use sustainable modes of transport (Alsaawy et al. 2022). Digitalization helps facilitate the proper communication between transportation providers and users. Enhanced information flow improves service delivery, thereby increasing customer satisfaction. For example, when people receive real-time schedules about public transport, their use increases manifold because of reduced navigation complexities. Increased usage reduces the number of cars that might have been used. Consequently, it reduces associated health risks caused by sedentary lifestyles (Fig. 11.5).

Limitation of Study

The limitations of the study are its dependence on secondary data, such as policy reports, case studies, and literature reviews, limiting context-specific analysis and real-time information on SUMP implementation. Furthermore, case studies considered in the study may not be universally applicable due to different socio-economic conditions and urban infrastructure. The lack of primary data constrains knowledge of SUMPs' direct impacts on local health outcomes and climate change mitigation. Future studies should be carried out considering primary data to determine the direct impact of SUMPs on urban population and their environment.

Fig. 11.5 Pathway to sustainable urban mobility. *Source* Author

Conclusion

In conclusion, urban mobility systems have a very significant impact on public health outcomes. Whether it is air quality and respiratory health, physical activity, mental well-being, or safety in traffic, the design and implementation of urban mobility systems strongly impact the health of citizens in cities. Sustainable mobility, active transport, clean public transport, and policies cutting down on air pollution in cities may create healthier and more resilient communities. The case studies are drawn from cities like Copenhagen, Barcelona, and Bogotá, highlighting the transformative potential of well-planned mobility reforms. These demonstrate health benefits and the greater social and economic advantages these initiatives can bring. However, the problems remain and are more complex, especially in rapidly growing cities experiencing traffic congestion, pollution, and limited resources for infrastructure development. As cities expand further, it is necessary to scale up sustainable urban mobility plans (SUMPs) through an integrated approach to meet local needs and respond to global challenges. By doing so, cities can have more efficient, sustainable, and inclusive urban transport systems.

References

Albrecher R, Curnier S, Kaufmann V (2023) Cycling and walking: literature review. Research notes, mobile lives forum

Alsaawy Y, Alkhodre A, Abi Sen A, Alshanqiti A, Bhat WA, Bahbouh NM (2022) A comprehensive and effective framework for traffic congestion problem based on the integration of IoT and data analytics. Appl Sci 12(4):2043

Angel S, Lamson-Hall P, Guerra B, Liu Y, Galarza N, Blei AM (2018) Our not-so-urban world. The Marron Institute of Urban Management, New York University, New York, NY, USA

Arsenio E, Martens K, Di Ciommo F (2016) Sustainable urban mobility plans: bridging climate change and equity targets? Res Transp Econ 55:30–39

Atmaca A (2024) Understanding carbon footprint: impact, assessment, and greenhouse gas emissions. In: Advances and technology development in greenhouse gases: emission, capture and conversion. Elsevier, pp 497–516

Attias D, Mira-Bonnardel S (2017) How public policies can pave the way for a new sustainable urban mobility? The automobile revolution: towards a new electro-mobility paradigm, pp 49–65

Bakker S, Zuidgeest M, De Coninck H, Huizenga C (2014) Transport, development and climate change mitigation: Towards an integrated approach. Transp Rev 34(3):335–355

Berger G, Feindt PH, Holden E, Rubik F (2014) Sustainable mobility—challenges for a complex transition, vol 16. Taylor & Francis, pp 303–320

Broberg A, Sarjala S (2015) School travel mode choice and the characteristics of the urban built environment: the case of Helsinki, Finland. Transp Policy 37:1–10

Buehler R, Pucher J (2011) Making public transport financially sustainable. Transp Policy 18(1):126–138

Butler L, Yigitcanlar T, Paz A (2020) Smart urban mobility innovations: a comprehensive review and evaluation. Ieee Access 8:196034–196049

Cavill N, Kahlmeier S, Rutter H, Racioppi F, Oja P (2008) Economic analyses of transport infrastructure and policies including health effects related to cycling and walking: a systematic review. Transp Policy 15(5):291–304

Conticelli E, Gobbi G, Saavedra Rosas PI, Tondelli S (2021) Assessing the performance of modal interchange for ensuring seamless and sustainable mobility in European cities. Sustainability 13(2):1001

Cyclist AJ (July 25, 2024) Does cycling reduce obesity? Comparing rates of obesity and cycling in 8 major countries. https://averagejoecyclist.com/cycling-and-obesity/

Des Enffans d'Avernas A, Paque B. What are the key success factors for Smart Mobility in Amsterdam?

Fernández-Aguilar C, Brosed-Lázaro M, Carmona-Derqui D (2023) Effectiveness of mobility and urban sustainability measures in improving citizen health: a scoping review. Int J Environ Res Public Health 20(3):2649

Finger M, Montero J, Petrozziello E (2024) Urban nodes along the trans-European transport network

Glazener A, Khreis H (2019) Transforming our cities: best practices towards clean air and active transportation. Curr Environ Health Reports 6:22–37

Ho CQ, Mulley C, Hensher DA (2020) Public preferences for mobility as a service: insights from stated preference surveys. Transp Res Part a: Policy Pract 131:70–90

Hristoski J, Jovanovic G, Petrovski A, Petrovska O (2020) Sustainable urban mobility plan for municipality of Veles 2019–2030. J Road Traffic Eng 66(1):19–24

Jaroudi I (2024) Mobility externalities and sustainable urban development: the case of Shared Automated Vehicles Université Paris-Saclay

Jordová R, Brůhová-Foltýnová H (2021) Rise of a new sustainable urban mobility planning paradigm in local governance: does the SUMP make a difference? Sustainability 13(11):5950

Joshi N, Gupta CK, Mangla Y, Chowdhuri A (2023) Green plants as a sustainable solution to air pollution. Int J Plant Environ 9(02):102–112

Kahlmeier S, Kasraina D, Murdock HE, Faghih Imani A, Yu Y, de Nazelle A, Stead D (2024) 129 Prioritising health in mobility planning: unlocking the health protential of sustainable urban mobility plans (SUMPs). European J Public Health 34(Suppl)

Kakderi C, Oikonomaki E, Papadaki I (2021) Smart and resilient urban futures for sustainability in the post COVID-19 era: a review of policy responses on urban mobility. Sustainability 13(11):6486

Kang B (2019) Identifying street design elements associated with vehicle-to-pedestrian collision reduction at intersections in New York City. Accid Anal Prev 122:308–317

Kasraian D, Murdock HE, Imani AF, Yu Y, de Nazelle A, Stead D, Kahlmeier S (2024) Health in mobility planning: an assessment of how health is considered in sustainable urban mobility plans. J Transp Health 39:101919

Lah O (2019) Sustainable urban mobility in action. In: Sustainable urban mobility pathways. Elsevier, pp 133–282

Lewis E (2024) Stepping towards sustainability: analyzing walkability in urban environments

López I, Ortega J, Pardo M (2020) Mobility infrastructures in cities and climate change: an analysis through the superblocks in Barcelona. Atmosphere 11(4):410

Maji S, Ahmed S, Kaur-Sidhu M, Mor S, Ravindra K (2023) Health risks of major air pollutants, their drivers and mitigation strategies: a review. Air, Soil Water Res 16:11786221231154660

Mallett WJ (2020) Transportation infrastructure investment as economic stimulus: lessons from the American recovery and reinvestment act of 2009. Congressional Res Serv 14

Márquez HRG, Bracho RL, Ramirez-Nafarrate A (2021) A simulation-optimization study of the inventory of a bike-sharing system: the case of Mexico City Ecobici's system. Case Stud Transport Policy 9(3):1059–1072

Meerow S, Pajouhesh P, Miller TR (2019) Social equity in urban resilience planning. Local Environ 24(9):793–808

Miltiadou M, Taxiltaris C, Mintsis G, Basbas S, Tsoukala A, Fylaktakis A, Panousi E (2021) Case studies of sustainable urban mobility plans and measures implemented in the framework of the SUMPORT project. WIT Trans Built Environ 204:259–273

Monardo B, Ravagnan C (2022) Sustainable strategic mobility plans towards the resilient metropolis. In: International symposium: new metropolitan perspectives

Montero S (2019) Worlding Bogotá's Ciclovía: from urban experiment to international "best practice" 1. In: Latin America and policy diffusion. Routledge, pp 115–137
Morales EB (2010) Promoting the right to the city through a transport system? The case of Transmilenio, the BRT of Bogota thesis submitted as part of the requirements for an MSc in Urban Development
Mueller N, Rojas-Rueda D, Khreis H, Cirach M, Andrés D, Ballester J, Bartoll X, Daher C, Deluca A, Echave C (2020) Changing the urban design of cities for health: the superblock model. Environ Int 134:105132
Nieuwenhuijsen MJ, Khreis H (2016) Car free cities: pathway to healthy urban living. Environ Int 94:251–262
Oviedo D (2021) Making the links between accessibility, social and spatial inequality, and social exclusion: a framework for cities in Latin America. In: Advances in transport policy and planning, vol 8. Elsevier, pp 135–172
Paiva S, Ahad MA, Tripathi G, Feroz N, Casalino G (2021) Enabling technologies for urban smart mobility: recent trends, opportunities and challenges. Sensors 21(6):2143
Pathak R, Wyczalkowski CK, Huang X (2017) Public transit access and the changing spatial distribution of poverty. Reg Sci Urban Econ 66:198–212
Qiu L-Y, He L-Y (2018) Bike sharing and the economy, the environment, and health-related externalities. Sustainability 10(4):1145
Rojas-Rueda D (2021) Health impacts of urban bicycling in Mexico. Int J Environ Res Public Health 18(5):2300
Rojas-Rueda D, Morales-Zamora E, Alsufyani WA, Herbst CH, AlBalawi SM, Alsukait R, Alomran M (2021) Environmental risk factors and health: an umbrella review of meta-analyses. Int J Environ Res Public Health 18(2):704
Rosan CD (2016) Governing the fragmented metropolis: planning for regional sustainability. University of Pennsylvania Press
Rutka A, Michnej M, Zwoliński T, Gabory M (2024) Core indicators for monitoring the sustainable urban mobility plan: a case study. Sustainability 16(13):5632
Scott AJ (2017) Urbanization, work and community: the logic of city life in the contemporary world. Qual Innov Prosperity 21(1):9-30–39-30
Soliman AM (2021) Urbanization and urban informality in the era of globalization. In: Urban informality: experiences and urban sustainability transitions in middle east cities. Springer, pp 85–120
Sperling D (2018) Three revolutions: steering automated, shared, and electric vehicles to a better future. Island Press
Sundling C, Jakobsson M (2023) How do urban walking environments impact pedestrians' experience and psychological health? A systematic review. Sustainability 15(14):10817
Tiboni M, Rossetti S, Vetturi D, Torrisi V, Botticini F, Schaefer MD (2021) Urban policies and planning approaches for a safer and climate friendlier mobility in cities: strategies, initiatives and some analysis. Sustainability 13(4):1778
Tirachini A (2020) Ride-hailing, travel behaviour and sustainable mobility: an international review. Transportation 47(4):2011–2047
Torrisi V, Garau C, Ignaccolo M, Inturri G (2020) Sustainable urban mobility plans: key concepts and a critical revision on SUMPs guidelines. In: International conference on computational science and its applications
Velez LXH (2018) Master plan contribution evaluation in the construction of urban resilience and sustainability in the city of Bogota. Respuestas 23(1):72–77
Wefering F, Rupprecht S, Bührmann S, Böhler-Baedeker S (2013) Guidelines. developing and implementing a sustainable urban mobility plan. Workshop
World Health Organization (2019) Noncommunicable diseases and air pollution: WHO European high-level conference on noncommunicable diseases: time to deliver–meeting NCD targets to achieve Sustainable Development Goals in Europe: 9–10 April 2019, Ashgabat, Turkmenistan.

Zhang H, Song X, Xia T, Zheng J, Haung D, Shibasaki R, Yan Y, Liang Y (2018) MaaS in bike-sharing: smart phone GPS data based layout optimization and emission reduction potential analysis. Energy Procedia 152:649–654

Part III
Tools, Strategies, and Innovations for Sustainable and Resilient Urban Futures

Chapter 12
Developing a Framework for Microclimate Assessment of Indian Cities

Meet Kumar and Sairam Dasari

Abstract Urbanization is a defining trend in global population dynamics, that creates challenges like harsh and varying urban microclimates. Researchers highlight the impact of rapid urbanization on microclimates as a global concern. Defined by multiple meteorological conditions within city limits, urban microclimate studies are crucial for enhancing well-being and comfort of urban residents. This study develops a comprehensive framework for assessing urban microclimate in Indian cities, involving Delhi as the case study area to analyse and model these influences. In contrast to most existing studies that tend to investigate microclimate in relation to a single, isolated parameter, this study examines the combined influence of multiple parameters, thereby offering a more holistic and nuanced understanding of microclimatic dynamics. The study computes a Microclimate Index, specifically for Delhi, that offers a valuable method to identify microclimate distress zones and project the benefits of proposed interventions. This chapter contributes to the broader discourse on urban resilience by presenting a replicable framework for microclimate assessment in cities, supporting urban planners and policymakers in making evidence-based and informed decisions and developing strategies that address the complex challenges posed by rapid urbanization in the face of climate change.

Keywords Urban microclimate · Urbanization · Influencing factors · Built environment · Land use · Local climatic zones (LCZs)

M. Kumar (✉) · S. Dasari
Department of Planning, School of Planning and Architecture, New Delhi, India
e-mail: meetkumar.spa@gmail.com

S. Dasari
e-mail: sairam.dasari@spa.ac.in

© The Author(s), under exclusive license to Springer Nature Switzerland AG 2025

A. I. Almulhim et al. (eds.), *Urbanization, Climate Change, and Health*, Climate Change Management, https://doi.org/10.1007/978-3-031-99132-5_12

Introduction

Urbanization is the dominant factor shaping the global population, with 55% of the world's population living in cities as of 2017, a number estimated to rise to 68% by 2050. India accounts for around 11% of the world's urban population, owing to robust economic development and growing population. While urbanization creates hubs of economic activities and opportunities, it also causes complicated issues in urban areas. One of the phenomena it affects significantly is Urban Microclimate.

Microclimates are climate differences in a small area of the earth's surface (Williams 1991), while Urban microclimate (UMC) is a unique climate within a small urban area that is mostly shaped by the built environment around it (Yang et al. 2018; Makropoulou 2017). Authors and researchers believe that the impact of rapid urbanization in cities on their microclimate is at present a great cause of global concern (Swain et al. 2016). Over the years, due to urbanization, the urban morphology and geometry have become more complex because of which one can witness large spatial temperature gradients within the city (Mohan et al. 2012a).

Microclimatic conditions in urban areas vary significantly across different locations. It is a multifaceted phenomenon influenced by factors such as building density, environmental characteristics, thermal response of buildings, presence of plants and water bodies among others. As cities grow, the replacement of natural surfaces by concrete, asphalt and other man-made elements alters both visual character and climatic behaviour of a region. Poorer surface albedo in urban areas increases heat absorption, while declining vegetation limits natural cooling through evapotranspiration, raising temperatures. Increased air pollution from vehicles and industries further degrades air quality and contributes to climate change, impacting local and long-term climate patterns. As a result, research into urban climates is critical for the healthy development of our living spaces.

Research Gaps

For a rapidly developing country like India, a study of urban microclimate for cities is more important for research since it directly impacts people's living conditions. It could potentially be revised through technological interventions to improve cities' resilience in the face of climate change and disasters (Zhou et al. 2022). According to a research by Hashemi et al. (2023), urban areas typically experience yearly mean temperatures that are 1 to 2 degrees Celsius higher than those in rural areas. These temperature variations have serious consequences for human health, as revealed by Shi et al. (2015) who discovered that mere 1 °C increase in daily mean temperature is associated with a significant 1.8% increase in mortality rates. Furthermore, Cetin et al. (2023) found a strong positive correlation between land surface temperature, sleep deprivation, and heat stress, emphasizing the complex connection between microclimate and human well-being.

A comprehensive study of the intracity microclimate diversification is necessary in order to determine specific conditions that are most relevant for the comfort, health, and general well-being of urban residents. Hashemi et al. (2023) adds to this narrative by pointing out the limited assessment of microclimate conditions in multiple urban blocks of a city, hampering a meaningful comparison of impacts caused by various urban parameters. A quantitative understanding of spatial distribution of microclimate is essential for improving urban planning decision-making and design practices (Shi et al. 2015). The quantitative models have become of significant interest to planners, architects, and environmental assessors to simulate urban microclimate as individual assessments are not very feasible. The limited feasibility is owed to differences in urban fabrics among cities, leading to substantial efforts and resources required for field surveys and observations.

A multi-parameter analysis is a significantly evident void in urban microclimate research. There is lack of a methodology for a comprehensive analysis of all factors shaping the microclimate of a city (Giyasova 2021). This was also pointed out by Bherwani et al. (2020) in their research where they highlighted the need of a 'Multi parameter analysis for robust assessment' in Urban microclimate studies. This gap is compounded by the absence of clarity in target results that can be practically used by cities and hinder the development of holistic strategies for environmental management.

The aim of this research is to address these gaps (Fig. 12.1) by devising a way to quantify urban microclimates of Indian cities. The eventual goal of the study would be to facilitate informed decision making so as to make urban areas more resilient to climate change and comfortable to live in. To achieve this, 4 objectives of the research were outlined—

1. To identify the critical factors affecting the microclimatic conditions in the context of urban areas from existing literature.
2. To critically analyse and develop the intra and inter relationships between planning parameters and microclimatic conditions for a case study area.
3. To develop an Index, specific to the case study area, for assessing its microclimatic conditions.
4. To recommend an analytical framework for Microclimate assessment.

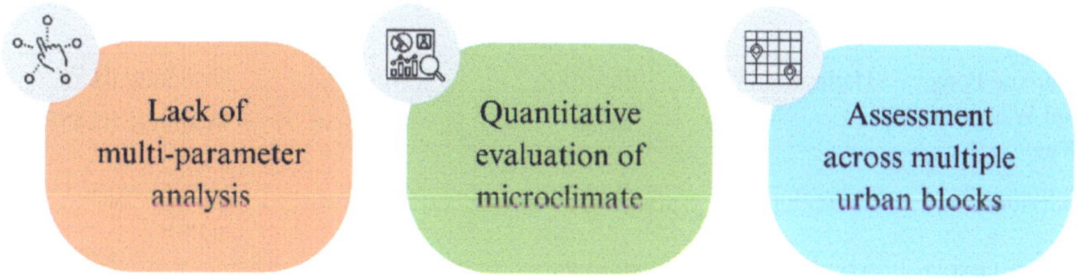

Fig. 12.1 Key research gaps addressed

Literature Review: Built Environment and Microclimate

To better understand the relationships between built and microclimatic parameters, an extensive literature review was conducted to find established relationships as quantified or stated by authors in existing research. This descriptive study intended to build a baseline understanding of expected relationships, which would then drive the analysis. Established relationships from researched literature has been compiled in Table 12.1.

Table 12.1 Relationship of temperature parameters with built parameters established in literature

Author	Study sites	Parameters	Type of correlation	Strength of correlation
R.B. Singh and Akriti Grover	Delhi	LULC change-Built	Positive	Strong
	Delhi and Mumbai	NDVI	Negative	Moderate for Delhi, Strong for Mumbai
Megha Jain	Bhopal	Distance from water body	Negative	Moderate
Javed Mallick and Atiqur Rehman	Delhi	Population density	Positive	Strong (R^2 = 0.748) for Delhi
Farzad Hashemi	Philadelphia	Vegetation with air temperature	Negative	Strong
		Albedo with MRT	Negative	Moderate (– 0.60)
		Sky View factor *with MRT*	Positive	Strong (0.61)
Daniela Maroni	Vila Rodrigues neighbourhood, Brazil	NDVI	Negative	Moderate to Strong (0.42 to 0.8)
Nguyen Thanh Hoan	Hanoi, Vietnam	Water	Negative	Strong (– 0.7)
		Vegetation	Negative	Strong (0.67)
		Urban built cover	Positive	Moderate (– 0.43)
		Air temperature	Positive	Strong
Jinchao Song	China	Built density	Positive	Moderate (0.55 to 0.65)
Xiaoxue Peng and Wenyuan Wu	Hangzhou, China	Elevation	Negative	Moderate (0.5)
		Shade	Negative	Moderate (0.45)
G Suthar	Bangalore	PM2.5 Conc	Positive	Variable (0.5 approx.)
Kamyar Fuladlu	Tehran	PM2.5 Conc	Positive	Moderate (0.47)
Sanjoy Garai	West Bengal	Rainfall	Negative	Strong (0.839)

It is evident from the literature that the relationship between built characteristics and microclimatic conditions is well established. Temperature and allied microclimatic indicators are strongly influenced by parameters like built density, vegetation, water bodies, population density, albedo etc. However, as identified in the research gaps, most of these focus on individual parameters in isolation.

This review provides a useful starting point for the study by identifying key built environment parameters that affect microclimate. It supports the development of the study's methodology and guides the analysis.

Research Methodology

The methodology of this research employs a comprehensive approach to assess the microclimatic conditions in Delhi.

Study Area Profile

Delhi is the largest metropolitan city in India. As per census 2011, Delhi saw 4.1% rise in Urban population thus witnessing the fastest rate of urbanization in the world. Consequently, it goes without saying that Delhi is undergoing dramatic changes in its climatic landscape due to urbanization. Despite the fact that Delhi is one of the most rapidly growing cities of the world, there is a scarcity of research on its land cover change and the creation of microclimates (Grover and Singh 2014). Microclimatic variation in Delhi is quite evident and its urban heat island phenomena is comparable to other major cities of the world such as London, Tokyo and Beijing and calls for mitigation action plans (Mohan et al. 2012a, b). Thus, the research takes Delhi as the case study area in order to develop and at later stages, apply the Microclimate index.

Delhi has a very diverse urban landscape. To facilitate the research, the city is divided into grids measuring one kilometer by one kilometer. The horizontal range of 1 km is the most commonly employed scale for microclimatic studies. The sites are selected majorly on the basis of homogeneity of Local Climatic Zones (LCZs). As there is a mix of multiple types of morphologies within an area, I.D. Stewart and T.R. Oke aimed to create a climate-relevant classification system for both urban and rural areas that would allow for meaningful inter-comparison of observations. They developed the Local Climatic Zones (LCZ) framework in 2012, defining LCZs as, "regions of uniform surface cover, structure, material, and human activity that span hundreds of meters to several kilometres in horizontal scale." (Stewart and Oke 2012) The 17 LCZs are divided into 10 "built" zones representing urbanized areas (LCZs 1–10) and 7 "land cover" zones representing non-urban or natural landscapes (LCZs A-G). A total of 10 grids are selected throughout Delhi representing 9 out of 10 Urban LCZs, as shown in Fig. 12.2.

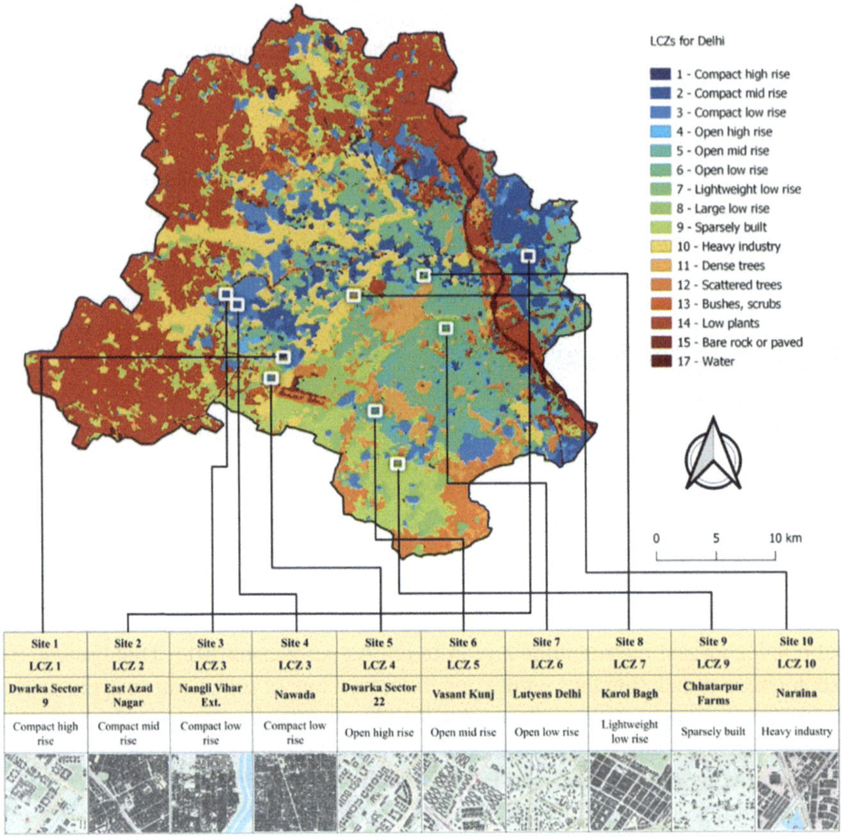

Fig. 12.2 Study sites marked on the local climatic zones (LCZs) map of Delhi

Data Collection

The data collection process in this study was grounded in a comprehensive multi-source approach, combining remote sensing, on-site measurements, and survey-based assessments to gather data for both built environment and microclimatic parameters. These parameters were selected based on a literature review, identifying them as critical representatives and influencers of urban microclimate. The selection process considered four key criteria:

(1) *Influence on microclimate*—The parameters which have identified impacts on urban microclimates as per the literature are chosen.
(2) *Relevance at micro level*—Emphasis was placed on parameters which are directly relevant at the micro level. Ensuring that the parameters align with the localized dynamics of urban environment.

Table 12.2 Parameters for the research

Parameter	Data collection method	Measurable indicator
Population	Estimation	Number of people
Population density	Delhi election commission ward wise population report	Population per hectare
Land surface temperature	Land surface temperature through Landsat data	Temperature value
Air temperature	AQI monitor	Temperature value
Precipitation	Data from centre for hydrometeorology and remote sensing (CHRS) at the University of California, Irvine	Average annual rainfall (in mm)
Humidity	AQI monitor	Absolute humidity
Vegetation cover	NDVI analysis	Value of NDVI
Air quality and pollution	AQI monitor	PM2.5 value
Land use and land cover	LULC mapping through supervised classification	–
Surface albedo	Averaging the albedo values of different surfaces in the site	Albedo per hectare
Built to open space ratio	Open buildings data from Google earth engine	Built/Open area
Presence of water bodies	Open street maps	Extent of waterbody in sq.m
Shaded areas	Mapping of areas under the shade during daytime	Shaded areas in sq.m

(3) *Availability of data*—Accessibility of reliable and comprehensive data is one factor.
(4) *Feasibility of analysis*—Parameters that could be effectively studied and integrated into the research framework were prioritized.

Based on these criteria, 13 parameters are chosen for the study, as given in Table 12.2.

Primary Data Dollection

The primary data collection for the study consisted of Resident surveys, Expert surveys, and on-site measurements. The Resident Survey is designed to collect insights into various aspects, including the socio-economic profile of residents, their travel patterns, and perceptions related to microclimate. On the other hand, the Expert Survey employed the Analytic Hierarchy Process (AHP) to collect expert opinions and derive weightages for microclimate parameters. Additionally, experts also offered qualitative inputs on microclimate management and improvement. Concurrently, on-site measurements were carried out to collect empirical data for key climatic characteristics of the sites. This included determining the concentration of PM2.5, estimating humidity and air temperature, and making observations to estimate areas under continuous shade by vegetation and the surface albedo. These extensive data gathering efforts were critical in improving the understanding of microclimate dynamics in the study sites.

Secondary Data Collection

The secondary data collection for the study extensively relied on online data portals and software tools. Landsat data sourced from the United States Geological Survey (USGS) was instrumental in obtaining crucial information such as Normalized Difference Vegetation Index (NDVI), Land Surface Temperature (LST) assessments, mapping of water bodies, and Land Use Land Cover (LULC) mapping. The LULC mapping data was further leveraged for demographic estimations and albedo calculations, enriching the study's analytical depth. Additionally, data from the Centre for Hydrometeorology and Remote Sensing (CHRS) was utilized for estimating annual rainfall patterns, providing valuable insights into the climatic conditions of the study area. Google Earth Engine emerged as a valuable resource, particularly for its open buildings dataset, facilitating the estimation of population distribution and built density across the study region. Data from the Delhi Election Commission's ward-wise reports was also integrated to access population density data specific to different wards in Delhi. Open Street Map data was also harnessed for extracting essential shapefiles related to roads, green spaces, water bodies etc.

Data Processing and Analysis Methods

As microclimate is not a tangible parameter, an Analytical Hierarchy Process (AHP) was performed through expert surveys to derive the criteria weightages for 5 proxy parameters of microclimate—Land surface temperature, air temperature, humidity, rainfall and PM2.5 concentrations. AHP is a decision-making tool used to assign weights to multiple criteria based on their relative importance. The final Microclimate Value for each study site was computed through weighted aggregation of the normalized indicators.

Subsequently, single variable regression was done among the identified parameters to assess the relationship between them. Multiple regression equations were further developed to model the relationships between the Microclimate and identified parameters as independent variables. The regression coefficients quantified the influence of each parameter on microclimate and enabled the identification of significant parameters for the case study area. Using the regression equation as the microclimate index, sensitivity analysis was performed to understand the sensitivity of microclimate towards the independent variables. The developed Microclimate Index for Delhi quantifies the microclimatic conditions of an area, enabling the identification of locations in distress and the projection of benefits from proposed interventions.

Analysis and Results

Assessing Parameter Relationships Using Single Variable Regression

The application of single variable regression analysis offers clarity on the relationships between different built and microclimatic parameters in Delhi. Multiple single variable regressions were performed among the study parameters to understand the relationship between them. Some of the key relationships are described in Table 12.3 and Fig. 12.3.

Table 12.3 Results obtained from Single variable regression analysis

S. No.	Dependent variable	Independent variable	Type of regression	Equation	R^2 value	Relationship type
1	LST	Net density	Linear	$y = 0.0037x + 34.058$	0.624	Positive
2	LST	NDVI	Polynomial	$y = -10.004x^2 + 0.4925x + 37.346$	0.737	Negative
3	Air temperature	Net density	Polynomial	$y = 6E{-}06x^2 - 0.0032x + 24.79$	0.305	Positive
4	LST	Rainfall	Linear	$y = 0.0159x + 22.415$	0.044	Positive
5	Air temperature	Rainfall	Exponential	$y = 2.0314e^{0.0029x}$	0.577	Positive
6	NDVI	Population	Logarithmic	$y = -0.108\ln(x) + 1.3722$	0.874	Negative
7	B to O ratio	Net density	Exponential	$y = 0.0903e^{0.0029x}$	0.656	Positive
8	Shaded area	NDVI	Polynomial	$y = -42.767x^2 + 89.755x - 11.922$	0.761	Positive
9	NDVI	Built to open ratio	Power	$y = 0.2221x^{-0.453}$	0.959	Negative

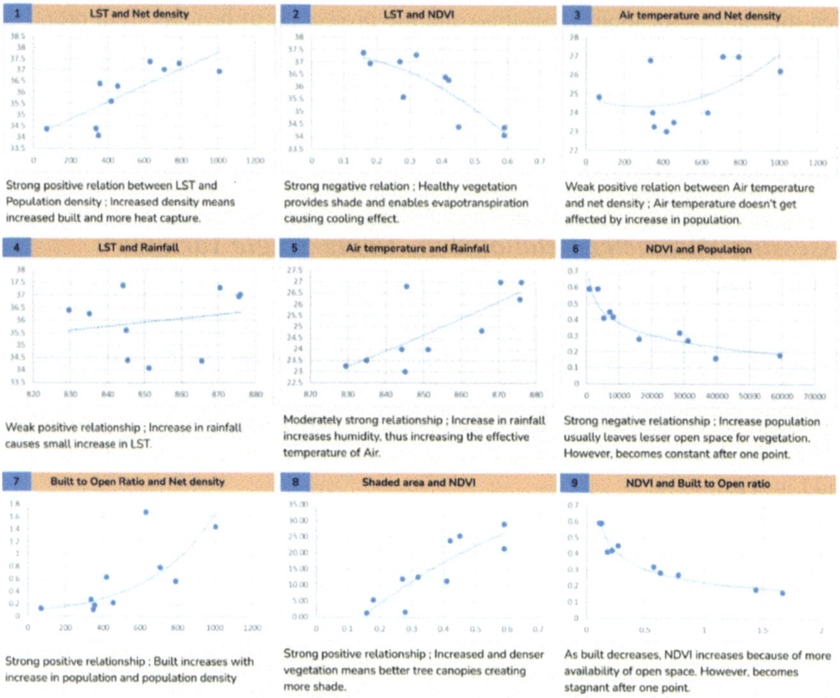

Fig. 12.3 Single variable regression graphs between the study parameters

The analysis shows that higher population density is strongly linked to increased land surface temperature (LST) due to more built-up areas, while vegetation (NDVI) significantly reduces LST through cooling effects. Air temperature, however, shows only weak correlations with both density and vegetation. Rainfall slightly increases LST but has a stronger impact on air temperature. Dense areas tend to have less vegetation, though this effect stabilizes at very high densities. Built-to-open space ratio increases with population density, and a strong inverse relationship exists between vegetation and built-up areas, which also plateaus beyond a point. These results highlight the complex relationships between microclimate parameters and urbanization, highlighting the vital role that built environments and vegetation play in affecting microclimate conditions.

Quantification of Microclimate

As an intangible value, microclimate was quantified using an Analytical Hierarchy Process (AHP). Expert opinion surveys were initially collected and consistency check was done. The consistent responses were used to estimate the criteria weightages for the five proxy parameters—air temperature, land surface temperature (LST), rainfall, humidity, and pollution, which served as the microclimate representatives. The methodology of quantification of microclimate is—

(1) **Identification of microclimate representatives**: Initial selection of representative indicators of microclimate through literature study.
(2) **Experts survey**: Experts performed pairwise comparisons of the five indicators based on their relative influence on microclimate, using the following matrix (Table 12.4) and key (Table 12.5).

(3) **Normalization of the matrix**: The pairwise comparison matrices filled by the experts were compiled. Each element in the matrix, A_{ij} represents the relative importance of parameter i over parameter j. Each element of the matrix was normalized by dividing it by the sum of its column, using the formula—

$$N_{ij} = A_{ij} / \sum_{1}^{n} A_{ij}$$

(4) **Derivation of criteria weights**: For all expert responses, the normalized values in each row were averaged to get the criteria weight for each parameter.

$$w_i = \left(\sum_{j=1}^{n} N_{ij} \right) / n$$

(5) **Consistency check**: A consistency check was done to ensure the reliability of expert judgments. Only responses with a consistency ratio < 0.1 were considered valid.

Table 12.4 Pair-wise comparison matrix for AHP

	Air temperature	*Land surface temperature*	*Precipitation*	*Humidity*	*Air pollution*
Air temperature	1				
Land surface temperature		1			
Precipitation			1		
Humidity				1	
Air Pollution					1

Table 12.5 Marking scheme for the pair-wise comparison

Marking key	Strongly **more** important		Moderately **more** important		**Equal** importance		Moderately **less** important		Strongly **less** important
	5	4	3	2	1	1/2	1/3	1/4	1/5

(6) **Criteria weight averaging**: The final criteria weights for respective parameters were obtained by averaging the weights from the consistent responses.

LST received the highest weightage of 30.4%, showing its importance as the most crucial parameter in microclimate assessment. Following LST, humidity, air temperature, and pollution were weighted at 18.8%, 18.6%, and 17.1%, respectively. Notably, rainfall was rated the least efficient representation of microclimate, as evidenced by its weightage of only 15.1%.

(7) **Parameter data normalization**: Raw data for influencing parameters corresponding to each site is normalized to a common scale ranging from 0 to 1, by Min–Max normalization, using the formula:

$$X_{norm} = (X - X_{min})/(X_{max} - X_{min})$$

where X is the original value, and X_{min} and X_{max} are the minimum and maximum values in the dataset.

(8) **Weighted aggregation**: Normalized data is then subjected to weighted aggregation, where the final criteria weights are applied to the respective parameter data.

(9) **Score summation**: The weighted scores obtained from the aggregation process are added to give the Microclimate value for each site, which serves as a quantitative measure of the prevailing microclimatic conditions based on the selected indicators.

The results, given in Table 12.6, indicated that LCZ 6 had the best microclimate conditions, followed by LCZ 9. This can be explained in terms of better vegetation and less residents in these areas. In contrast, LCZ 10 had the least favorable microclimate, with LCZ 7 following close behind in terms of poorer conditions. After reviewing specific sites, it was discovered that, while being in the common LCZ 3 of compact low rise, Site 3 had a superior microclimate compared to Site 4, owing to its lower air temperature and Land Surface Temperature (LST), which might be influenced by the presence of a water body. Furthermore, an association between vegetation and microclimate quality is observed, as the sites with more and healthier vegetation generally have better microclimate conditions.

Table 12.6 Quantification of microclimate values by application of criteria weights from AHP

Final criteria weightages from AHP				0.186	0.304	0.151	0.188	0.171	1
Basic site characteristics				Air temperature	Land surface temperature	Rainfall	Humidity	Pollution	Weighted microclimate value (0 to 1)
Site 1	LCZ 1	Dwarka Sector 9	Compact high rise	0.94	0.3	0	0.16	1	0.466
Site 2	LCZ 2	East Azad Nagar	Compact mid rise	0.19	0.13	0.99	0	0.94	0.383
Site 3	LCZ 3	Nangli Vihar Ext	Compact low rise	1	0.54	0.34	0.24	0.8	0.581
Site 4	LCZ 3	Nawada	Compact low rise	0.75	0	0.32	0.17	0.8	0.355
Site 5	LCZ 4	Dwarka Sector 22	Open high rise	0.88	0.34	0.12	0.14	0.93	0.469
Site 6	LCZ 5	Vasant Kunj	Open mid rise	0.05	0.9	0.34	0.04	0.85	0.489
Site 7	LCZ 6	Lutyens Delhi	Open low rise	0.54	0.91	0.77	1	0	0.683
Site 8	LCZ 7	Karol Bagh	Lightweight low rise	0	0.1	1	0.11	0.79	0.338
Site 9	LCZ 9	Chhatarpur Farms	Sparsely built	0.75	1	0.47	0.06	0.71	0.648
Site 10	LCZ 10	Naraina	Heavy industry	0	0.02	0.88	0.11	0.86	0.307

Multiple Regression Analysis for Microclimate Index

After quantifying the intangible parameter of Microclimate for the study sites, multiple regression analysis was conducted to establish the Microclimate Index, incorporating key built and demographic parameters. Through regression equations derived from these variables, the Microclimate Index was formulated specifically for the case of Delhi to quantitatively assess and evaluate the microclimatic conditions across the study sites.

In the initial step of regression analysis, the dependent variable (Y) was defined as the Microclimate Index, while the independent variables (X) included population, net population density, shaded area, NDVI, water body extent, albedo, and built-to-open space ratio (B to O Ratio). Upon analyzing the regression results, it was observed that the coefficients for population, net density, and shaded area were statistically insignificant. This implies that these variables may not have a significant impact on the Microclimate Index as compared to other factors with higher coefficients considered in the regression model.

1. Regression analysis between microclimate and built parameters

Parameters	Intercept	Population	Net density	Shaded area	Water body	B to O ratio	NDVI	Albedo per HA
Coefficients	– 0.26985	**0.00001**	**– 0.00031**	**0.00308**	0.01901	0.06412	0.82076	1.73468

To improve upon the equation, the insignificant parameters were removed from the regression analysis in the second step so as to get effective results. For the second step of regression analysis, the dependent variable (Y), that is, the Microclimate index was analysed against the independent variables (X) consisting of Water body, B to O Ratio, NDVI and Albedo. The results obtained gave the coefficients of the dependent variables to be considered in the Microclimate index.

2. Regression analysis between microclimate and significant parameters

Parameters	Intercept	Water body	B to O ratio	NDVI	Albedo per HA
Coefficients	– 0.385	0.019	0.192	1.342	0.987

The R^2 value of the regression was also significant, with a value of 0.912. The regression equation formed with the significant parameters gives us the Microclimate index for Delhi, that is—

$$\begin{aligned}\text{Microclimate Index} = & -0.385 + 0.019 * \text{Water Body Cover} \\ & + 0.192 * \text{B to O Ratio} + 1.342 * \text{NDVI} \\ & + 0.987 * \text{Albedo}\end{aligned}$$

The established Microclimate Index solves the identified research gaps by providing a consistent, scalable, and replicable tool for quantifying microclimatic conditions across an urban area. It allows for the spatial mapping of microclimate variation as well as comparisons between multiple urban blocks within a city, which is critical for understanding intra-city variation of temperature and comfort. By converting complicated environmental data for various influencing factors into a single measurable value, the index can facilitate evidence-based planning and assist cities in prioritizing measures that improve the comfort and overall well-being of individuals living in every part of a city.

Discussion

The relationship between the Independent variables and the dependent variable, Microclimate can be examined through Fig. 12.4. Minor increases in Albedo and NDVI result in significant improvements in microclimate conditions, owing to the steeper slope seen in the graphs for these variables. In contrast, increasing the Built-to-Open ratio reduces microclimate value, albeit in a small degree. Furthermore, although the presence of water bodies exhibits a positive correlation with microclimate conditions, the effect is relatively minor, as indicated by the low regression coefficient. Additionally, the preliminary plots revealed an insignificant gradient, and therefore, the variable was not included in the graphical representation.

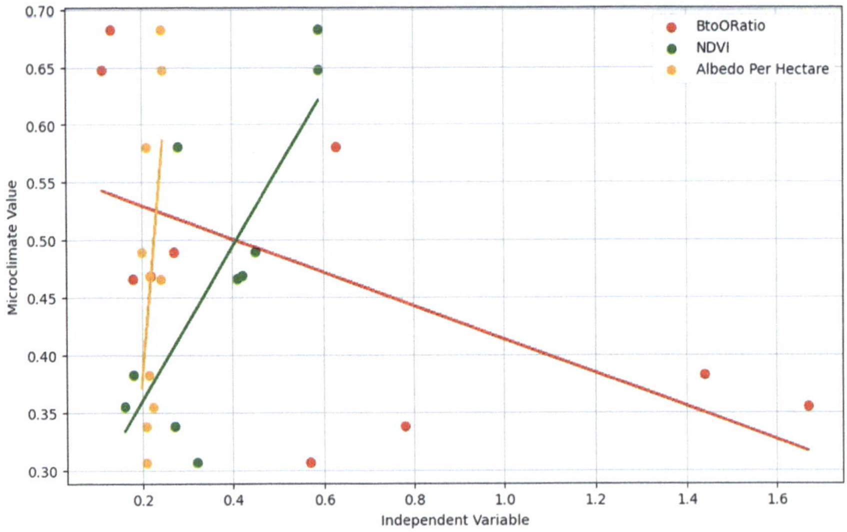

Fig. 12.4 Scatter Plot graph depicting relationship between microclimate and independent variables

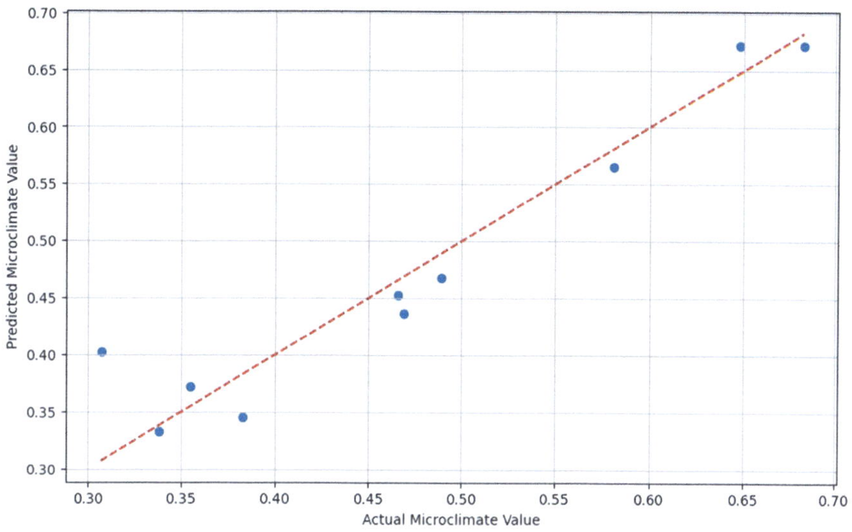

Fig. 12.5 Actual values versus predicted values of microclimate

To evaluate the accuracy of the model, the GEH (Geoffrey E. Havers) statistic was calculated. Commonly used in transport planning, the GEH statistic quantifies the difference between model-predicted values and actual observed values. The formula for the GEH statistic is:

$$GEH = \sqrt{\frac{2(M - C)^2}{M + C}}$$

Where M is the value obtained from the model (or new count) and C is the original value (or old count).

For the microclimate index, the average GEH value computed was 0.042, indicating high model accuracy. Typically, a GEH value below 5 is considered acceptable, signifying a close agreement between predicted and actual values. This can also be represented through a graph between the Predicted values and Actual values of Microclimate, as shown in Fig. 12.5.

Sensitivity Analysis

To understand the sensitivity of Microclimate and Well-being towards the independent variables, a sensitivity analysis was also performed. As per the equations formed for microclimate in the previous steps, NDVI (1.342) and Albedo (0.987) are the two independent variables with the highest coefficient values, and thus the highest impact on the effective value of the index. Thus, for sensitivity analysis, the two parameters of NDVI and Albedo are systematically increased so as to quantify their effect on the Microclimate, as shown in Tables 12.7 and 12.8.

The sensitivity analysis investigated the influence of increased Normalized Difference Vegetation Index (NDVI) and surface albedo on microclimate. A 10% increase in NDVI resulted in a 1.104-fold increase in the microclimate value, indicating a significant positive impact. Conversely, a 10% increase in surface albedo led to a smaller 1.046-fold increase. Another notable observation is that increasing albedo

Table 12.7 Sensitivity analysis for microclimate against NDVI

Increasing NDVI by 10%							
Water body	Built to Open Ratio	Old NDVI	New NDVI	Albedo Per Hectare	Old Microclimate value	New Microclimate Value	Improvement
0.79	0.18	0.41	0.451	0.241	0.466	0.508	0.055
0	1.44	0.18	0.198	0.215	0.383	0.369	0.024
13	0.63	0.28	0.308	0.21	0.581	0.604	0.038
0	1.67	0.16	0.176	0.225	0.355	0.394	0.021
0	0.22	0.42	0.462	0.218	0.469	0.492	0.056
0	0.27	0.45	0.495	0.199	0.489	0.528	0.060
0	0.13	0.59	0.649	0.243	0.683	0.751	0.079
0	0.78	0.27	0.297	0.208	0.338	0.369	0.036
0	0.11	0.59	0.649	0.247	0.648	0.751	0.079
2.26	0.57	0.32	0.352	0.208	0.307	0.445	0.043
				Average		0.521	
				Improvement		**1.104 times**	

Table 12.8 Sensitivity analysis for microclimate against Albedo

Increasing Albedo by 10%							
Water body	Built to Open Ratio	NDVI	Old Albedo	New Albedo	Old Microclimate value	New Microclimate Value	Improvement
0.79	0.18	0.41	0.241	0.2651	0.466	0.476	0.024
0	1.44	0.18	0.215	0.2365	0.383	0.366	0.021
13	0.63	0.28	0.21	0.231	0.581	0.587	0.021
0	1.67	0.16	0.225	0.2475	0.355	0.395	0.022
0	0.22	0.42	0.218	0.2398	0.469	0.458	0.022
0	0.27	0.45	0.199	0.2189	0.489	0.487	0.020
0	0.13	0.59	0.243	0.2673	0.683	0.696	0.024
0	0.78	0.27	0.208	0.2288	0.338	0.353	0.021
0	0.11	0.59	0.247	0.2717	0.648	0.696	0.024
2.26	0.57	0.32	0.208	0.2288	0.307	0.423	0.021
				Average		0.494	
				Improvement		**1.046 times**	

causes a greater positive effect on the microclimate of the sites which have a greater built density. This is because in high-density built-up areas, the concentration of impervious and heat-absorbing surfaces is significantly higher, leading to greater heat retention. Increasing the surface albedo in such environments enhances the reflection of incoming solar radiation, thereby reducing surface temperatures more effectively compared to areas with lower built density, where natural surfaces already contribute to cooling.

Limitations of the Study

While the study provides a comprehensive framework for microclimate assessment, certain limitations must be acknowledged. Due to constraints in time and resources, the scope of the research was limited to a selected set of variables deemed most critical and feasible. Additionally, the complex nature of urban microclimates means that data variability and availability can impact precision. Minor data gaps or inconsistencies, especially in remotely sensed or survey-based inputs, may introduce a degree of inaccuracy in the results. Nonetheless, the framework offers a robust foundation for future research and refinement.

Conclusion

Summary of Findings

This study revealed a clear connection between microclimate and built environment parameters, suggesting significant potential for urban planning to influence this relationship. The research identified a differential sensitivity of microclimate to specific parameters as microclimate appeared more responsive to changes in Normalized Difference Vegetation Index (NDVI), with a significant positive correlation of 0.78.

As per the experts in the field as well as in the literature study, current urban planning practices often fail to adequately address microclimatic concerns, with existing plans neglecting essential measures to improve comfort and mitigate extreme heat. 78.57% of surveyed experts believe existing plans inadequately address the challenges of harsh microclimates and their variations. For instance, the Delhi Heat Wave Action Plan for 2023 focuses solely on managing heat waves through water delivery systems, early warning systems, training and awareness programs etc. rather than incorporating heat mitigation strategies.

The developed microclimate index for Delhi identified NDVI and surface albedo as the most influential parameters affecting microclimate. The index also revealed a positive coefficient for the Building to Open-Space Ratio (B to O Ratio). This finding suggests that while a higher building density might typically lead to a negative impact on microclimate, incorporating increased albedo can counteract this effect, ultimately leading to a positive influence on microclimate due to its benefits. This analysis also identified several critical issues related to microclimate and well-being in urban environments. Areas with higher building density and less vegetation consistently exhibited the poorest microclimate conditions. Furthermore, a concerning disparity emerged, with sites with residents having lower average income levels experiencing significantly poorer microclimate and well-being compared to wealthier areas. These findings highlight a major concern of climate equity in our cities, where lower-income residents may be disproportionately burdened by harsh microclimates.

Proposed Conceptual Framework for Microclimate Assessment

The methodology of this research is a replicable and data-driven framework that could be used for microclimate assessment of other Indian cities. This research presents a first-of-its-kind framework for an Indian city that quantifies microclimatic conditions at a granular level by considering multiple influencing factors. The framework integrates diverse data sources including remote sensing datasets, open-source information, and primary data collection, to capture microclimatic, morphological, and demographic characteristics comprehensively. This process involves majorly 5 steps: Site selection, Data preparation, Parameter estimation, Modelling and analysis, and finally Projection of benefits, as shown in Fig. 12.6. This allows the establishment and quantification of relationship between key built parameters and microclimate in the form of the Microclimate index.

This microclimate index can serve dual purposes—Firstly, it can act as a diagnostic tool to help the urban planners and policymakers identify the locations within urban areas experiencing microclimate distress. Identification of these locations can help in targeted allocation of resources and implementation of strategic interventions to address their microclimate concerns. Secondly, the framework empowers informed decision-making. The microclimate index not only predicts the impact of various built and planning parameters on microclimate, but also projects the potential benefits planning interventions can bring. This allows planners and policymakers to strategically choose between alternatives and prioritize projects with the greatest positive impact on resident comfort and well-being.

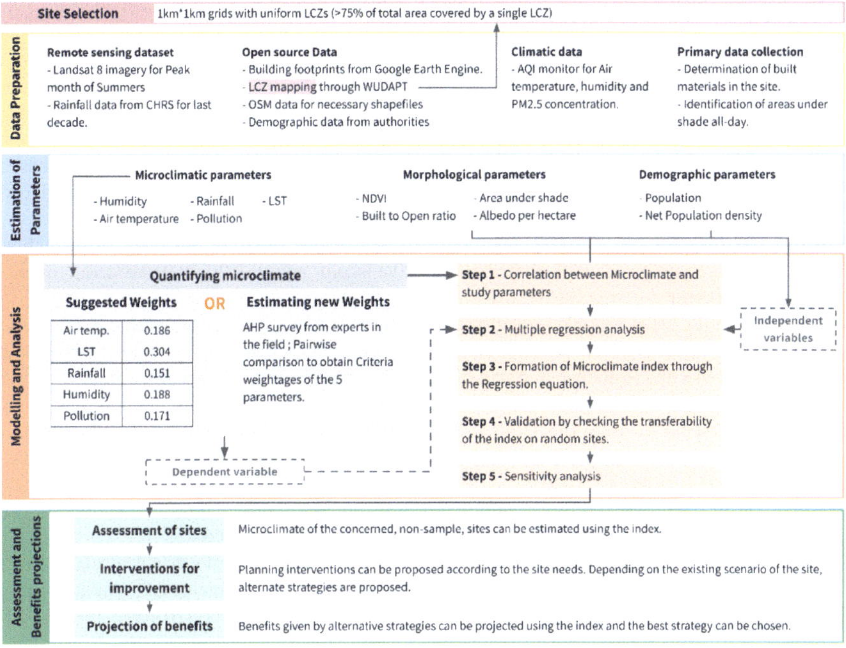

Fig. 12.6 Proposed framework for microclimate assessment

Way Forward

The research paves the way for further research in microclimate assessment. The framework developed in this study can be extended to encompass other Indian cities, allowing for the creation of microclimate indices made for their specific contexts. The core framework can be adapted, with adjustments to index parameters and weightages to reflect local contexts. This expansion will facilitate comparative analyses and the development of region-specific strategies for urban microclimate management.

The perception of microclimate can vary significantly based on income levels. Wealthier people may possess greater resources to mitigate discomfort from harsh microclimates, such as air conditioning. In contrast, underprivileged communities often lack these means of adaptation, experiencing a disproportionate burden from harsh microclimate. Future research can include these considerations so as to refine the assessment framework to account for these wealth disparities. This can ensure that microclimate improvements benefit all the residents equitably.

Overall, the utility of this research extends beyond academia, being particularly relevant for urban planners, researchers, urban local bodies involved in development, real estate developers and stakeholders engaged in shaping the built environment. It provides practical tools, insights, and evidence-based interventions that help to develop more livable, resilient, and equitable cities in India and around the world.

References

Bherwani H, Singh A, Kumar R (2020) Assessment methods of urban microclimate and its parameters: a critical review to take the research from lab to land. Urban Climate. https://www.sciencedirect.com/science/article/abs/pii/S2212095520302364

Cetin IZ, Varol T, Ozel HB (2023) A geographic information systems and remote sensing–based approach to assess urban micro-climate change and its impact on human health in Bartin, Turkey. Environ Monit Assess. https://doi.org/10.1007/s10661-023-11105-z

Giyasova I (2021) Factors affecting microclimatic conditions in. EDP Sciences. https://www.e3s-conferences.org/articles/e3sconf/abs/2021/20/e3sconf_emmft2020_06010/e3sconf_emmft2020_06010.html

Grover A, Singh R (2014) Spatial correlations of changing land use, surface temperature (UHI) and NDVI in Delhi using landsat satellite images. In: Grover A, Singh R (eds) Urban development challenges, risks and resilience in Asian megacities. Springer, Tokyo, pp 83–97. https://doi.org/10.1007/978-4-431-55043-3_5

Hashemi F, Poerschke U, Lulo LD, Chi G (2023) Urban microclimate, outdoor thermal comfort, and socio-economic mapping: a case study of Philadelphia, PA. MDPI Build 13(4). https://www.mdpi.com/2075-5309/13/4/1040

Makropoulou M (2017) Microclimate improvement of inner-city urban areas in a mediterranean coastal city. MDPI Sustain 9(6). https://www.mdpi.com/2071-1050/9/6/882

Mohan M, Kikegawa Y, Gurjar B, Bhati S, Kolli NR (2012) Assessment of urban heat island effect for different land use–land cover from micrometeorological measurements and remote sensing data for megacity Delhi. Theor Appl Climatol:647–658. https://doi.org/10.1007/s00704-012-0758-z

Mohan M, Kikegawa Y, Gurjar B, Bhati S, Kandya A, Ogawa K (2012b) Urban heat island assessment for a tropical urban airshed in India. Atmos Clim Sci. https://www.scirp.org/html/1-4700073_18813.htm

Shi Y, Ren C, Zheng Y, Ng E (2015) Mapping the urban microclimatic spatial distribution in a sub-tropical high-density urban environment. Arch Sci Rev:370–384. https://doi.org/10.1080/00038628.2015.1105195

Stewart ID, Oke TR (2012) Local climatic zones for urban temperature studies. Bulletin of American Meteorological Society. https://journals.ametsoc.org/view/journals/bams/93/12/bams-d-11-00019.1.xml

Swain D, Roberts G, Dash J, Vinoj V, Lekshmi K, Tripathy S (2016) Impact of rapid urbanization on the microclimate of Indian cities: a case study for the city of Bhubaneswar. In: Land surface and cryosphere remote sensing III. SPIE, Bhubaneswar. https://doi.org/10.1117/12.2228111.short

Williams TB (1991) Microclimatic temperature relationships over different surfaces. J Geogr 90(6):285–291. https://doi.org/10.1080/00221349108979321

Yang W, Lin Y, Qing Li C (2018) Effects of landscape design on urban microclimate and thermal comfort in tropical climate. Adv Meteorol. https://doi.org/10.1155/2018/2809649

Zhou Y, An N, Yao J (2022) Characteristics, progress and trends of urban microclimate research: a systematic literature review and bibliometric analysis. MDPI Build 12(7). https://www.mdpi.com/2075-5309/12/7/877

Chapter 13
Evaluating the Impact of Land Use and Green Space on Urban Heat Mitigation: Insights from Mirpur Botanical Garden and Surrounding Areas

Anika Tahsin Odri, Md. Asfaque Hossain, Md. Anamul Haque, and Ummeh Saika

Abstract Accelerated urbanization and climate change have made many cities vulnerable to high summer temperatures, putting residents' lives at risk. Rapid urban growth alters microclimates and ecosystems by replacing permeable, green spaces with impermeable surfaces. These surfaces, more efficient at heat transfer than water, absorb and store solar radiation, affecting surface energy balance. LST (Land Surface Temperature) influences an urban resident's comfort in extreme heat. This study examines how changes in LULC impact midsummer LST and analyzes the botanical garden in the Mirpur neighborhood and its 800-m buffer zone for green space cooling efficiency (GCE), park cooling intensity, and green space cooling gradient (GCG). We compute the LST of the Botanical Garden and its surrounding buffer zone, correlating it with land use metrics like NDVI, NDBI, and NDWI, using Landsat 8 data from summer 2024. ArcGIS is employed to analyze the correlation between LST and relevant indicators. NDVI showed a negative correlation with LST, while NDBI had a strong positive correlation. NDWI displayed a weak correlation. Results indicate that LST increases within the buffer zones, peaking at the farthest point. The park's cooling intensity shows that the botanical garden significantly cools its nearby area, with the effect decreasing as distance increases, though it remains noticeable. The GCE value of 6.72 m^2 suggests that each square meter of green space cools

A. T. Odri (✉) · Md. A. Hossain · U. Saika
Department of Geography and Environment, Jahangirnagar University, Savar, Bangladesh
e-mail: odri.47@geography-juniv.edu.bd

Md. A. Hossain
e-mail: asfaque.47@geography-juniv.edu.bd

U. Saika
e-mail: usaikau@juniv.edu

Md. A. Haque
Center for Environmental and Geographic Information Services (CEGIS), Dhaka, Bangladesh
e-mail: anamul.geo46.ju@gmail.com

© The Author(s), under exclusive license to Springer Nature Switzerland AG 2025

A. I. Almulhim et al. (eds.), *Urbanization, Climate Change, and Health*, Climate Change Management, https://doi.org/10.1007/978-3-031-99132-5_13

6.72 square meters of the surrounding area. The GCG value of 0.04 °C/m^2 indicates a relatively slow decrease in cooling effect, meaning the botanical garden has a notable cooling influence extending across its buffer area. This study emphasizes the importance of green spaces in mitigating urban heat and create a healthy and comfortable thermal environment.

Keywords Urban heat island · Park cooling intensity · Park cooling efficiency · Park cooling gradient

Introduction

Urban green spaces (UGS) are widely acknowledged as vital components of contemporary urban environments. They aid in the conservation of biodiversity, enhance human well-being, and provide other more benefits that are too numerous to detail. UGSs in cities refer to perfectly maintained parks and gardens that may be in their natural state or have been significantly transformed. The landscape is enriched by distributed vegetation clusters connected to roads and occasional green spaces (Sangwan et al. 2022). Unbuilt areas in metropolitan contexts are critical for mitigating the impact of urban heat islands, which are formed when heat accumulates in densely populated areas, as well as lowering pollution levels. The metropolitan Heat Island (UHI) effect, a relatively recent phenomenon, has exacerbated this issue in metropolitan regions (Taha et al. 1988; Oke 1982). Research by Dimoudi et al. (2013) indicates that the urban heat island effect can result in cities being 1.0 to 6.0 °C warmer than adjacent non-urban regions. The intensity of the urban heat island can be accurately estimated by the temperature differential between the city center and its periphery. Projections from the United Nations from 1997 suggest that by 2030, 61% of the global population will inhabit metropolitan areas or municipal jurisdictions. Many studies have linked high temperatures to health problems and death (Liu et al. 2022). Studies link higher temperatures to more health hazards, diseases, and deaths. It initiates a cycle of "high temperature—energy consumption—"greenhouse effect"—high temperature," which rapidly degrades urban ecological and environmental health (Hirano and Fujita 2012). Many people lived in small settlements until recent centuries and decades. Metropolitan areas will house 50% of the world population by 2030. Urbanization will transform Africa and Asia socially, economically, and environmentally. Migration from rural to urban is crucial. Cities are warmer than rural locations. This difference is due to urban heat islands. Dhakal and Hanaki (2002) found that vegetation around buildings lowered air conditioning needs by 25–80%, surface temperatures by 17 °C, and midday indoor air temperatures by 0.47 °C. Outdoor temperatures can drop two to eight degrees Celsius in green regions, according to Taha et al. (1988).

Akbari and Kolokotsa (2016) examined how urban green spaces affect urban heat island phenomenon. Givoni (1991) claims that urban parks possess a greater capacity to substantially affect thermal comfort and cooling compared to smaller green spaces, due to their expansive area and varied vegetation cover. The pleasant climate in city parks is known as "Park Cool Island" (PCI) in functions on urban studies and sustainability (Cao et al. 2010; Spronken-Smith and Oke 1998), and it's been called "Green Space Cool Island" (GCI) more recently. In modern studies of growth planning, the UGS cooling effect is seen as a good way to deal with thermal areas. The air temperature can drop by two to four degrees Celsius in city parks. The quality of the open space and the area around an urban park affect how cool it is. As Lin et al. (2019) said, many things have an effect on urban parks.The percentage of vegetation cover could be increased to substantially reduce air temperature, while the percentage of impervious surface area could be increased to significantly increase it (Yan et al., 2017).

Dhaka city is one of the few cities with urban population increase above 2.4% between 1975 and 2005. It was a top-10 megacity in 2011. Dhaka frequently ranks high in hostility (Alam 2019). A recent study found that Dhaka's summer and monsoon temperatures are 2.0 degrees Celsius higher than rural areas. The rapid urbanization, increased motor activity, and loss of flora and water have raised temperatures (Issue-I 2020). As indicated in topic-I, 2020, policy texts have paid little attention to suburban heat island effects. Dhaka's urban heat island issues include new vectors and diseases' survival. Infections including dengue, chikungunya, and respiratory issues are becoming widespread. Animals, plants, and humans suffer in urban heat islands effects (Issue-I 2018a, b).

Dhaka, an expeditiously developing metropolis experiencing increasing temperatures, emphasizes public comfort and well-being through extensive urban parks and green spaces. Vegetation moderates the temperature of the atmosphere. This impacts the park and its surroundings enhances thermal comfort Liu et al. (2022). High temperatures, intense solar radiation, and humidity induce heat stress and discomfort among urban residents of Dhaka. The summer season in Bangladesh extends from mid-April to July, encompassing the months of Baishakh and Jaishtha. Summer now commences and concludes in a distinct manner. National seasonal patterns in recent decades have prolonged summer. Recent observations encompass the initial fortnight of the Bangla monsoon Asharh. The prolonged heat induces numerous challenges for the inhabitants of Dhaka. Summer in Dhaka is a struggle for survival, extending beyond mere heat. Celsius may attain 43 (Zaman 2024). Since 1989, Dhaka has experienced a significant decline in vegetation and waterways, accompanied by an increase in high-rise buildings and infrastructure evidence substantiates this. The temperature in Dhaka surged dramatically. Humidity exacerbated discomfort. Dhaka requires 20% green space. If that is the case, the temperature will be sustained. The Forestry Department reports that 14% of the nation is covered in greenery. Nine percent in Dhaka. 'Global Forest Watch' reports that Dhaka possesses 5.9% green spaces (Mahmud and Mahmud 2024).

Bangladesh's biggest heat wave in 50 years shut down schools, stretched electricity, and disrupted livelihoods from mid-April to mid-June 2024 (Report 2024). The Meteorological Department of Bangladesh reported it hasn't had a heatwave this long since 1971 (Mahmud 2023). In ten days, the hottest temperature rose from 32 to 41 °C (105.8 F). According to Bangladesh Meteorological Department, the heatwave continues (Islam, 2024b). Scientists warn summer heat waves are getting worse, more frequent, and longer. Dense forests lower air temperatures by 1 °C on sunny days and 0.5 °C on cloudy days (Kim et al., 2023). This study analyzes vegetated and non-vegetated land surface temperatures to determine how parks and other green spaces affect local temperatures and how UGSs reduce the city heat island effect.

The aim of this investigation is to evaluate the impact of land use and green space on urban heat mitigation.

(1) to quantify and obtain spatial distribution of LST, NDVI, NDBI and NDWI of the study area
(2) to examine the correlation between the LST with NDVI, NDBI and NDWI and quantify the Park Cooling Intensity (PCI), Green space Cooling Efficiency (GCE) and Green Space cooling Gradient (GCG) of the study area.

Study Area

The National Botanical Garden of Bangladesh is Bangladesh's largest plant conservation facility, covering 84 hectares (210 acres). It is near the Dhaka Zoo in Mirpur and the garden has 56,000 tree, herb, and shrub species plus a large aquatic plant collection. Many exotic plant species have been placed into the garden to acclimate and multiply according to local weather patterns (Shoab 2018).

- **Park Dimensions**: 210 Acres
- **City Corporation**: Dhaka North City Corporation (DNCC)
- **Relative Location**: Mirpur, a northern suburb of Dhaka, adjacent to the Dhaka Zoo.
- **Absolute Location**: **23.8127° N, 90.3476° E**
- **Park Status**: Design quality and well maintained
- **Park Type**: Extra- large park**.**

In a homogeneous setting, temperature and thermal comfort are more likely to influence green space than other factors. The study might compare the eastern heterogeneous zone to the more uniform surrounding areas to see how various urban landscapes affect green spaces' heat mitigation and thermal comfort. Despite its homogeneity, the Mirpur Botanical Garden and its buffer zones were good choices. The green space's homogeneity makes it easy to analyze its cooling effects, while the eastern section's diversity adds contrast. Understanding how urban green spaces improve outdoor thermal comfort and guiding future urban planning efforts to produce more resilient and livable communities is the goal of this strategy. Figure 13.1 shows the urban park and its four buffer zones.

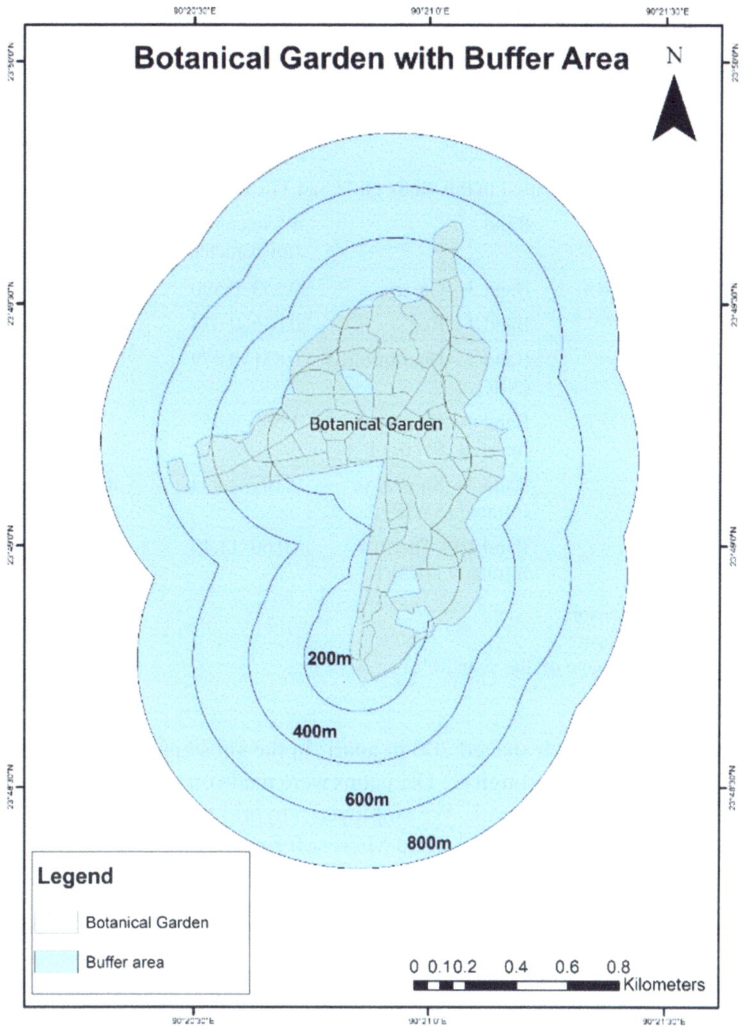

Fig. 13.1 Study area

Data Sources and Methods

It aimed to collect remote sensing imageries from Landsat 8 TM captured in 2024 of 30 m resolution and less than 5% of cloud coverage images from April 2024 during the summer months. All accessible LST pictures during the study period were combined for analysis to produce a more comprehensive conclusion and mitigate the impact of specific temporal weather conditions. An 800-m buffer zone was established around the parks to determine the optimal distance for cooling as green spaces and various infrastructures produce significant interference. The radius was to be divided into

Table 13.1 Information of Landsat satellite images used in this study

Date of acquisition	Sensor and satellite	Time of year	Resolution (m)	Cloud coverage (%)
April 2024	Landsat 8 OLI/TIRS	Summer	30 m	Less than 5%

Table 13.2 Detail of Landsat used in this study OLI* and TIRS**

Satellite	Band	Wavelength (micrometers)	Resolution (m)
Landsat-8 OLI and TIRS	Band 3—Green	0.533–0.590	30
	Band 4—Red	0.636–0.673	30
	Band 5—Near Infrared (NIR)	0.851–0.879	30
	Band 6—Shortwave Infrared (SWIR) 1	1.566–1.651	
	Band 7—Shortwave Infrared (SWIR) 2	2.107–2.294	30
	Band 10—Thermal Infrared (TIRS) 1	10.60–11.19	100 (30)

**Thermal Infrared Sensor
*Operational Land Imager
Source (Survey n.d.)(Image of the year 2024)

four equal segments, each spaced 200 m apart. In the subsequent sections, we will refer to them as "zones 1 through 4." Decisions were made on the average LST values for each zone, including the park. We employed a hybrid pixel reduction method to achieve this objective. ArcMap 10.8 and Microsoft Excel office 2021 was utilized to conduct further analytical operations, employing various tools for processing. The Landsat used in this study is illustrated in Tables 13.1 and 13.2. Additionally, a detailed description of the Landsat used in this study is provided in Table 13.2.

LST Mapping

Determining the (TOA) spectral radiance

$$L\lambda = ML \times Qcal + AL \tag{13.1}$$

Here, ML is the multiplicative rescaling factor for a specific band, Qcal corresponds to Band10 of the image, and AL represents the additive rescaling factor of the metadata file attached with Landsat images (Misbari & Hashim, 2016).

TOA to brightness temperature conversion

$$BT = \frac{K2}{\ln\left(\frac{K1}{L\lambda} + 1\right)} - 273.15 \tag{13.2}$$

NDVI calculation for emissivity correction

$$NDVI = \frac{NIR(Band5) - Red(Band4)}{NIR(Band5) + Red(Band4)} \tag{13.3}$$

Proportion of vegetation (PV)

$$PV = \left[\frac{NDVI - NDVI_{\min}}{NDVI_{\max} - NDVI_{\min}}\right]^2 \tag{13.4}$$

Land surface temperature calculation (LST)

$$LST = \frac{BT}{\left\{1 + \left[\left(\frac{\lambda BT}{\rho}\right)\ln \varepsilon_\lambda\right]\right\}} \tag{13.5}$$

$$\rho = h\frac{c}{\sigma} = 1.438 \times 10^{-2}\,\text{mK} \tag{13.6}$$

NDBI & NDWI Calculation

$$NDBI = \frac{SWIR - NIR}{SWIR + NIR} = \frac{BAND6 - BAND5}{BAND6 + BAND5} \tag{13.7}$$

$$NDWI = \frac{NIR - SWIR}{NIR + SWIR} = \frac{BAND5 - BAND6}{BAND5 + BAND6} \tag{13.8}$$

PCI (Park Cooling Intensity)

The aim of this measurement is to quantify the degree to which a green space is less severe than nearby parts that are impervious (for example, concrete or asphalt) (Liu et al. 2022). In the context of a verdant environment, such as a park, it is a representation of the temperature differential that exists between the space itself and its environs. It offers a numerical assessment of how much a park contributes to the improvement of local environmental conditions and surface temperature reduction. The PCI is a measurement that shows the extent to which the park area is cooler than the areas that are immediately surrounding it. An additional analysis was performed on the LST readings by determining the PCI intensity (ΔT) by utilizing the equation that is presented below.

$$\Delta T = T_u - T_P \tag{13.9}$$

Tu signifies the land temperature at the surface (LST) in the buffer zones, whereas Tp denotes the LST of the park. Yu et al. (2018) state that the cooling effect can only be observed when the intensity of the PCI reaches a plateau or starts to decrease. Therefore, the regression analysis only evaluated the buffer zones leading up to a potential stagnation point. This is because it is necessary to identify the cooling effect prior to the onset of diminishing PCI intensity.

GCE (Green Space Cooling Efficiency)

Green spaces, such as urban woodlands, gardens, and parks, have the capacity to decrease the surface temperatures of their surrounds, hence reducing the total temperature of the environment. The measure assesses the degree to which green areas contribute to creating a thermal climate that is both healthier and more acceptable, while also reducing the influence of the urban heat island (UHI). The effectiveness of these green locations is assessed by measuring the extent to which they may reduce temperatures in the surrounding area through shade, evapotranspiration, and cooling from the soil and plant surfaces. The GCE was calculated by utilizing the following equation to perform a comprehensive examination of the LST data.

$$GCE = \frac{GCA}{GA} \tag{13.10}$$

Here,

GCA (Green Space Cooling Area): In terms of square meters, this is the area that is able to reap the benefits of the cooling impact that the green space provides.
GA (Green Space Area): This is the actual area of the green space (also in square meters).

Green Space Cooling Gradient (GCG)

The Green area Cooling Gradient (GCG) is a term that denotes the spatial variation in cooling intensity that occurs within urban natural spaces. It quantifies the extent to which the chilling effect varies in the vicinity of green space. The spatial variation in cooling effects provided by green spaces, which is typically quantified as the rate of temperature decrease when one passes farther from the focal point of a natural space, is referred to as GCG. The GCG was calculated using the following equation

$$GCG = \frac{GCI}{L_{\max}} \tag{13.11}$$

Here,

GCI (Green Space Cooling Intensity): This is the difference in heat among the green area's mean temperature and the warmest nearby impermeable surface.
Lmax: Up until it reaches this maximum distance, which is the maximum distance, the green area imparts a cooling impact on the surrounding environment

Result and Discussions

Quantify and Obtain Spatial Distribution of LST

In the botanical park's core, LST values are low (dark green and green), indicating colder temperatures. Temperatures range from 25.34 to 29.62 °C. The buffer zones' LST values have increased from yellow to red, indicating a temperature spike from 29.62 to 40.27 °C. The middle of the botanical park has dark and light green sections with temperatures from 25.34 to 29.62 °C. The lush vegetation, which provides shade and cools via evapotranspiration, may explain this cooler location. Yellow areas in the core cool zone have slightly higher temperatures (29.62 to 31.49 °C). These places have diversified vegetation and wide spaces, cooling somewhat. The orange and red outlying parts of the chart have the highest temperatures, 31.49 to 40.27 °C. These areas are near urban growth and developed areas, raising surface temperatures. Less vegetation and more impermeable surfaces absorb and retain heat. Plants significantly affect cooling processes, as shown by the temperature gradient. Green areas with low temperatures have lots of plants, while yellow and red areas with high temperatures don't have many plants. The picture clearly shows how land surface temperature affects vegetation cover in the botanical garden and buffer zone. This shows how thick forest cools and how sparse vegetation or urbanization raises temperatures. Urban development, ecological conservation, and green infrastructure in the region depend on this data. Understanding these temperature swings can help urban planners create more green spaces to reduce urban heat islands. Green spaces may reduce heat and improve well-being for residents and tourists. The map that is shown in Fig. 13.2 includes an illustration of the distribution of Land Surface Temperature (LST) throughout the botanical garden and the buffer region that surrounds it.

Built-Up Area, Vegetation Cover and Water Bodies of the Study Area

The Normalized Difference Vegetation Index (NDVI) map provides a clear visual representation of the vegetation density and health of the botanical garden and its surrounding regions. The primary regions of the botanical garden display high NDVI values, indicating dense and vigorous vegetation. The existence of well-maintained vegetation improves air quality and sustains biodiversity in these areas. Regions with

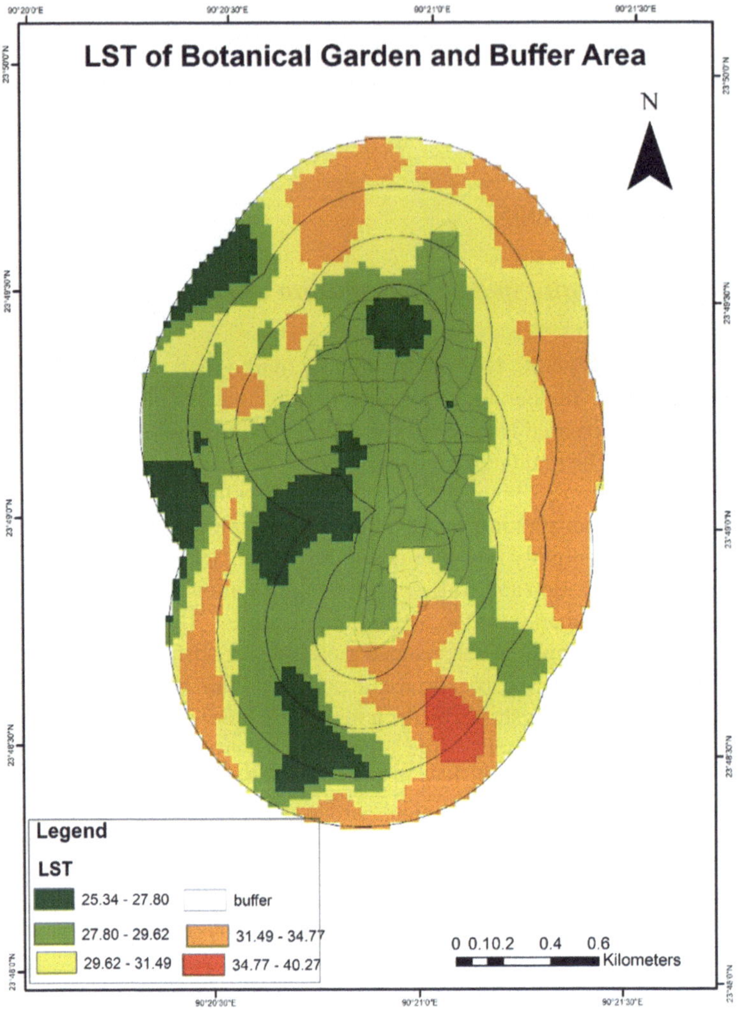

Fig. 13.2 LST of botanical garden and the buffer area

intermediate NDVI values are identifiable adjacent to the dense vegetation pattern. The vegetation in these areas primarily comprises grasses, shrubs, and tiny trees. The boundaries of the buffer zone and its elements exhibit low NDVI values, indicating that these areas have little vegetative cover. These may include urban regions, roadways, or untouched environments. These locations do not enhance the cooling benefits of natural ecosystems, making them more sensitive to heat island effects and increased temperatures. The botanical garden is part of the buffer zone as a way to lessen the damage that urban growth does to natural ecosystems. It is very important to keep the ecosystem in balance and get the most out of green areas by managing this buffer well. NDVI map depicts the variation in vegetation density between the botanical garden and the surrounding buffer zone. The abundant flora in central regions markedly enhances the ecological viability of the area. The scarcity of vegetation in the periphery zones suggests sites requiring rapid action. The buffer zone is essential for safeguarding the botanical garden against urban encroachment. This function guarantees the green space remains resilient and sustainable over a prolonged period. This research's findings may provide substantial direction for future conservation and urban planning initiatives focused on improving and safeguarding the ecological benefits of the botanical garden. The NDBI map represents the Normalized Difference Built up Index, is an essential tool for evaluating the degree of development in the botanical garden and its adjacent buffer zone. This evaluation differentiates between urbanized (developed) and non-built-up (natural or vegetative) environments. An extensive examination of green space allocation, conservation efforts, and urban planning is essential for evaluating the impacts of urbanization. The subsequent map offers a comprehensive visual depiction of developed and underdeveloped areas within the examined regions. The map utilizes a color scheme to represent different land cover classifications. The map illustrates urban regions highlighted in red. These regions are predominantly situated at the map's periphery.

The central zone of the botanical garden exhibits a diminished presence of built environments (denoted by the color red), indicating a lower level of urbanization and an increased amount of green space in this area. The park's boundaries both allocate and centralize all these items. The eastern and northeastern regions display heightened concentration levels. The yellow zones, predominantly situated in the central area of the botanical park, denote the non-urbanized regions. It illustrates the lush terrain in the vicinity. The color yellow clearly indicates the abundance of green landscapes in the western and southern areas. It conveys an impression of lush vegetation. The NDBI-generated map depicts the existing built environment in the region. The NDBI map shows the equilibrium between cultivated and undeveloped regions, showcasing the attributes of the botanical garden. It underscores the vital significance of green spaces in urban environments, especially for ecological health, biodiversity, and air circulation. Conserving these undeveloped regions is crucial for fostering a sustainable urban ecosystem and alleviating the urban heat island phenomenon. The natural parts of the park are very important for keeping the temperature stable, which cools down nearby cities. The Urban Heat Island effect has a lot of different effects, and this map shows them in a unique way. The NDWI stands for Normalized Difference Water Index, which is crucial for evaluating the quantity and distribution of aquatic features within the botanical garden and its surrounding buffer zone.

It is essential for identifying water bodies and assessing vegetation moisture content, hence providing critical information on the region's hydrological conditions. Comprehending the distribution of water is crucial for the effective management of water resources, environmental regulation, and conservation efforts. The map shows how the botanical park and the buffer zone around it are distributed in terms of water features and levels of moisture. The picture clearly shows how green the area is. The green areas on the map show places with lots of plants. There is more greenery in these places, which means that the plants are doing well and may even be making the environment better. The middle, western, and southern parts are mostly made up of plants. The carefully placed botanical park is known for its rich plants. This part mostly talks about the region's urban features. The red areas show places that don't have a lot of water in them. These could be places with little vegetation or dry areas. This common phenomenon occurs in buffer zones when human activities or natural elements diminish soil moisture. This region is primarily composed of the eastern and northeastern parts. This section aims to identify which region has the greatest temperature. This section is frequently used to represent an area that includes aquatic bodies. The blue patches indicate the presence of aquatic bodies. These locations exhibit the greatest density of water and encompass features such as lakes, ponds, or other notable aquatic formations. They serve an essential function in the local environment, offering habitats for several aquatic species and enhancing biodiversity. The water bodies are also visible within the lush landscape. Botanical gardens typically include aquatic elements, such as lakes. Furthermore, the western and southern sections have bodies of water. Although the proportion of aquatic settings is reduced relative to the total green space and constructed areas, the map clearly illustrates the water bodies within the study area, highlighting the differing moisture content levels throughout the landscape. The botanical garden mostly consists of lush, humid sections, whereas the buffer zones mainly feature arid regions with diminished moisture levels. Minor aquatic bodies are distinctly observable. The NDWI map is crucial for developing conservation strategies, managing water resources, and improving the sustainability of the botanical garden and its vicinity. Figures 13.3, 13.4 and 13.5 illustrate the NDVI, NDBI, and NDWI of the research area.

Relationship of the LST with NDVI, NDBI and NDWI

Relationship Between LST and NDVI

Graph 13.1 indicates the relationship of Land Surface Temperature (LST) to Normalized Vegetation Index (NDVI). The vertical axis depicts LST from 25 to 40 degrees Celsius, meaning that as the number becomes higher, so does the temperature; while on the horizontal axis are NDVI values from 0 to 0.5, with larger values signifying denser vegetation. This negative relationship clearly illustrates that higher NDVI values are associated with cooler LST, wherein vegetation is attributed with cooling. What can be noticed is that for NDVI between 0.2 and 0.35 and LST between 30 and 35 degrees, the vegetation played a substantial influence in reducing surface temperatures. Such a relationship emphasizes vegetation's importance in urban planning. The botanical garden, with the largest green area in the study, has the lowest temperatures. The botanical garden is in a central position in the other zones, providing the greenest space out of all of the zones and being the coolest among the others.

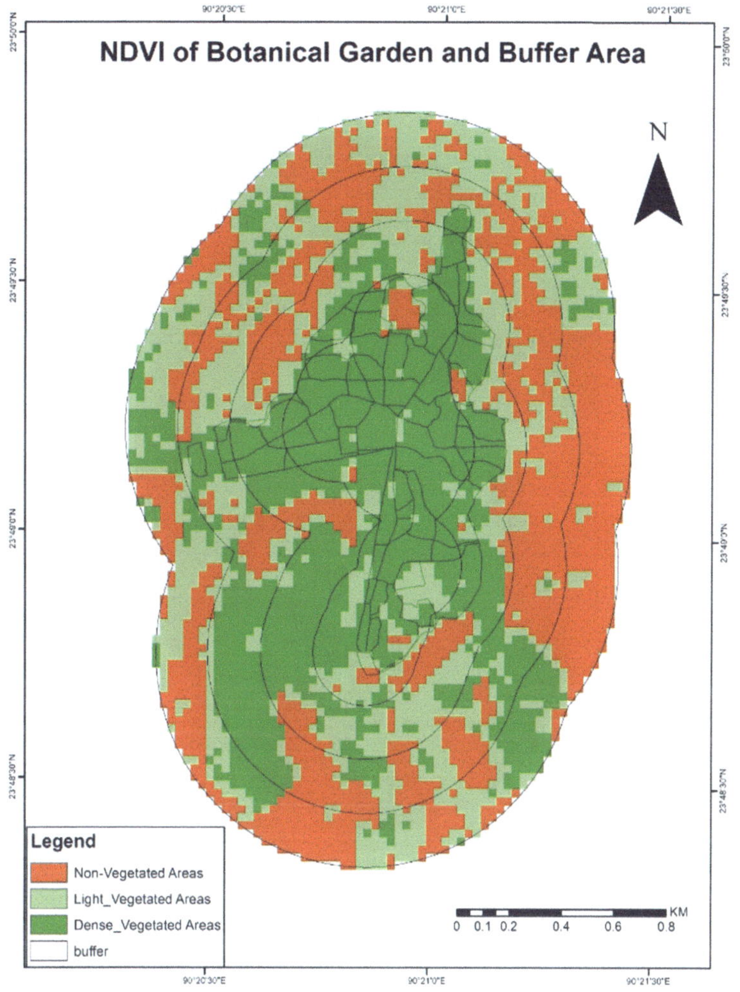

Fig. 13.3 NDVI of botanical and its buffer area

Relationship Between LST and NDBI

Graph 13.2 shows the relationship between the Land Surface Temperature (LST) and the Normalized Difference Built-up Index (NDBI). The vertical axis represents the Land Surface Temperature (ranging between 25 to 40 degrees). Higher values entail higher temperature readings. The horizontal axis represents NDBI, which ranges from − 0.25 to 0.15. The higher the value, the more built-up areas there are and less foliage. A positive correlation- higher NDBI, higher LST- reflects the heat island effect. High temperature readings are found in areas with lots of buildings and not much vegetation. Low temperature readings are found in areas with vegetation. This

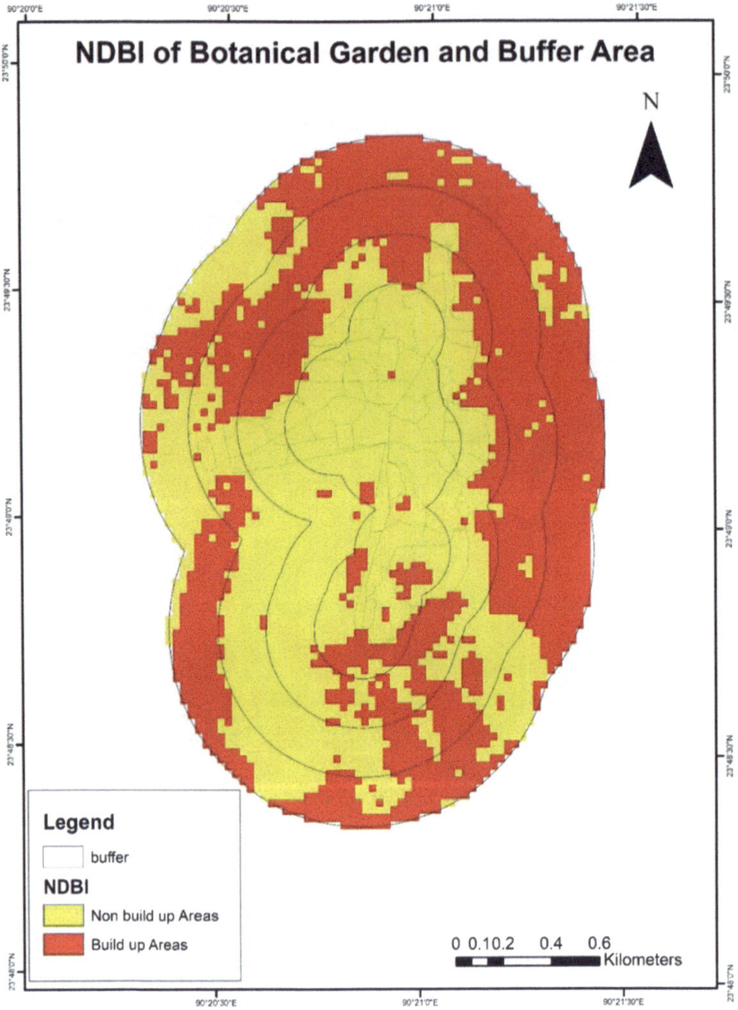

Fig. 13.4 NDBI of botanical and its buffer area

shows that plants cool the air. Compared to the central zone, the buffer zone has higher temperatures due to more built-up space; thus, urbanization is considered a means of increasing the land surface temperature.

Relationship Between LST and NDWI

Graph 13.3 demonstrates that Land Surface Temperature (LST) of 25–40 °C has a reverse relationship with the Normalized Difference Water Index (NDWI) of − 0.3

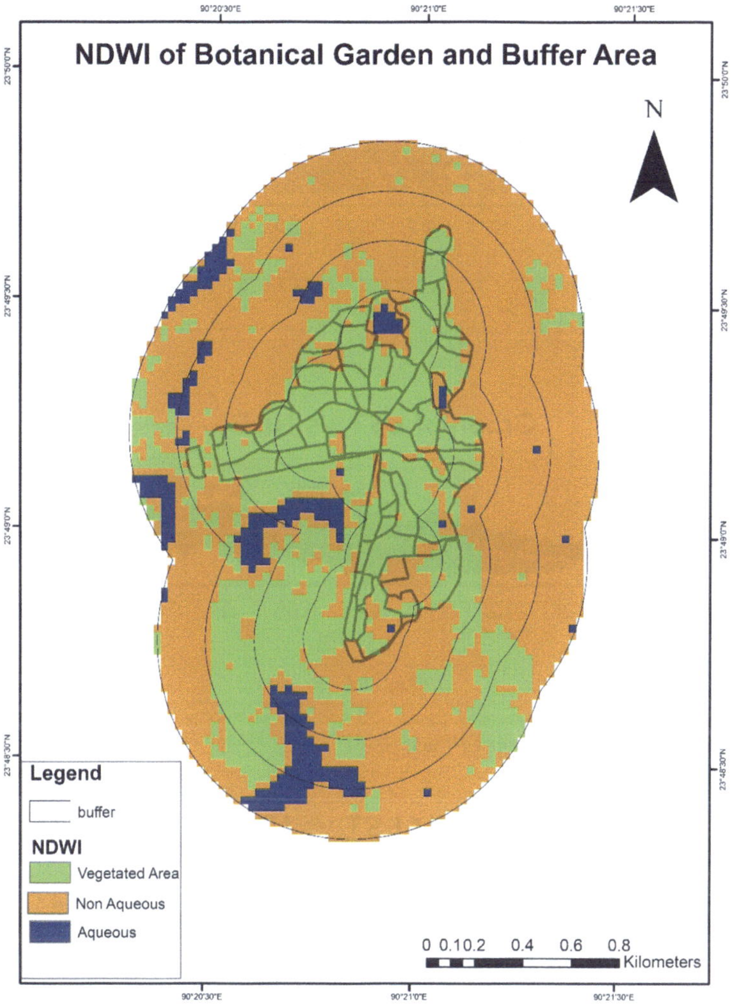

Fig. 13.5 NDWI of botanical and its buffer area

to − 0.05. A small degree of negative association has been observed, which means that places with more moisture content (as represented by NDWI being greater) generally have lower LST, thus verifying the effect of water and well-set vegetation on reducing temperature. Water bodies and irrigated plants can cool the urban heat island effect, but it is not significant. The areas that are more enriched with water bodies, especially in the central to western part of the study area, have less elevated temperatures compared to the more urbanized counterparts. This particular effect is, however, far lesser than that by the green spaces.

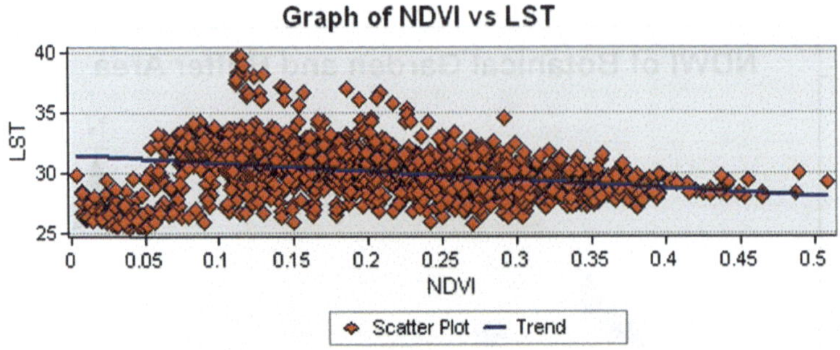

Graph 13.1 Relationship between LST and NDVI

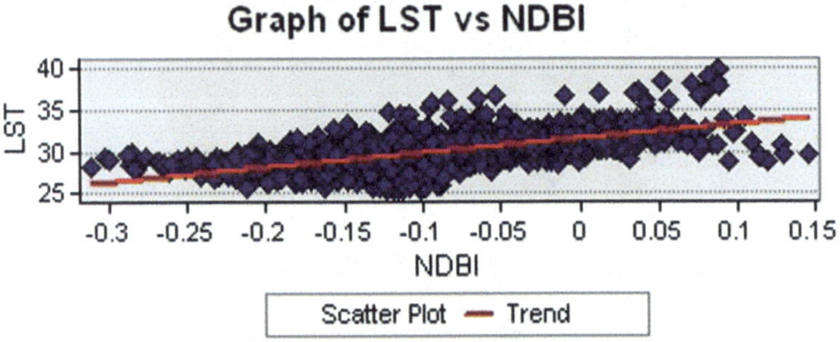

Graph 13.2 Relationship between LST and NDVI

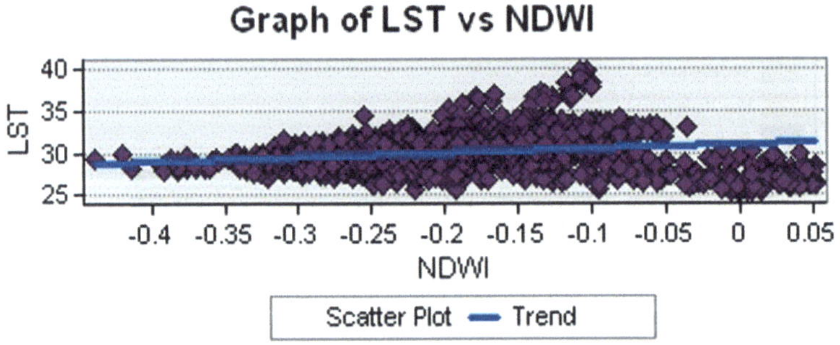

Graph 13.3 Relationship between LST and NDWI

Green Space/Park Cooling Intensity, Green Space Cooling Gradient and Green Space Cooling Efficiency

Park Cooling Intensity

This metric measures the extent to which vegetation lowers the ambient temperature. PCI measures the extent to which a park or green space contributes to the cooling of its urban environment.

The measurement assesses the temperature difference between the park area and the neighboring urban zone. If the park's temperature is inferior than that of the city, the PCI will be positive. A more significant cooling impact correlates with an elevated PCI value.

The PCI is determined using Eq. 13.9. PCI is computed for each zone. The graph is depicted to provide a clear representation (Fig. 13.6; Table 13.3).

The term, Park Cooling Intensity, refers to the temperature difference in degrees of reduction between an open green space and adjacent impervious surfaces such as asphalt or concrete. The X-axis shows the buffer zone distance from the botanical garden. Buffer Zone 1 lies adjacent to the garden, while with each farther zone, these will have a greater distance from it. The y-axis shows the temperature difference between the green space (botanical garden) and impervious surfaces. As the y-axis value increases, the cooling effect of the botanical garden becomes more

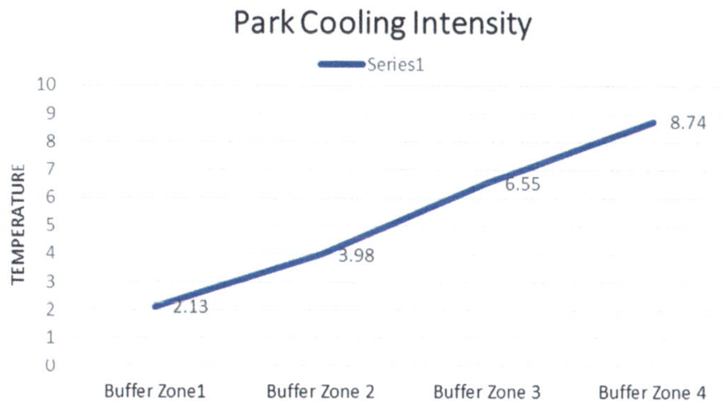

Fig. 13.6 PCI of different buffer zones. *Source* Author 2024

Table 13.3 Buffer zones with PCI

Buffer zone	PCI
1 (200 m)	2.13
2(400 m)	3.98
3(600 m)	6.55
4(800 m)	8.74

Source Author 2024

pronounced. In Buffer Zone 1, temperature variation is about 2.13 degrees, which reflects a cooling. Buffer Zone 2 sees it rise to roughly 3.98. Buffer Zone 3, gets a rise to approximately 6.55. Buffer Zone 4 sees the highest temperature differential, around 8.74 degrees. The botanical park reduces temperature significantly at the surrounding area, Buffer Zone 1. The botanical garden significantly lowers the temperature of the adjacent environment (Buffer Zone 1). As the distance grows, the cooling effect progressively diminishes yet remains substantial. Green spaces provide relief from high heat in urban settings, benefiting both persons and the ecosystem.

Green Space Cooling Efficiency (GCE)

This metric quantifies the degree to which a green space reduces the temperature of the surrounding area. Urban cooling is the ability of green spaces, including urban forests, gardens, and parks, to reduce surface temperatures in their immediate proximity.

This assesses the effectiveness of green spaces in lessening the urban heat island (UHI) impact and their role in creating a healthier and also suitable environment which has optimum thermal satisfaction. Equation 13.10 is used to ascertain the thermal efficiency of the verdant space.

$$\boldsymbol{GCE} = \frac{\boldsymbol{GCA}}{\boldsymbol{GA}} = \frac{(285.84)\,\mathrm{m}^2}{(42.48)\,\mathrm{m}^2}$$

[GCA = 285.84m^2, GA = 42.48m^2]. So, the GCE is 6.72 m^2

A GCE value of 6.72 m^2 for the botanical garden and its buffer means that there is appreciable cooling for each square meter of green space that extends into the surrounding area to a distance of 6.72 m^2. This GCE assesses the effectiveness of the botanical garden and its adjacent buffer zone in mitigating urban heat islands and providing cooler air to the surrounding urban area. The GCE score of 6.72 indicates that approximately 6.72 m^2 of weather are cooled for every square meter of total green space (which includes the botanical garden and buffer zone).

This action lowers the temperature around the place and makes a cooler microclimate. With each increase in GCE, the cooling efficiency goes up even more. This clearly indicates that this particular green space design does allow for substantial cooling of the surrounding area. The effectiveness of cooling is strongly influenced by vegetation density, shade, and spatial arrangement.

Green Space Cooling Gradient (GCG)

This parameter is indicative of a decrease in the cooling power of a green area as one moves away from it. Urban microclimatology refers to the spatial variability of cooling within green areas in the city. It describes the variability of cooling effects across different parts of a park or green space.

The GCG shows how the cooling effect varies as we go from the botanical garden into its adjacent buffer zones. It specifically measures from the garden the rate of temperature change—that is, cooling—per unit distance (in meter). Equation 13.11 is used to calculate the GCG of the botanical garden.

$$GCG = \frac{\boldsymbol{GCI}}{\boldsymbol{L}_{\max}} = \frac{(6.11)^{\circ}\mathrm{C}}{(131.94)\,\mathrm{m}^2}$$

[GCI = 6.11 °C, Lmax = 131.94 m^2].

So, the GCG is **0.04** °C/m^2.

This finding indicates that the cooling impact diminishes by 0.04 °C for each square meter distance from the botanical garden. The temperature in the buffer zone surrounding the botanical garden would progressively rise with increasing distance from the garden. The GCG value of 0.04 °C/m^2 signifies a gradual reduction in cooling impact, indicating that the botanical garden has a significant cooling influence throughout its buffer zone. Due to the closeness of tall buildings and large, impermeable surfaces like roads and parking lots, high urban density can make it harder for cooling effects to spread. To address these challenges, we need integrated future urban planning for sustained maintenance and significant investments in sustainable infrastructure.

The Land Surface Temperature (LST) is positively and even negatively correlated with NDVI NDBI and NDWI. The land surface temperature is reduced in vegetative areas, as LST and NDVI exhibit a negative correlation. Similar to NDVI, NDWI lacks the depth of NDVI. Urbanisation leads to an increase in land surface temperature, which is positively influenced by NDBI. This is especially apparent in the peripheral regions. The significant difference is represented by a departure from the verdant landscape.

By distances from green areas, it indicates how with the increasing distance from park areas, the temperature differences and the urban heat island effects increase. The LST is positively correlated with NDVI; this cooling of the green space is supposed to provide it for green land. Nevertheless, that is something that is actually influenced by urbanized areas. A green cooling gradient reflects a cool value; yet, it does not fully accommodate urban growth. The creation of impermeable surfaces and the urban heat island effect, specifically the deterioration of surrounding areas due to inadequate management of green spaces, increasingly threatens us as we distance ourselves from these regions.

Conclusion and Recommendations

The research has produced comprehensive insights into the several advantages of urban greenery. This study highlights the significance of green areas in alleviating the urban heat island effect, improving thermal comfort, and promoting environmental sustainability in urban environments. Urban green spaces markedly lower adjacent temperatures, improving the experience of outdoor areas during the sweltering summer months. The cooling benefits of green spaces reach beyond their immediate vicinity, offering advantages to neighboring urban areas. This phenomenon is especially evident in highly populated areas such as Mirpur, where the built environment can exacerbate heat retention. The cooling effect is evident within a specific range; however, it decreases in areas with more infrastructure and building density. The study underscores the significance of judicious urban planning and policy initiatives that prioritize the creation and conservation of natural areas. Continuous research and assessment are crucial to guarantee the optimal design and efficacy of urban green areas. The Mirpur, Dhaka case study exemplifies a model for other urban regions facing similar challenges. This illustrates the beneficial effect of thoughtfully designed green spaces on residents' quality of life by mitigating heat stress and enhancing urban livability. The findings of this study may enable the subsequent actions: The development of green spaces in urban environments can facilitate the establishment of sustainable urban regions and improve community well-being. This may mitigate the effects of Urban Heat Island (UHI) and improve thermal comfort, thereby making the city more livable and secure for future generations.

Recommendations—

(1) A rise in the cooling efficiency and a fall in land surface temperatures will enhance thermal comfort in very densely populated areas through the expansion of parks, botanical gardens, and urban forests.
(2) The maintenance and enlargement of vegetation within the buffers can allow further cooling from the parks and counter the urban heat island effects in the various areas in their vicinity.
(3) By integrating green corridors, water bodies, and permeable surfaces, nature-based solutions can enhance microclimatic conditions and advance urban sustainability.

References

Alam D (2019, February 18) The dark side of Dhaka's urbanisation. The Daily Star. https://www.thedailystar.net/supplements/28th-anniversary-supplements/avoiding-urban-nightmare-time-get-planning-right/news/the-dark-side-dhakas-urbanisation-1703425

Akbari H, Kolokotsa D (2016) Three decades of urban heat islands and mitigation technologies research. Energy Build 133:834–842

Cao X, Onishi A, Chen J, Imura H (2010) Quantifying the cool island intensity of urban parks using ASTER and IKONOS data. Landsc Urban Plan 96(4):224–231. https://doi.org/10.1016/j.landurbplan.2010.03.008

Dhakal S, Hanaki K (2002) Improvement of urban thermal environment by managing heat discharge sources and surface modification in Tokyo. Energy Build 34(1):13–23

Dimoudi A, Kantzioura A, Zoras S, Pallas C, Kosmopoulos P (2013) Investigation of urban microclimate parameters in an urban center. Energy Build 64:1–9. https://doi.org/10.1016/j.enbuild.2013.04.014

Givoni B (1991) Impact of planted areas on urban environmental quality: a review. Atmos Environ Part b. Urban Atmos 25(3):289–299

Hirano Y, Fujita T (2012) Evaluation of the impact of the urban heat island on residential and commercial energy consumption in Tokyo. Energy 37(1):371–383

Islam MJ (2024b, April 15) Bangladesh faces extreme weather as longer heatwave looms. The Business Standard. https://www.tbsnews.net/bangladesh/environment/bangladesh-faces-extreme-weather-longer-heatwave-looms-828771

Issue-I TA (2018a, March 24) Dhaka's summertime woes. The Financial Express. https://thefinancialexpress.com.bd/views/dhakas-summertime-woes-1521906111

Issue-I TA (2018b, December 12) Dhaka: the growing urban heat island. The Financial Express. https://thefinancialexpress.com.bd/views/dhaka-the-growing-urban-heat-island-1544627993.

Issue-I SA (2020, January 20) Dhaka's temp stays 2.0 degrees higher. The Financial Express. https://thefinancialexpress.com.bd/national/dhakas-temp-stays-20-degrees-higher-1579496377

Kim J, Khouakhi A, Corstanje R, Johnston AS (2023) Greater local cooling effects of trees across globally distributed urban green spaces. Sci Total Environ 911:168494. https://doi.org/10.1016/j.scitotenv.2023.168494

Lin W, Chen Q, Jiang M, Zhang X, Liu Z, Tao J, ... Zeng Q (2019) The effect of green space behaviour and per capita area in small urban green spaces on psychophysiological responses. Landsc Urban Plann 192:103637

Liu W, Zhao H, Sun S, Xu X, Huang T, Zhu J (2022) Green space cooling effect and contribution to mitigate heat island effect of surrounding communities in Beijing metropolitan area. Front Public Health 10:870403

Mahmud I, Mahmud I (2023, May 1) Dhaka turning into 'heat island': study. Prothomalo. https://en.prothomalo.com/bangladesh/city/erywbwtbnu

Mahmud I, Mahmud I (2024, April 25) Dhaka temperature 3 degrees higher due to lack of necessary trees, water bodies. Prothomalo. https://en.prothomalo.com/environment/5xh2t3wyrx

Misbari S, Hashim M (2016) Change detection of submerged seagrass biomass in shallow coastal water. Remote Sens 8(3):200. https://doi.org/10.3390/rs8030200

Oke TR (1982) The energetic basis of the urban heat island. Q J R Meteorol Soc 108(455):1–24

Report T (2024, April 28) Heatwave: schools, colleges in Dhaka, 4 other districts to remain closed tomorrow. Business Standard. https://www.tbsnews.net/bangladesh/education/heatwave-schools-colleges-dhaka-4-other-districts-remain-closed-tomorrow-838516

Sangwan A, Saraswat A, Kumar N, Pipralia S, Kumar A (2022) Urban green spaces prospects and retrospect's. IntechOpen eBooks. https://doi.org/10.5772/intechopen.102857

Shoab N (2018) A brief study on botanical garden of Bangladesh. Phyjnu. https://www.academia.edu/37718636/A_brief_study_on_botanical_garden_of_bangladesh

Spronken-Smith RA, Oke TR (1998) The thermal regime of urban parks in two cities with different summer climates. Int J Remote Sens 19(11):2085–2104

Survey U-UG (n.d.) EarthExplorer. None. https://earthexplorer.usgs.gov/

Taha H, Akbari H, Rosenfeld A, Huang J (1988) Residential cooling loads and the urban heat island—the effects of albedo. Build Environ 23(4):271–283

Yan H, Wu F, Dong L (2017) Influence of a large urban park on the local urban thermal environment. Sci Total Environ 622–623:882–891. https://doi.org/10.1016/j.scitotenv.2017.11.327

Zaman KT (2024, April 29) Surviving Dhaka's summer: a guide to keeping you cool. The Daily Star. https://www.thedailystar.net/life-living/lifehacks/news/surviving-dhakas-summer-guide-keeping-you-cool-3597951

Chapter 14
Building for Tomorrow: Climate-Resilient Infrastructure Strategies for Urban Sustainability

Mohammad Salman, S. V. P. Najid, Umair Ali, Zulnoorain Khairoowala, and Tahir Abdul Rahman Siddiquee

Abstract Urban areas are increasingly exposed to the impacts of climate change, including extreme heat, flooding, and deteriorating air quality, all of which pose significant threats to public health and urban sustainability. Traditional infrastructure systems, often rigid and outdated, are ill-equipped to respond to these dynamic challenges. This chapter explores climate-resilient infrastructure strategies that enhance urban sustainability by integrating green infrastructure, adaptive urban planning, and public health considerations. Using a qualitative research approach, the study synthesis data from peer-reviewed literature, global policy frameworks, and comparative case studies of cities including Ahmedabad, Chennai, Rotterdam, New York, Singapore, and Copenhagen. The four core dimensions were identified: technological innovation, governance structures, health outcomes, and community participation. The results highlight that cities integrating nature-based solutions, such as permeable surfaces, green roofs, and urban forests, alongside early warning systems and adaptive water infrastructure, have significantly reduced health risks and improved resilience to climate-related stresses. However, challenges remain, especially in low- and middle-income cities, where financial, institutional, and policy constraints hinder large-scale implementation. The chapter argues that climate-resilient urban infrastructure must be inclusive, health-centred, and adaptable to varying urban contexts. It

M. Salman (✉) · U. Ali · Z. Khairoowala
Faculty of Architecture, Planning and Design, Integral University, Lucknow, Uttar Pradesh, India
e-mail: mdsalman@iul.ac.in

U. Ali
e-mail: ar.umair.ali@gmail.com

Z. Khairoowala
e-mail: khairoowala5291@gmail.com

S. V. P. Najid
Beary's Enviro Architecture Design School, Mangalore, India
e-mail: najid@beads.edu.in

T. A. R. Siddiquee
Architecture Section, University Polytechnic, Aligarh Muslim University, Aligarh, India
e-mail: tarsiddiquee.co@amu.ac.in

© The Author(s), under exclusive license to Springer Nature Switzerland AG 2025
A. I. Almulhim et al. (eds.), *Urbanization, Climate Change, and Health*, Climate Change Management, https://doi.org/10.1007/978-3-031-99132-5_14

concludes by recommending mandatory health impact assessments, stronger policy integration, and community co-production in resilience planning. These strategies, if systematically adopted, can help urban areas not only withstand climate-related shocks but also enhance long-term well-being, equity, and sustainability for all residents.

Keywords Climate resilient · Urban sustainability · Urbanization

Introduction

Urbanization and climate change are converging to create unprecedented challenges for urban health, making climate-resilient infrastructure a critical priority. Rising temperatures, worsening air pollution, and the spread of vector-borne diseases are putting millions of urban residents at risk, with vulnerable populations such as the elderly, children and low-income communities disproportionately affected (Sharifi et al. 2021). The urban heat island (UHI) effect, caused by dense concrete and asphalt absorbing and retaining heat, intensifies heat-related illnesses like heatstroke and cardiovascular stress (Piracha and Chaudhary 2022) Additionally, climate change exacerbates air pollution, increasing ground-level ozone and fine particulate matter, which contribute to respiratory diseases such as asthma and bronchitis (Susca and Pomponi 2020). Furthermore, warming climates are expanding the range of disease-carrying vectors, such as mosquitoes, leading to a rise in malaria, dengue, and other infectious diseases (Heaviside et al. 2017).

Poor urban planning and inadequate infrastructure further compound these issues, leaving cities ill-prepared to manage the growing public health crisis. As climate change accelerates, its impacts are felt most severely by marginalized and economically disadvantaged populations. Many low-income communities live in overcrowded, poorly ventilated housing with limited access to air conditioning, clean water, and quality healthcare, increasing their vulnerability to extreme heat, pollution, and disease outbreaks (Hoeben et al. 2023). This integrated perspective on health and infrastructure is illustrated in Fig. 14.1, which outlines the urban resilience approach adopted in this study. Heatwaves disproportionately affect those who cannot afford cooling systems, while pollution exposure is highest in densely populated urban areas with heavy traffic and industrial activity (Sheehan et al. 2021). Similarly, inadequate drainage and sanitation systems in informal settlements create breeding grounds for mosquitoes, worsening the spread of vector-borne diseases (Puntub and Greiving 2022). Addressing these challenges requires a holistic urban planning approach that prioritizes both climate resilience and public health. Green infrastructure, such as urban forests, green roofs, and permeable pavements, helps lower urban temperatures, while air quality regulations, clean energy initiatives, and improved public transit reduce pollution levels. Resilient water and sanitation systems are essential to preventing disease outbreaks, and investing in early warning systems and climate responsive healthcare can further safeguard public health (Semenza 2021).

Fig. 14.1 Urban resilience approach

This research explores how climate-resilient infrastructure can serve as a crucial intervention to mitigate health risks and enhance urban sustainability. By analyzing global case studies and best practices, this study examines how cities can integrate nature-based solutions, smart urban planning, and policy frameworks to protect public health while adapting to climate change. The key questions explored include how infrastructure strategies can reduce heat exposure, improve air quality, and limit disease transmission while ensuring that resilience efforts benefit all segments of society. Additionally, the study investigates the role of governance, funding mechanisms, and interdisciplinary collaboration in fostering climate resilience in urban infrastructure. As climate change continues to reshape urban landscapes, prioritizing public health within resilience strategies is not only a necessity but also an opportunity to create sustainable, equitable, and future-ready cities that can thrive despite environmental uncertainties.

Literature Review

Climate change has emerged as a major challenge to urban sustainability, particularly in its impact on public health. As global temperatures rise, cities face increased risks of heatwaves, declining air quality, and extreme weather events such as flooding, all of which threaten the health and well-being of urban populations. Research has consistently shown that these challenges disproportionately affect vulnerable communities, including low-income groups, children, and the elderly. In response, climate-resilient infrastructure has been identified as a key strategy for mitigating these health risks while promoting urban sustainability. This literature review examines existing research on the health risks of climate change in urban areas, the role of green infrastructure in improving public health, and international policy frameworks that support the integration of health considerations into urban resilience planning.

Health Risks of Climate Change in Urban Areas

Climate change has emerged as a defining challenge of the twenty-first century, with urban areas bearing a disproportionate share of its adverse impacts. Rapid urbanization, coupled with inadequate infrastructure and socio-economic disparities, has heightened the vulnerability of urban populations to climate-induced health risks. Rising temperatures, declining air quality, increased frequency of extreme weather events, and deteriorating water sanitation systems have all contributed to a complex web of public health threats. Understanding these risks is essential for informing infrastructure planning and resilience strategies. This section examines the major health hazards associated with climate change in urban settings, with a focus on heat-related mortality, air pollution, and waterborne diseases.

Heat-Related Mortality and the Urban Heat Island Effect

One of the most well-documented health impacts of climate change in urban settings is heat-related mortality and morbidity. The urban heat island (UHI) effect exacerbates high temperatures in cities by trapping heat due to impervious surfaces like concrete, asphalt, and metal (Huang et al. 2025). Studies show that urban areas can be up to 7 °C hotter than surrounding rural areas, significantly increasing the risk of heatstroke, dehydration, and cardiovascular illnesses (Ho et al. 2023). A study conducted in Hong Kong from 2010 to 2019 found that heat-related mortality was significantly higher in urban areas compared to rural counterparts, especially under extreme heat scenarios (Huang et al. 2025). Furthermore, economic valuation of heat-related mortality in European cities indicated that the economic burden of UHI effects is substantial, further underscoring the need for climate-adaptive strategies (Macintyre et al. 2021). Heat-adaptive urban infrastructure, including green roofs, reflective materials, and increased vegetation, has been proposed as a crucial intervention to mitigate UHI effects and protect public health (Heaviside et al. 2017). These strategies not only reduce surface and ambient temperatures but also improve air quality and overall urban liveability.

Air Quality Degradation and Respiratory Health

Air pollution is another significant urban health challenge linked to climate change. Vehicle emissions, industrial pollution, and the burning of fossil fuels contribute to high concentrations of particulate matter (PM2.5), nitrogen dioxide (NO_2), sulphur dioxide (SO_2), and ground-level ozone (O_3) (Lu et al. 2021). These pollutants are known to cause chronic respiratory diseases, cardiovascular illnesses, and premature mortality (Rovira et al. 2020). A study analysing global PM2.5 levels found that urbanization and meteorological changes have exacerbated air pollution-related health burdens, particularly in densely populated cities (Lu et al. 2021). Additionally, exposure to PM2.5 has been linked to significant reductions in life expectancy, with long-term inhalation contributing to severe respiratory and cardiovascular conditions (Yang et al. 2019). A case study in China confirmed that air pollution is a major risk factor for stroke and other life-threatening diseases, further emphasizing the urgency of effective mitigation strategies (Wu et al. 2017). Urban areas with high vehicle congestion and industrial activity experience the worst air quality, disproportionately affecting low-income populations who lack access to healthcare and live in areas with poor air circulation. Solutions such as low-emission zones, clean energy public transport, and urban greening have been found to effectively reduce air pollution and its health impacts. Studies highlight that integrating these measures into urban planning can significantly lower pollution-related mortality and improve overall public health (Garcia et al. 2016).

Flooding and Waterborne Diseases

Flooding, exacerbated by climate change-induced extreme rainfall events, significantly increases the risk of waterborne diseases. Contaminated floodwaters can spread pathogens that cause cholera, typhoid, leptospirosis, and dysentery, particularly in informal settlements with inadequate sanitation infrastructure (Noureen et al. 2022). A study found that climate change-related extreme weather events, such as floods, are increasing waterborne disease risks, particularly in low-income regions where access to clean water and sanitation is limited (Cissé 2019). Moreover, research indicates that post-flood outbreaks of diarrheal and vector-borne diseases disproportionately affect children under five and elderly populations (Walker 2018). In regions such as Bangladesh, India, and Nigeria, flood-prone urban areas have seen significant increases in malaria and dengue transmission due to prolonged water stagnation, which creates breeding grounds for mosquitoes (Levy et al. 2018). Climate-resilient infrastructure solutions, such as improved drainage systems, flood-resistant housing, and integrated water management strategies, are critical for mitigating these risks. A review suggests that addressing the links between climate change and waterborne diseases through targeted interventions can reduce public health risks and improve urban resilience (Davies et al. 2014).

Green Infrastructure and Health

Green infrastructure has gained prominence as a vital component of climate-resilient urban development due to its multifaceted benefits for environmental sustainability and human well-being. Beyond its ecological value, green infrastructure contributes significantly to public health by mitigating climate-related hazards such as extreme heat and air pollution. Urban vegetation, parks, green roofs, and bioswales not only regulate microclimates but also enhance respiratory health through pollutant filtration and improved air circulation. This section explores the health co-benefits of green infrastructure, particularly its roles in reducing urban heat exposure and improving air quality, thereby supporting healthier and more liveable cities.

Urban Green Spaces and Heat Reduction

Green infrastructure, such as urban parks, green roofs, and street trees, has been extensively studied for its ability to reduce urban temperatures and improve public health. Research shows that green spaces can effectively mitigate the urban heat island effect by lowering ambient temperatures and reducing the risks associated with extreme heat events (Azmeer et al. 2024). A study evaluating green infrastructure in European cities found that strategic placement of urban greenery significantly reduces the temperature burden in densely populated areas, directly benefiting public health (Iungman et al. 2023). Additionally, studies have demonstrated that green spaces can reduce ambient temperatures by 2–5 °C, significantly lowering the risk of heat-related illnesses (Gobatti et al. 2023). Tree-lined streets and urban parks provide cooling effects that improve thermal comfort and reduce mortality during heatwaves (Budzik et al. 2024). These findings support the integration of green infrastructure into urban planning as an essential tool for climate resilience and public health improvement.

Air Quality Improvements and Respiratory Health

Green spaces also play a crucial role in filtering pollutants and improving air quality. Studies have shown that urban green infrastructure, such as urban forests and green walls, can effectively mitigate airborne pollutants, reducing exposure to harmful particles like PM2.5 and nitrogen oxides (Venter et al. 2024). A large-scale assessment found that green spaces help regulate air quality by absorbing pollutants and enhancing ventilation, leading to better respiratory health outcomes (Diener and Mudu 2021). Moreover, urban parks and tree-lined streets have been linked to lower rates of respiratory diseases such as asthma and bronchitis, particularly in densely populated areas (Tomson et al. 2021). Research has further demonstrated that green corridors and bioswales can serve as effective air filtration systems, improving local air quality and reducing stress levels among urban residents (Moradpour and Hosseini 2020). Urban planners increasingly advocate for tree-planting programs and sustainable vegetation management as cost-effective strategies for enhancing air quality and climate resilience.

International Frameworks Supporting Climate-Resilient Infrastructure

In response to the growing intersection of climate change and public health, several international frameworks have emerged to guide urban policy and infrastructure development. These frameworks emphasise the need for integrated, cross-sectoral strategies that align climate resilience with health equity and sustainable urbanization. By embedding health considerations within broader planning and governance processes, global initiatives such as the World Health Organization's Health in All Policies (HiAP) and the United Nations Sustainable Development Goals (SDGs) provide essential guidance for cities aiming to build climate-resilient infrastructure. This section explores how these frameworks support local governments in translating global objectives into actionable urban strategies.

WHO's Health in All Policies (HiAP) Approach

The Health in All Policies (HiAP) approach, developed by the World Health Organization (WHO), emphasizes integrating health considerations into urban planning, transportation, housing, and climate policies. This framework promotes cross-sector collaboration to ensure that health equity and sustainability are central to decision-making (Greer et al. 2022). Cities that implement HiAP principles benefit from multi-sectoral cooperation, leading to reduced health disparities and enhanced climate adaptation efforts. A study highlighted that HiAP fosters healthier urban environments by aligning public health with sustainable development goals (Ramirez-Rubio et al. 2019). Additionally, local governments that have adopted HiAP report improvements in community well-being and policy coherence. For example, a review of HiAP implementation at the municipal level found that integrating health into urban policies enhances resilience against climate-related health risks while also promoting equity in public service delivery (Guglielmin et al. 2018).

United Nations Sustainable Development Goals (SDGs)

The United Nations Sustainable Development Goals (SDGs) provide a global framework for addressing urban health and sustainability. SDG 3 (Good Health and Well-Being) and SDG 11 (Sustainable Cities and Communities) specifically promote climate-resilient infrastructure, access to clean water, and sustainable urban environments (Barcellos-Paula et al. 2025). Cities that align their policies with SDG targets have reported improvements in air quality, heat resilience, and disease prevention. A study found that urban regeneration efforts incorporating SDG 11 principles foster climate resilience and enhance urban sustainability (Allan et al. 2024). Additionally, SDG-aligned policies play a crucial role in promoting health equity and climate adaptation in urban settings. Research has demonstrated that integrating sustainability strategies into urban planning can mitigate climate-related hazards and reduce environmental health risks (Sebestyén et al. 2023). Cities that incorporate these principles into their governance frameworks have seen measurable improvements in public health outcomes and environmental resilience (Orsetti et al. 2022).

Methodology

Research Design

This study employs a qualitative, multi-method research design to investigate how climate-resilient infrastructure contributes to urban sustainability and public health. The approach integrates an extensive literature review with comparative case study analysis to identify effective adaptation strategies used across diverse urban contexts. Figure 14.2 presents a schematic representation of the research design employed. This design allows for in-depth exploration of policy, planning, and implementation dynamics in both high-income and resource-constrained cities.

Data Sources and Literature Review Protocol

The literature review was conducted using a systematic approach across major academic databases, including Scopus, Web of Science, and Google Scholar. The search covered literature published between 2010 and 2024 using keywords such as "climate-resilient infrastructure," "urban heat island," "green infrastructure," "urban flooding," "air quality and health," and "sustainable urban planning." Government reports, urban resilience strategies, and policy documents were also reviewed to contextualise findings.

Inclusion criteria required sources to:

i. Address urban resilience strategies related to climate and health.
ii. Focus on infrastructure or policy-level interventions.
iii. Provide empirical or conceptual analysis.

Sources focusing exclusively on rural areas, speculative design without implementation, or lacking health-related dimensions were excluded. In total, over 150 documents were screened, of which 48 met the inclusion criteria for analysis.

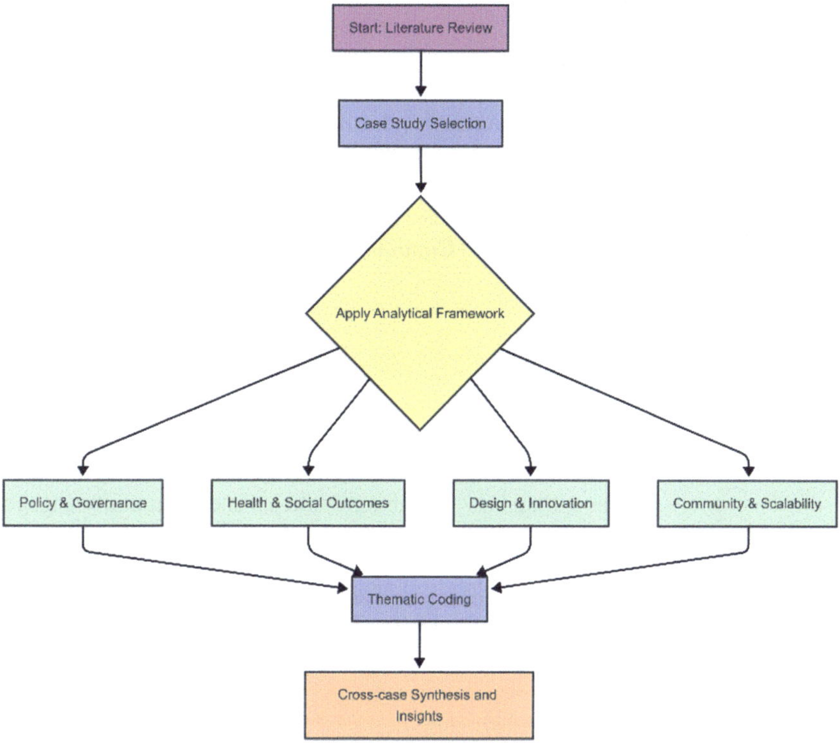

Fig. 14.2 Research design

Analytical Framework for Case Studies

To standardise comparison and minimise subjectivity across diverse case studies, a four-dimensional analytical framework was developed. Each case study was analysed according to the following axes:

i. Policy and Governance Integration: Degree to which resilience strategies are embedded in urban policy and institutional frameworks.
ii. Health and Social Outcomes: Impact on public health metrics (e.g., heat mortality, respiratory illness, waterborne disease) and equity outcomes.
iii. Design and Technological Innovation: Use of adaptive, nature-based, or smart infrastructure strategies.
iv. Community Engagement and Scalability: Extent of local participation and potential for replication in different urban contexts.

This framework guided data extraction and synthesis from secondary sources for each case, ensuring comparability despite contextual variation.

Analytical Approach

The analysis was conducted through a structured, interpretive synthesis of secondary sources, primarily scholarly literature, policy documents, and urban planning reports. Instead of formal coding software or grounded theory procedures, the study applied a deductive framework based on four dimensions identified during the literature review: (1) Policy and Governance Integration, (2) Health and Social Outcomes, (3) Design and Technological Innovation, and (4) Community Engagement and Scalability.

Case Studies

The case studies highlight the success of climate-resilient infrastructure strategies in both India and abroad. The Ahmedabad Heat Action Plan and Chennai's flood resilience initiatives demonstrate how data-driven policies and community engagement can protect vulnerable populations. Research shows that optimizing critical infrastructure projects enhances urban resilience to climate change, allowing cities to better withstand extreme weather events and environmental stressors. Meanwhile, international examples such as Rotterdam's climate adaptation and Singapore's green infrastructure model illustrate how integrating nature-based solutions, technology, and sustainable urban planning can create climate-resilient, livable cities. Studies indicate that hybrid infrastructure designs, which combine traditional and green infrastructure elements, are crucial for urban climate adaptation and disaster resilience. Moreover, cities that adopt comprehensive resilience strategies improve their ability to recover from climate-induced disruptions and safeguard public health. As climate change continues to challenge urban environments, lessons from these case studies can help other cities develop comprehensive, adaptive infrastructure strategies to safeguard both human health and urban sustainability.

Case Studies in Context: Indian Cities

Ahmedabad Heat Action Plan (HAP), India

Ahmedabad, a major city in western India, has been significantly impacted by rising temperatures and extreme heat events, exacerbated by the urban heat island (UHI) effect. In 2010, a severe heatwave caused over 1344 deaths, prompting the Ahmedabad Municipal Corporation (AMC) to develop India's first Heat Action Plan (HAP) in collaboration with health and climate experts. The plan, implemented in 2013, introduced an early warning system to alert residents about impending heatwaves through SMS, media, and community networks (Limaye et al. 2023).

Additionally, the initiative promoted cool roofs, encouraging the use of white reflective paint and green roofs to lower indoor temperatures in vulnerable communities. The city also established public cooling centres, distributed water kiosks, and trained healthcare workers to treat heat-related illnesses effectively. Research highlights that Ahmedabad's HAP has significantly reduced heat-related mortality and improved urban resilience (Golechha et al. 2021). Over the years, the HAP has proven successful in reducing heat-related mortality, with similar models now being implemented in other Indian cities such as Hyderabad, Bhubaneswar, and Nagpur. A broader analysis of India's heat adaptation strategies found that Ahmedabad's HAP has been instrumental in scaling up heat risk governance nationwide (Singh et al. 2024). Recognised internationally by the Natural Resources Defence Council (NRDC) and the World Health Organization (WHO), Ahmedabad's initiative serves as a scalable model for other heat-vulnerable cities worldwide.

Chennai's Flood Resilience Program, India

Chennai, a coastal megacity in southern India, has experienced severe flooding, particularly during the devastating 2015 floods that displaced thousands and caused significant economic losses. Unregulated urban expansion, poor drainage infrastructure, and the destruction of natural wetlands contributed to the crisis. In response, the Chennai Flood Resilience Program adopted sponge city principles, focusing on reviving water bodies, improving drainage systems, and integrating permeable surfaces into urban planning. The master plan for Chennai's sponge city model aims to enhance flood resilience by incorporating nature-based solutions (Mishra et al. 2023). The restoration of over 200 lakes and reservoirs improved the city's natural flood absorption capacity while enhancing groundwater recharge. Research highlights that nature-based solutions, including bioswales, porous pavements, and rain gardens, have been instrumental in reducing the intensity of surface runoff and improving Chennai's flood resilience (Ahmad and Hassan 2025). Additionally, blue-green infrastructure approaches, inspired by successful implementations in China and the Netherlands, have proven effective in mitigating flood risks while enhancing urban climate adaptation (García Sánchez and Govindarajulu 2023). These measures are now being replicated in other Indian cities like Mumbai, Kolkata, and Bengaluru, where flooding remains a major urban challenge. Chennai's initiative serves as a model for sustainable urban flood management, demonstrating how climate-responsive planning can enhance resilience in rapidly urbanizing environments (Banga et al. 2023).

Case Studies in Context of Various Cities Across the Globe

Rotterdam Climate Adaptation Strategy, Netherlands

Rotterdam, one of Europe's largest port cities, faces increasing threats from sea-level rise, heavy rainfall, and urban heat stress. Given that much of the city lies below sea level, it has become a global leader in climate adaptation and flood resilience. The Rotterdam Climate Adaptation Strategy integrates multiple innovative solutions, including floating infrastructure, green roofs, and water plazas. Urban planning in Rotterdam prioritizes sustainable land use and responsible water management, making the city a model for adaptive strategies (van der Berg 2023). The city has developed floating houses, farms, and offices, allowing urban functions to continue despite rising water levels. Additionally, over 300,000 m^2 of green roofs have been installed across buildings, significantly reducing temperatures and improving air quality (Khader 2021). Another innovative measure is the introduction of water plazas, which serve as recreational spaces during dry weather and transform into temporary water storage basins during heavy rainfall. These multifunctional spaces enhance public engagement while strengthening urban resilience. Rotterdam has also strengthened its flood protection through storm surge barriers like the Maeslantkering, one of the world's most advanced movable barriers. This structure, along with other large-scale interventions, provides long-term flood risk management while integrating nature-based solutions (Restemeyer et al. 2018). These interventions have successfully enhanced the city's urban resilience, serving as a model for flood-prone cities such as New Orleans, Jakarta, and Bangkok.

Singapore's Green Urban Planning Model

Singapore, a high-density island city-state, has pioneered climate-resilient urban planning by integrating green infrastructure and water-sensitive design into its urban development. The Skyrise Greenery Program has transformed rooftops and vertical facades into lush green spaces, reducing urban heat and improving air quality. Research shows that green space policies play a crucial role in enhancing urban resilience to climate change by reducing surface temperatures and mitigating air pollution (Sabri et al. 2023). Additionally, the Active, Beautiful, Clean (ABC) Waters Program has revolutionised stormwater management by incorporating rain gardens, bioswales, and constructed wetlands, which naturally filter pollutants and enhance flood resilience. Singapore has adopted an integrated approach to drainage and wastewater management, reinforcing its commitment to sustainable water-sensitive urban design (Goh et al. 2024).

The city has also implemented strict building codes mandating energy-efficient construction and cooling systems to combat rising temperatures. Singapore's green infrastructure strategies, including the use of nature-based solutions, have helped improve urban sustainability and mitigate the adverse impacts of climate change.

Additionally, the city has advanced waste-to-energy initiatives, reducing landfill dependency and promoting a circular economy. Due to these efforts, urban temperatures in Singapore have been reduced by 2 °C in certain areas, significantly improving thermal comfort and public health (Xu and Fricker 2023). Recognised as a global leader in sustainable urban planning, Singapore's strategies have influenced other cities, including Dubai, Hong Kong, and Shanghai, to adopt similar approaches. Its innovative policies serve as a model for climate-responsive urban development worldwide.

New York City's Urban Heat Island Initiative, USA

New York City (NYC), one of the largest urban centers in the world, faces significant climate challenges, especially from rising temperatures and the urban heat island (UHI) effect. The city's dense development, minimal green spaces, and extensive concrete surfaces result in temperatures that are often 5–10 °F higher than surrounding rural areas, exacerbating heat-related illnesses and energy demand. In response, NYC launched the Urban Heat Island Initiative (UHII) in 2007, which includes a comprehensive range of climate-resilient measures aimed at cooling down the city. The initiative is part of broader climate adaptation efforts developed by the New York City Panel on Climate Change (NPCC) to track and mitigate UHI effects (Solecki and Rosenzweig 2020). One of the key strategies of this initiative is the Cool Roofs program, which encourages building owners to apply reflective coatings to their rooftops, reducing the absorption of heat. Studies have shown that cool roof applications and light-coloured surfaces significantly lower indoor and outdoor temperatures in dense urban environments (Akbari et al. 2016). The initiative also focuses on expanding urban green spaces and increasing tree canopy coverage through programs like Million Trees NYC, which aims to plant and maintain a million trees across the five boroughs. These green infrastructure projects contribute to air quality improvements and heat stress reduction (Leal Filho et al. 2018).

The city has also introduced green walls and green infrastructure solutions, including bioswales and permeable pavements, to reduce surface temperatures and manage stormwater. The effectiveness of such measures has been widely recognised, with similar UHI mitigation strategies being adopted in cities such as Los Angeles and Chicago (Parihar and Birman 2024). Additionally, the NYC Department of Environmental Protection (DEP) collaborates with local communities to ensure that heatwave response plans are in place, particularly for vulnerable populations. These measures have led to a 2 °C reduction in temperature in certain neighbourhoods and a noticeable improvement in air quality and public health. The success of these initiatives highlights NYC's leadership in climate resilience planning and its potential as a model for other heat-vulnerable cities globally.

Copenhagen's Climate Adaptation Plan, Denmark

Copenhagen, Denmark's capital, has long been recognized as a leader in climate resilience and sustainable urban development. The city faces a variety of climate risks, particularly from rising sea levels, increased rainfall, and urban flooding. In 2011, Copenhagen unveiled its Climate Adaptation Plan with the ambitious goal of becoming carbon–neutral by 2025 and ensuring the city is prepared for climate impacts. A key feature of the plan is its integrated approach to stormwater management, which has successfully merged climate resilience with broader sustainability targets (Xu et al. 2021). Copenhagen has developed an innovative system of green roofs, urban wetlands, and underground water storage that can hold excess stormwater during heavy rainfall and release it gradually into the city's sewer system. Additionally, the city has transformed several public spaces into water plazas, designed to temporarily store water during extreme rainfall events while providing recreational space during dry conditions. Research highlights that these nature-based adaptation projects serve as a model for urban resilience and sustainable planning (Jørgensen et al. 2022).

Beyond flood management, the city has invested heavily in energy-efficient buildings, sustainable mobility infrastructure (e.g., bike lanes and electric public transport), and urban greening initiatives. Copenhagen's climate strategy is deeply rooted in public–private partnerships, fostering co-creation of ambitious climate mitigation and adaptation goals (Sørensen and Torfing 2022). The city's Climate Adaptation Plan is part of Copenhagen's broader sustainability vision, aligned with global frameworks such as the UN Sustainable Development Goals (SDGs), particularly SDG 13 (Climate Action) and SDG 11 (Sustainable Cities and Communities). Due to these efforts, Copenhagen has become a model for other European cities like Amsterdam and Stockholm and is frequently cited as one of the most climate-resilient cities in the world (Andreucci et al. 2021).

Discussion

A comparative summary of the six case studies across four resilience dimensions is provided in Table 14.1. The analysis of climate-resilient infrastructure strategies across six diverse cities highlights several critical patterns and actionable insights. While the specific interventions varied by geography, governance capacity, and socio-economic context, key success factors emerged across all cases. This discussion synthesizes these findings under four central themes: Comprehensive Planning, Policy Integration, Community Engagement, and Scalability, with attention to public health co-benefits and cross-context applicability.

Table 14.1 Comparative analysis of climate-resilient urban infrastructure strategies across global cities

City	1. Policy and governance integration	2. Health and social outcomes	3. Design and technological innovation	4. Community engagement and scalability
Ahmedabad heat action plan (HAP), India	Municipal-led Heat Action Plan integrated post-crisis; now nationally recognized	Direct health benefits; significant reduction in heat-related mortality	Cool roofs, early warning systems, low-cost interventions	High community outreach via NGOs and health workers; scaled to 30 + Indian cities
Chennai's flood resilience program, India	Multi-agency flood planning emerging; sponge city ideas adopted but fragmented	Moderate reduction in waterborne disease risks	Wetland restoration, rain gardens, permeable surfaces	Community awareness through NGOs; moderate scalability under Smart Cities Mission
Rotterdam climate adaptation strategy, Netherlands	Deeply institutionalized adaptation strategy ("Rotterdam Climate Proof")	Improved public health via flood mitigation and cooling infrastructure	Floating buildings, green roofs, multifunctional water plazas	Strong co-design processes with residents; model studied globally
Singapore's green urban planning model	Urban Redevelopment Authority mandates greening and resilience standards	Measurable improvements in heat and air pollution-related outcomes	Vertical greening, bioswales, stormwater management	Moderate engagement through public campaigns; high replication potential in dense cities
New York City's urban heat island initiative, USA	Climate policies embedded within municipal and state frameworks (NPCC)	Targeted programs for low-income populations reduce heat vulnerability	Green roofs, bioswales, "Cool Neighborhoods NYC" initiative	High civic participation in greening efforts; model expanded to other U.S. cities
Copenhagen's climate adaptation plan, Denmark	Resilience mainstreamed into long-term climate and land use planning	Co-benefits include reduced air pollution, improved mental well-being	Green mobility, stormwater tunnels, energy-efficient design	Very strong citizen co-design; widely emulated in European urban contexts

Policy and Governance Integration

A consistent finding across cases is that strong institutional alignment and cross-sectoral governance significantly enhance resilience outcomes. Rotterdam, Singapore, and Copenhagen demonstrated a high level of policy integration, where climate adaptation is institutionalized through regulatory mandates and supported by dedicated agencies or interdepartmental cooperation. In contrast, cities like Chennai and

Ahmedabad showed emerging models of policy innovation, often driven by crisis events (e.g., floods, heatwaves), but lacked enduring governance integration. Moreover, cities with top-down but coordinated governance (e.g., Singapore) were able to enforce resilience mandates through urban redevelopment authorities, while those with fragmented municipal systems (e.g., many Indian cities) faced challenges in sustaining intersectoral collaboration. Institutionalized planning, particularly where climate objectives are embedded in land use and infrastructure regulation, proved critical for delivering long-term resilience benefits.

Health and Social Outcomes

All six cities demonstrated recognition of the intersection between climate risks and public health, but their ability to operationalise this understanding varied. Ahmedabad's Heat Action Plan stands out for directly targeting health outcomes—reducing heat-related mortality through early warning systems, medical training, and public awareness. Similarly, New York's green roof and cooling programs prioritised heat stress mitigation in low-income neighbourhoods. European cities, while not always health-centric in language, delivered strong co-benefits through flood control, pollution reduction, and access to green space. Singapore's vertical greening and Copenhagen's active mobility systems reduced respiratory illness and improved mental health through enhanced liveability. Yet, the findings reveal that health outcomes are often indirect byproducts of resilience strategies rather than core metrics of planning success. Few cities mandated Health Impact Assessments (HIAs), and equity-focused health data were seldom central to decision-making. This suggests a need for more explicit inclusion of health indicators in climate-resilient infrastructure design.

Design and Technological Innovation

Innovation was evident in both high- and middle-income cities, though scale and sophistication differed. Rotterdam and Copenhagen pioneered multifunctional infrastructure—such as water plazas and underground retention systems—that combine urban utility with public space and ecological benefits. Singapore's Skyrise Greenery and ABC Waters programs exemplify integrated design at scale, merging aesthetics, resilience, and biodiversity. Ahmedabad and Chennai used more low-cost, context-specific strategies—like reflective roofs, restored wetlands, and urban lakes—that are affordable yet impactful. New York's hybrid gray-green infrastructure approach (e.g., bioswales, permeable pavements) illustrated how existing systems can be retrofitted for dual-use functions. This comparative view shows that technological innovation is not solely a function of resources, but of how well infrastructure strategies are tailored to local risks, cultural preferences, and spatial constraints. The flexibility and multifunctionality of infrastructure emerged as key success factors across contexts.

Community Engagement and Scalability

Public participation proved to be a strong enabler of both program uptake and long-term sustainability. In Ahmedabad and New York City, community health workers, civil society actors, and local greening initiatives played critical roles in implementing and maintaining resilience projects. Copenhagen's co-design of water plazas and urban green spaces fostered strong civic ownership, reinforcing public trust in climate governance. However, participatory practices were less institutionalized in some cases. While Singapore engaged communities through public campaigns, its model remains largely state-driven. Chennai's community engagement, though promising, was often project-dependent rather than structurally embedded. On scalability, several models demonstrated strong potential for replication. Ahmedabad's HAP has expanded to over 30 cities in India. Rotterdam's and Singapore's strategies have influenced similar delta or tropical cities globally. These examples underscore the importance of designing solutions that are not only innovative but also adaptable to diverse urban governance and resource conditions.

Case Study Inferences

Conclusion

As climate change accelerates and urban populations continue to grow, cities face the dual challenge of adapting to environmental risks while safeguarding public health. This study has shown that climate-resilient infrastructure—when integrated into comprehensive, health-centered urban planning—can significantly enhance both urban sustainability and population well-being. Case studies from Ahmedabad to Rotterdam reveal that cities can reduce the impacts of heatwaves, flooding, and air pollution through a combination of green infrastructure, early warning systems, and inclusive governance. A central finding of this research is the importance of embedding public health considerations within climate adaptation frameworks. Strategies that reduce urban heat islands, improve air quality, and mitigate flood risks offer direct and measurable co-benefits for human health, especially among vulnerable populations. However, realizing these benefits requires institutional support in the form of policy mandates, such as health impact assessments (HIAs), green building codes, and climate-integrated zoning regulations.

Moreover, community engagement emerged as a decisive factor in the success of climate-resilient interventions. Cities that fostered participatory planning—such as New York, Copenhagen, and Ahmedabad—achieved greater acceptance, contextual relevance, and sustainability of interventions. Equally, the challenge of scalability in resource-constrained cities underscores the need for adaptable and cost-effective solutions, including nature-based designs and locally-led resilience initiatives.

Moving forward, cities must adopt multidimensional resilience strategies that balance environmental imperatives with social equity and public health. National urban policies and international climate frameworks should prioritize urban resilience not merely as infrastructure upgrading, but as a transformative opportunity to design cities that are just, adaptive, and inclusive. Collaborative governance, sustained financing, and cross-sector innovation will be essential to realizing this vision. Only through such integrated efforts can urban areas navigate the complexity of climate change and emerge as healthier, more sustainable environments for all residents.

Acknowledgements This chapter is prepared with the assistance of AI tools (mainly ChatGPT) to refine the language and to maintain the flow of writing. The research analysis and arguments included in this chapter are the original contributions of the author, while the AI is just used to enhance the writing clarity and grammatical errors.

References

Ahmad N, Hassan Q (2025) Ecosystem services linked to nature-based solutions for resilient and sustainable cities in India. Front Water 6. https://doi.org/10.3389/frwa.2024.1504492

Akbari H, Cartalis C, Kolokotsa D, Muscio A, Pisello AL, Rossi F, Santamouris M, Synnefa A, Wong NH, Zinzi M (2016) Local climate change and urban heat island mitigation techniques—the state of the art. J Civ Eng Manag 22(1):1–16. https://doi.org/10.3846/13923730.2015.1111934

Allan M, Rajabifard A, Foliente G (2024) Climate resilient urban regeneration and SDG 11—stakeholders' view on pathways and digital infrastructures. Int J Digit Earth 17(1). https://doi.org/10.1080/17538947.2024.2385076

Andreucci MB, Marvuglia A, Baltov M, Hansen P (2021) Future city 15 rethinking sustainability towards a regenerative economy. http://www.springer.com/series/8178

Azmeer A, Tahir F, Al-Ghamdi SG (2024) Progress on green infrastructure for urban cooling: evaluating techniques, design strategies, and benefits. Urban Climate 56:102077. https://doi.org/10.1016/j.uclim.2024.102077

Banga JS, Ninsalam Y, Fien J (2023) Blue and green: hues of riverine flood risk reduction in Chennai, India. In: Water and environment for sustainability. Springer International Publishing, pp 17–44. https://doi.org/10.1007/978-3-031-27280-6_2

Barcellos-Paula L, Castro-Rezende A, Gil-Lafuente AM (2025) Understanding urban resilience and SDGs: a new approach in decision-making for sustainable cities. J Public Affairs 25(1). https://doi.org/10.1002/pa.70007

Budzik G, Sylla M, Kowalczyk T (2024) Understanding urban cooling of blue-green infrastructure: a review of spatial data and sustainable planning optimization methods for mitigating urban heat islands. Sustainability 17(1):142. https://doi.org/10.3390/su17010142

Cissé G (2019) Food-borne and water-borne diseases under climate change in low- and middle-income countries: Further efforts needed for reducing environmental health exposure risks. Acta Trop 194:181–188. https://doi.org/10.1016/j.actatropica.2019.03.012

Davies G, McIver L, Kim Y, Hashizume M, Iddings S, Chan V (2014) Water-borne diseases and extreme weather events in Cambodia: review of impacts and implications of climate change. Int J Environ Res Public Health 12(1):191–213. https://doi.org/10.3390/ijerph120100191

Diener A, Mudu P (2021) How can vegetation protect us from air pollution? A critical review on green spaces' mitigation abilities for air-borne particles from a public health perspective—with implications for urban planning. Sci Total Environ 796:148605. https://doi.org/10.1016/j.scitotenv.2021.148605
Garcia CA, Yap P-S, Park H-Y, Weller BL (2016) Association of long-term PM2.5 exposure with mortality using different air pollution exposure models: impacts in rural and urban California. Int J Environ Health Res 26(2):145–157. https://doi.org/10.1080/09603123.2015.1061113
García Sánchez F, Govindarajulu D (2023) Integrating blue-green infrastructure in urban planning for climate adaptation: lessons from Chennai and Kochi, India. Land Use Policy 124:106455. https://doi.org/10.1016/j.landusepol.2022.106455
Gobatti L, Bach PM, Scheidegger A, Leitão JP (2023) Using satellite imagery to investigate Blue-Green Infrastructure establishment time for urban cooling. Sustain Cities Soc 97:104768. https://doi.org/10.1016/j.scs.2023.104768
Goh KC, Kurniawan TA, Goh HH, Zhang D, Jiang M, Dai W, Khan MI, Othman MHD, Aziz F, Anouzla A, Meidiana C (2024) Strengthening infrastructure resilience for climate change mitigation: case studies from the southeast Asia region with a focus on wastewater treatment plants in addressing flooding challenges. ACS ES&T Water 4(9):3725–3740. https://doi.org/10.1021/acsestwater.4c00591
Golechha M, Mavalankar D, Bhan SC (2021) India: heat wave and action plan implementation in Indian cities, pp 285–308. https://doi.org/10.1007/978-3-030-87598-5_13
Greer SL, Falkenbach M, Siciliani L, McKee M, Wismar M, Figueras J (2022) From health in all policies to health for all policies. The Lancet Public Health 7(8):e718–e720. https://doi.org/10.1016/S2468-2667(22)00155-4
Guglielmin M, Muntaner C, O'Campo P, Shankardass K (2018) A scoping review of the implementation of health in all policies at the local level. Health Policy 122(3):284–292. https://doi.org/10.1016/j.healthpol.2017.12.005
Heaviside C, Macintyre H, Vardoulakis S (2017) The urban heat island: implications for health in a changing environment. Curr Environ Health Reports 4(3):296–305. https://doi.org/10.1007/s40572-017-0150-3
Ho JY, Shi Y, Lau KKL, Ng EYY, Ren C, Goggins WB (2023) Urban heat island effect-related mortality under extreme heat and non-extreme heat scenarios: a 2010–2019 case study in Hong Kong. Sci Total Environ 858:159791. https://doi.org/10.1016/j.scitotenv.2022.159791
Hoeben AD, Otto IM, Chersich MF (2023) Integrating public health in European climate change adaptation policy and planning. Climate Policy 23(5):609–622. https://doi.org/10.1080/14693062.2022.2143314
Huang Y, Mao F, Xu Y, Qi J, Zhang W, Li C, Zhou M (2025) Association between urban heat islands and heat-related mortality in 70 pairs of adjacent urban-rural counties among 8 topographic regions in China. Sustain Cities Society:106182. https://doi.org/10.1016/j.scs.2025.106182
Iungman T, Cirach M, Marando F, Pereira Barboza E, Khomenko S, Masselot P, Quijal-Zamorano M, Mueller N, Gasparrini A, Urquiza J, Heris M, Thondoo M, Nieuwenhuijsen M (2023) Cooling cities through urban green infrastructure: a health impact assessment of European cities. The Lancet 401(10376):577–589. https://doi.org/10.1016/S0140-6736(22)02585-5
Jørgensen G, Fryd O, Lund AA, Andersen PS, Herslund L (2022) Nature-based climate adaptation projects, their governance and transitional potential-cases from Copenhagen. Front Sustain Cities 4. https://doi.org/10.3389/frsc.2022.906960
Khader M (2021) Rotterdam resilience strategy, Rotterdam. In: Urban planning for transitions. Wiley, pp 1–18. https://doi.org/10.1002/9781119821670.ch1
Leal Filho W, Echevarria Icaza L, Neht A, Klavins M, Morgan EA (2018) Coping with the impacts of urban heat islands. A literature based study on understanding urban heat vulnerability and the need for resilience in cities in a global climate change context. J Clean Prod 171:1140–1149. https://doi.org/10.1016/j.jclepro.2017.10.086

Levy K, Smith SM, Carlton EJ (2018) Climate change impacts on waterborne diseases: moving toward designing interventions. Curr Environ Health Reports 5(2):272–282. https://doi.org/10.1007/s40572-018-0199-7

Limaye VS, Magal A, Joshi J, Maji S, Dutta P, Rajput P, Pingle S, Madan P, Mukerjee P, Bano S, Beig G, Mavalankar D, Jaiswal A, Knowlton K (2023) Air quality and health co-benefits of climate change mitigation and adaptation actions by 2030: an interdisciplinary modeling study in Ahmedabad, India. Environ Res: Health 1(2):021003. https://doi.org/10.1088/2752-5309/aca7d8

Lu X, Yuan D, Chen Y, Fung JCH (2021) Impacts of urbanization and long-term meteorological variations on global PM2.5 and its associated health burden. Environ Poll 270:116003. https://doi.org/10.1016/j.envpol.2020.116003

Macintyre HL, Heaviside C, Cai X, Phalkey R (2021) The winter urban heat island: impacts on cold-related mortality in a highly urbanized European region for present and future climate. Environ Int 154:106530. https://doi.org/10.1016/j.envint.2021.106530

Mishra RR, Verma J, Kanchan MR (2023) Water-sensitive urban design as a driver for accelerating sustainable urban development in India, pp 161–174. https://doi.org/10.1007/978-3-031-50132-6_12

Moradpour M, Hosseini V (2020) An investigation into the effects of green space on air quality of an urban area using CFD modeling. Urban Climate 34:100686. https://doi.org/10.1016/j.uclim.2020.100686

Noureen A, Aziz R, Ismail A, Trzcinski AP (2022) The impact of climate change on waterborne diseases in Pakistan. Sustain Clim Change 15(2):138–152. https://doi.org/10.1089/scc.2021.0070

Orsetti E, Tollin N, Lehmann M, Valderrama VA, Morató J (2022) Building resilient cities: climate change and health interlinkages in the planning of public spaces. Int J Environ Res Public Health 19(3):1355. https://doi.org/10.3390/ijerph19031355

Parihar J, Birman S (2024) Heat resilience in urban environments: strategies for sustainable city climate management. In: The climate-health-sustainability nexus. Springer Nature Switzerland, pp 305–324. https://doi.org/10.1007/978-3-031-56564-9_12

Piracha A, Chaudhary MT (2022) Urban air pollution, urban heat island and human health: a review of the literature. Sustainability 14(15):9234. https://doi.org/10.3390/su14159234

Puntub W, Greiving S (2022) Advanced operationalization framework for climate-resilient urban public health care services: composite indicators-based scenario assessment of Khon Kaen City, Thailand. Int J Environ Res Public Health 19(3):1283. https://doi.org/10.3390/ijerph19031283

Ramirez-Rubio O, Daher C, Fanjul G, Gascon M, Mueller N, Pajín L, Plasencia A, Rojas-Rueda D, Thondoo M, Nieuwenhuijsen MJ (2019) Urban health: an example of a "health in all policies" approach in the context of SDGs implementation. Glob Health 15(1):87. https://doi.org/10.1186/s12992-019-0529-z

Restemeyer B, van den Brink M, Woltjer J (2018) Resilience unpacked – framing of 'uncertainty' and 'adaptability' in long-term flood risk management strategies for London and Rotterdam. Eur Plan Stud 26(8):1559–1579. https://doi.org/10.1080/09654313.2018.1490393

Rovira J, Domingo JL, Schuhmacher M (2020) Air quality, health impacts and burden of disease due to air pollution (PM10, PM2.5, NO_2 and O_3): application of AirQ+ model to the Camp de Tarragona County (Catalonia, Spain). Sci Tot Environ 703:135538. https://doi.org/10.1016/j.scitotenv.2019.135538

Sabri SAM, Ponrahono Z, Bakar AA, Aziz FA (2023) Comparative analysis of open green spaces policies in enhancing urban resilience to climate change through small urban parks in Malaysia and Singapore. Chem Eng Trans 106(June):211–216. https://doi.org/10.3303/CET23106036

Sebestyén V, Dörgő G, Ipkovich Á, Abonyi J (2023) Identifying the links among urban climate hazards, mitigation and adaptation actions and sustainability for future resilient cities. Urban Climate 49:101557. https://doi.org/10.1016/j.uclim.2023.101557

Semenza JC (2021) Lateral public health: advancing systemic resilience to climate change. The Lancet Regional Health—Europe, 9. https://doi.org/10.1016/j.lanepe.2021.100231

Sharifi A, Pathak M, Joshi C, He B-J (2021) A systematic review of the health co-benefits of urban climate change adaptation. Sustain Cities Soc 74:103190. https://doi.org/10.1016/j.scs.2021.103190

Sheehan MC, Freire M, Martinez GS (2021) Piloting a city health adaptation typology with data from climate-engaged cities: toward identification of an urban health adaptation gap. Environ Res 196:110435. https://doi.org/10.1016/j.envres.2020.110435

Singh C, Vyas D, Patil S, Ranjit N, Poonacha P, Surampally S (2024) How are Indian cities adapting to extreme heat? Insights on heat risk governance and incremental adaptation from ten urban heat action plans. PLOS Climate 3(11):e0000484. https://doi.org/10.1371/journal.pclm.0000484

Solecki W, Rosenzweig C (2020) Indicators and monitoring systems for urban climate resiliency. Clim Change 163(4):1815–1837. https://doi.org/10.1007/s10584-020-02947-4

Sørensen E, Torfing J (2022) Co-creating ambitious climate change mitigation goals: the Copenhagen experience. Regulation Govern 16(2):572–587. https://doi.org/10.1111/rego.12374

Susca T, Pomponi F (2020) Heat island effects in urban life cycle assessment: novel insights to include the effects of the urban heat island and UHI-mitigation measures in LCA for effective policy making. J Ind Ecol 24(2):410–423. https://doi.org/10.1111/jiec.12980

Tomson M, Kumar P, Barwise Y, Perez P, Forehead H, French K, Morawska L, Watts JF (2021) Green infrastructure for air quality improvement in street canyons. Environ Int 146:106288. https://doi.org/10.1016/j.envint.2020.106288

van der Berg A (2023) Climate adaptation planning for resilient and sustainable cities: perspectives from the city of Rotterdam (Netherlands) and the City of Antwerp (Belgium). Eur J Risk Regulation 14(3):564–582. https://doi.org/10.1017/err.2022.17

Venter ZS, Hassani A, Stange E, Schneider P, Castell N (2024) Reassessing the role of urban green space in air pollution control. Proc Natl Acad Sci 121(6). https://doi.org/10.1073/pnas.2306200121

Walker J (2018) The influence of climate change on waterborne disease and Legionella: a review. Perspect Public Health 138(5):282–286. https://doi.org/10.1177/1757913918791198

Wu R, Dai H, Geng Y, Xie Y, Masui T, Liu Z, Qian Y (2017) Economic impacts from PM 2.5 pollution-related health effects: a case study in Shanghai. Environ Sci Technol 51(9):5035–5042. https://doi.org/10.1021/acs.est.7b00026

Xu M, Fricker P (2023) Urban agriculture: climate-responsive design strategies for blue infrastructure in the context of Singapore. J Digit Landscape Arch 2023(8):292–300. https://doi.org/10.14627/537740031

Xu H, Liu L, Ding P (2021) Building climate resilient city through multiple scale cooperative planning: experiences from Copenhagen. IOP Conf Ser: Mater Sci Eng 1203(3):032063. https://doi.org/10.1088/1757-899X/1203/3/032063

Yang S, Fang D, Chen B (2019) Human health impact and economic effect for PM2.5 exposure in typical cities. Appl Energy 249:316–325. https://doi.org/10.1016/j.apenergy.2019.04.173

Correction to: Genealogy of the Concept of Healthy Housing

Parsa Ahmadi Dehrashid, Ahmad Pourahmad, Hossein Mansourian, and Amir Reza Khavarian-Garmsir

Correction to: Chapter 9 in: A. I. Almulhim et al. (eds.), *Urbanization, Climate Change, and Health*, Climate Change Management, https://doi.org/10.1007/978-3-031-99132-5_9

In the original version of the book, the following belated correction have been incorporated: The author's "Parsa Ahmadi Dehrashid" given name and family name has been changed from "P. A. Dehrashid" to "Parsa Ahmadi Dehrashid" in the Front matter and in Chapter 9. The book and the correction chapter has been updated with the changes.

The updated version of this chapter can be found at
https://doi.org/10.1007/978-3-031-99132-5_9

© The Author(s), under exclusive license to Springer Nature Switzerland AG 2025

A. I. Almulhim et al. (eds.), *Urbanization, Climate Change, and Health*, Climate Change Management, https://doi.org/10.1007/978-3-031-99132-5_15

The manufacturer's authorised representative in the EU is Springer Nature Customer Service Centre GmbH, Europaplatz 3, 69115 Heidelberg, Germany. If you have any concerns regarding our products, please contact ProductSafety@springernature.com

Printed and bound by CPI Group (UK) Ltd, Croydon, CR0 4YY
01/10/2026
02225971-0010